Nonlinear Dynamics and the Beam-Beam Interaction
(Brookhaven National Laboratory, 1979)

AIP Conference Proceedings
Series Editor: Hugh C. Wolfe
Number 57

Nonlinear Dynamics and the Beam-Beam Interaction
(Brookhaven National Laboratory, 1979)

Editors
Melvin Month and John C. Herrera
Brookhaven National Laboratory

American Institute of Physics
New York   1979

Copying fees: The code at the bottom of the first page of each article in this volume gives the fee for each copy of the article made beyond the free copying permitted under the 1978 US Copyright Law. (See also the statement following "Copyright" below). This fee can be paid to the American Institute of Physics through the Copyright Clearance Center, Inc., Box 765, Schenectady, N.Y. 12301.

Copyright © 1980 American Institute of Physics

Individual readers of this volume and non-profit libraries, acting for them, are permitted to make fair use of the material in it, such as copying an article for use in teaching or research. Permission is granted to quote from this volume in scientific work with the customary acknowledgment of the source. To reprint a figure, table or other excerpt requires the consent of one of the original authors and notification to AIP. Republication or systematic or multiple reproduction of any material in this volume is permitted only under license from AIP. Address inquiries to Series Editor, AIP Conference Proceedings, AIP.

L.C. Catalog Card No. 79-57341
ISBN 0-88318-156-8
DOE CONF- 790318

SYMPOSIUM ON NONLINEARY DYNAMICS
AND
THE BEAM-BEAM INTERACTION

PREFACE

A symposium on Nonlinear Dynamics and the Beam-Beam Interaction, sponsored by the United States Department of Energy, High Energy Physics Division, was held at Brookhaven National Laboratory, March 19 through March 21, 1979. In bringing together researchers from the fields of accelerator physics and nonlinear theory, the symposium was an interdisciplinary approach to the study of the beam-beam interaction problem in high energy storage accelerators. Forty six scientists from universities and national laboratories in the United States and Western Europe attended. Twenty one papers were presented, thirteen of which were invited and eight of which were contributed. Their subject matter ranged from accelerator theory and experimental observations on beam-beam effects, through to the behavior of nonlinear oscillatory system.

The symposium was organized by a committee which met in September, 1978. These proceedings consist for the most part of manuscripts based on material presented orally at the symposium. An editorial committee which met in May, 1979 reviewed the papers submitted and approved most of them for publication. In some cases, considerable alterations to the oral presentation were incorporated by the author into the published version. In addition, some contributed papers stimulated by the exchange of ideas that took place at the symposium have been included in the proceedings.

During the last ten years, noteworthy success has been achieved in the operation of colliding beam systems. Due primarily to the large center-of-mass energies occurring in the particle collisions, experiment with counter-rotating colliding beams have already resulted in a greater understanding of the subnuclear interactions of elementary particles. But much still remains to be done, of course, and it is the hope of the high energy physics community that there will be more breakthroughs in the near future as more and higher energy colliding beams come into operation.

Since a stored beam has a much lower particle density than a material target, larger stored beam currents must be attained if the event rate required for successful experiments is to be achieved. For example, a high proton-proton luminosity, and therefore high beam current, is needed for studying rare processes, such as the production of a free W meson, the carrier of the weak force. However, an increase in current not only results in an increase in the luminosity (event rate) for particle interactions, but it is also accompanied by an increase in the strength of the classical electromagnetic interaction of the charge in one beam with the particles of the other beam. As it became apparent soon after the introduction of the first colliding beam storage ring, this "beam-beam interaction" is a significant limitation to the effectiveness of high energy physics experiments performed with colliding beams. And, in spite of the extensive experimental study of this two-beam interaction, as well as of computer simulation experiments and of other theoretical approaches, only limited success has been achieved in developing a detailed understanding of the mechanism causing this limitation.

Nonetheless, a worldwide expansion in the number of colliding beam facilities is about to take place. Rapid growth is occurring on many fronts. PETRA, a 15 GeV on 15 GeV electron-positron ring, recently began operation at DESY in Hamburg. A similar facility, PEP, at the Stanford Linear Accelerator Center, will be completed soon. And ISABELLE, a project at Brookhaven National Laboratory with high energy proton-proton colliding beams of energies up to 400 GeV per beam, is scheduled to become operational in late 1985 or early 1986. Also, a proton-antiproton system, potentially capable of achieving 540 GeV in the center-of-mass, is under construction at CERN in Geneva, while a similar type system, perhaps reaching a center-of-mass energy of as high as 2000 GeV (2 TeV), is being planned at Fermilab, near Chicago. And this is not all. Western Europe, in addition to the $\bar{p}p$ CERN project, is developing plans for major facilities capable of providing very high energy electron-positron (e^-e^+) and electron-proton (e-p) collisions. At CERN, a very large electron-positron colliding beam system, termed LEP, with energies in the range of 70-130 GeV per beam, is under consideration, the hope being to complete this facility by the late 1980's. At DESY, serious discussions have already begun on an electron-proton facility capable of producing 35 GeV electrons and 1 TeV protons, with a possible completion date also in the late 1980's. In the U.S., thought is being given to electron-proton systems as extensions to existing and planned facilities—by adding, for example, an electron ring in the Fermilab or ISABELLE tunnels, or by adding a proton ring in the PEP tunnel.

To translate this vision of high energy physics during the 80's and 90's into reality is a challenge of enormous magnitude. And it was with a recognition of this challenge, together with the realization that one of the major obstacles to be overcome is the beam-beam limitation, that the symposium was conceived. Its goals were twofold: first, to review the experimental and theoretical work that had already been accomplished on the beam-beam interaction; second, to study the relation of beam-beam effects to the larger class of phenomena associated with nonlinear forces.

An immediate benefit derived from the symposium was the introduction into accelerator physics of different approaches to nonlinear problems used for many years in statistical mechanics and celestial mechanics. These sophisticated mathematical techniques may indeed be the key needed to unravel the complex mechanisms involved in the strongly nonlinear beam-beam interaction.

Although the space charge beam-beam force constitutes a major limitation to the performance of colliding beams, it may, on the other hand, have a potential value in the general study of nonlinear systems. Existing facilities are capable of providing beams of particles oscillating with a lifetime greater than 10^{11} cycles, which is more than the number of planetary periods for our solar system. Since a variety of parameters of the storage rings and the colliding beams are under experimental control, including the strength of the nonlinear beam-beam force, it is clear that such systems may be of considerable importance in the continuing effort to understand the nature of perturbed oscillating systems. In particular, one can contemplate a study of the "long time" behavior of oscillating systems under weak nonlinear perturbations; or the "stochastic" behavior of such systems when strongly perturbed. Such experimental accelerator studies would enlighten and aid us in our understanding of these processes, and the knowledge thus gleaned would have ramifications in the many fields of physics where the dynamics are governed by nonlinear forces. The present symposium has allowed us to make a start.

        M. Month
        J.C. Herrera

ACKNOWLEDGMENTS

The organizers of the symposium take this opportunity to acknowledge the contributions of the many people toward the success of the symposium. Particular recognition is accorded to the following for their special efforts: to J.R. Sanford, Head of ISABELLE Project (BNL), R.R. Rau, Associate Director for High Energy Physics (BNL), and J. Rees, Head of PEP Project (SLAC) for their continued assistance during the planning of the symposium; to G.H. Vineyard, Director (BNL), W.F.K. Panofsky, Director (SLAC), and W.A. Wallenmeyer, Director, High Energy Physics Division, Department of Energy, for their intellectual support of the enterprise; and to J. Moser, Courant Institute, NYU, for his advice and encouragement.

The editors would like to thank Mrs. Paula Hughes for her painstaking efforts in converting the initial manuscripts to the final proceedings.

ORGANIZING COMMITTEE

M. Month, Brookhaven National Laboratory (Co-chairman)
H. Wiedemann, Stanford Linear Accelerator Center (Co-chairman)
E.D. Courant, Brookhaven National Laboratory
J. Ford, Georgia Institute of Technology
R.H.G. Helleman, Georgia Institute of Technology
L.J. Laslett, Lawrence Berkeley Laboratory
D. Sutter, Department of Energy
L.C. Teng, Fermi National Accelerator Laboratory

EDITORIAL COMMITTEE

M. Month, Brookhaven National Laboratory (Chairman)
T. Bountis, California Institute of Technology
R.H.G. Helleman, Georgia Institute of Technology
J.C. Herrera, Brookhaven National Laboratory
M. Schiff, City University of New York at Staten Island
J. Tennyson, University of California at Berkeley
H. Wiedemann, Stanford Linear Accelerator Center

PARTICIPANTS

M.Q. Barton, Brookhaven National Laboratory
M. Bassetti, Laboratori Nazionali di Frascati
T. Bountis, California Institute of Technology
M. Braun, Queens College
P. Bryant, Centre Europeene de Recherche Nucleaire (CERN)
A. Chao, Stanford Linear Accelerator Center
J. Claus, Brookhaven National Laboratory
E. Close, Lawrence Berkeley Laboratory
M. Cornacchia, Brookhaven National Laboratory
E.D. Courant, Brookhaven National Laboratory
D. Deacon, Stanford University
A.J. Dragt, Los Alamos Scientific Laboratory
C. Eminihizer, Physical Dynamics
J. Ford, Georgia Institute of Technology
J. Gareyte, Centre Europeene de Recherche Nucleaire (CERN)
J.M. Greene, Princeton University
S. Guiducci, Laboratori Nazionali di Frascati
I. Haber, Naval Research Laboratory
R.H.G. Helleman, Georgia Institute of Technology
J.C. Herrera, Brookhaven National Laboratory
S. Kheifets, Stanford Linear Accelerator Center
T. Khoe, Argonne National Laboratory
L.J. Laslett, Lawrence Berkeley Laboratory
C. Leemann, Lawrence Berkeley Laboratory
H. Leidecker, American University
M. Levi, Northwestern University
A. Lichtenberg, University of California at Berkeley
M.A. Lieberman, University of California at Berkeley
F. Mills, Fermi National Accelerator Laboratory
M. Month, Brookhaven National Laboratory
J. Moser, New York University
S. Ohnuma, Fermi National Accelerator Laboratory
R.F. Peierls, Brookhaven National Laboratory
C. Pellegrini, Brookhaven National Laboratory
I. Percival, Queen Mary College
A. Piwinski, Deutsches Elektronen Synchrotron
A.G. Ruggiero, Fermi National Accelerator Laboratory
D.F. Sutter, U.S. Department of Energy
S. Tazzari, Laboratori Nazionali di Frascati
L.C. Teng, Fermi National Accelerator Laboratory
J. Tennyson, University of California at Berkeley
P.A. Vuillermot, Georgia Institute of Technology
A. Weinstein, Rice University
W.T. Weng, Brookhaven National Laboratory
H. Wiedemann, Stanford Linear Accelerator Center
H. Zyngier, Institute de Physique Nucleaire, Orsay

TABLE OF CONTENTS

Overview of Colliding Beam Facilities
    J.C. Herrera & M. Month 1

Electromagnetic Interaction of Colliding Beams in
Storage Rings
    J.C. Herrera 29

A Summary of Some Beam-Beam Models
    A.W. Chao.. 42

Review of the Investigations on the Beam-Beam
Interactions at the ISR
    G. Guignard.. 69

Experiments on the Beam-Beam Effect in e^+e^-
Storage Rings
    H. Wiedemann... 84

Report on Some Beam-Beam Functional Dependencies
in SPEAR
    M. Cornacchia.. 99

Recent Results from DORIS and PETRA
    A. Piwinski...115

Beam-Beam Effects at the 1.5 GeV e^+e^- Storage Ring ADONE
    S. Tazzari..128

Beam-Beam Effect - A Review of the Observations made at Orsay
    H. Zyngier..136

Transfer Map Approach to the Beam-Beam Interaction
    A.J. Dragt..143

The Instability Threshold for Bunched Beams in ISABELLE
    J. Tennyson...158

The Beam-Beam Interaction as a Nonlinear Problem:
A Numerical Study
   E.D. Courant 194

Simple Computer Model for the Nonlinear Beam-Beam
Interaction in ISABELLE
   J.C. Herrera, M. Month, R.F. Peierls 202

Beam-Beam Limit Simulation of SPEAR 1, A Summary
of Results
   E. Close 210

Stable and Unstable Motion in Dynamical Systems
   J. Moser 222

Exact Results for Some Linear and Nonlinear
Beam-Beam Effects
   R.H.G. Helleman 236

KAM Surfaces Computed from the Hénon-Heiles Hamiltonian
   J.M. Greene 257

Diffusion in Near-Integrable Hamiltonian Systems with
Three Degrees of Freedom
   J.L. Tennyson, M.A. Lieberman, A.J. Lichtenberg 272

Variational Principles for Invariant Tori and Cantori
   I.C. Percival 302

On an Application of Perturbation Methods to the Beam-Beam
Interaction
   T. Bountis, E. Coutsias 311

Some Numerical Studies of Arnold Diffusion In a Simple
Model
   B.V. Chirikov, J. Ford, F. Vivaldi 323

OVERVIEW OF COLLIDING BEAM FACILITIES

J.C. Herrera and M. Month
Brookhaven National Laboratory, Upton, New York 11973

I. INTRODUCTION

The high energy physics community worldwide is on the threshold of a great expansion in colliding beam facilities. The promising future that can be foreseen for the field is the result of a number of auspicious events. First, there are the breakthroughs that have been made in the technology of superconducting magnets, which have made it possible to envisage very large accelerators with energies in the hundreds of GeV; second, there are the recent experimental discoveries and theoretical advances that have been made in elementary particle physics, which suggest that new and significant elementary processes take place in this higher energy regime; and third, there is the continuing desire of society through its government agencies to support this fundamental research into the ultimate nature of matter.

An overall view of how these high energy facilities have progressed in recent years can be had by considering Fig. 1. Storage rings, utilizing two counter rotating interacting beams, are clearly the better choice for achieving higher center-of-mass energies. That this fact is recognized is evidenced by the increase in the number of colliding beam projects being constructed and proposed. Note the steeper slope for the storage ring lines which graphically expresses this idea.

Accompanying this trend in the direction of colliding beams is the need for reaching high luminosities. As can be seen from Fig. 2, the fixed target synchrotrons have mean luminosities about 4 to 5 orders of magnitude higher than storage rings. Unfortunately, increasing the useful or c.m. energy does not appear promising since the size of the synchrotron rises essentially as the square of the c.m. energy. Thus, the main thrust for future machines is in the colliding beam direction and the principal effort for performance gains is in reaching higher luminosities in colliding beam systems.

In section II, we describe the major high energy physics facilities around the world: those existing, those under construction and those being proposed. The emphasis here is on the nature and size of the facilities and on the major technologies required to support the worldwide high energy physics effort. In section III, we attempt to present a view of the beam collisions, describing briefly the instruments used to make the beams collide and those used to detect the products of particle interactions in the beam overlap region. Section IV is brief summary.

II. FACILITIES

In Table I we present a list of the major colliding beam facilities operating in the 1970's and those projected for the 1980's and

1990's. One can see that the large and costly projects are either in the construction phase or in the design phase. The only existing proton-proton facility is the one kilometer circumference Intersecting Storage Ring (ISR) at CERN in Geneva. A schematic of the various machines on the CERN site, including the ISR, is shown in Fig. 3; while a picture of the Geneva area with the PS, ISR, and SPS machines superimposed is given in Fig. 4. In contrast to the number of pp colliding rings, there are in existence many electron-positron facilities, which were developed in the 1960's and early 1970's. The last of the "small" storage rings is SPEAR (Stanford Positron and Electron Accelerator Ring) at the Stanford Linear Accelerator Center (SLAC). Figure 5 is an aerial view of this facility with the 250 meter ring (SPEAR) on the lower right. The two mile long linear accelerator, which is used as the electron and positron injector can be seen, passing under the Freeway.

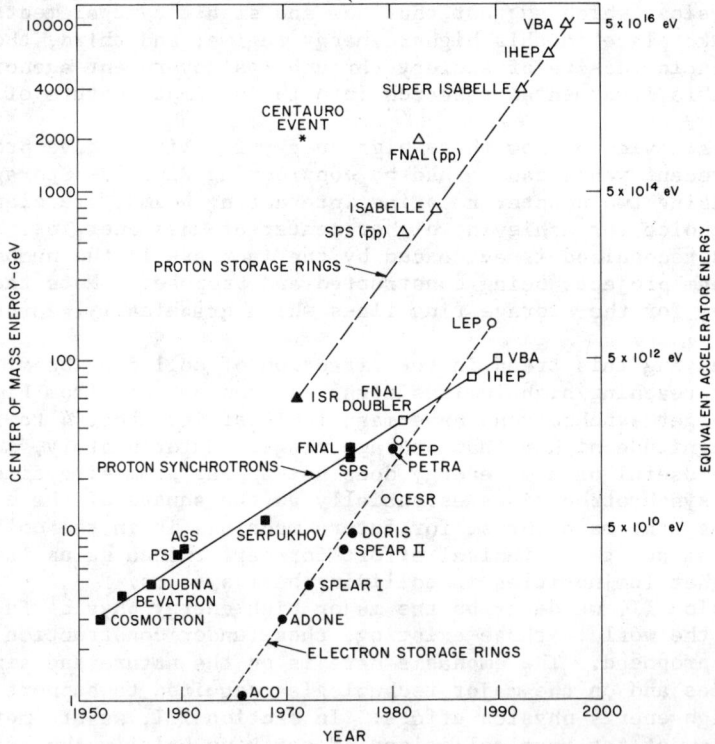

Fig. 1. Center-of-mass energy for high energy accelerators and storage rings over recent decades.

The first of the new storage rings to be available is the 2.3 kilometer PETRA ring at the DESY laboratory in Hamburg. This ring was completed in mid 1978 and became operational in October, 1978. A birdseye view of the DESY site is given in Fig. 6. In addition to the large PETRA ring circumscribing the site, one can also make out the smaller DORIS storage ring and, as well, the DESY synchrotron, the

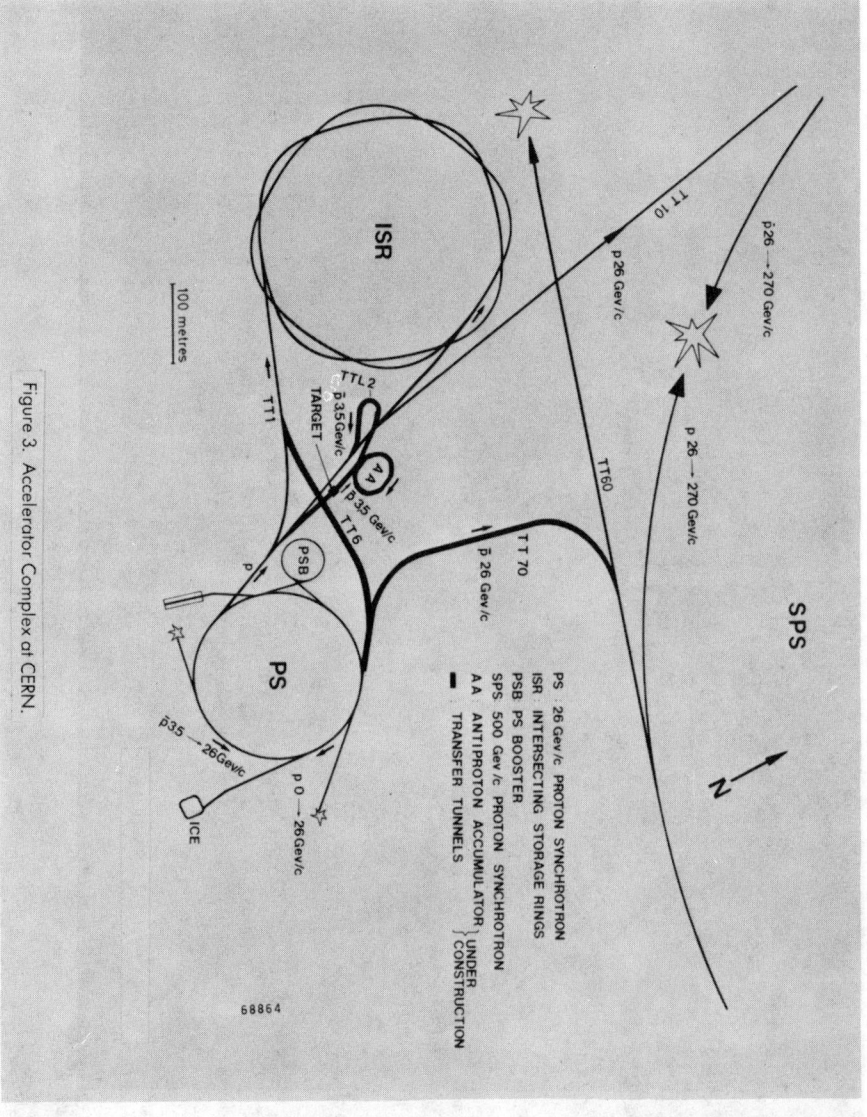

Figure 3. Accelerator Complex at CERN.

Figure 4. Aerial View of the CERN Site.

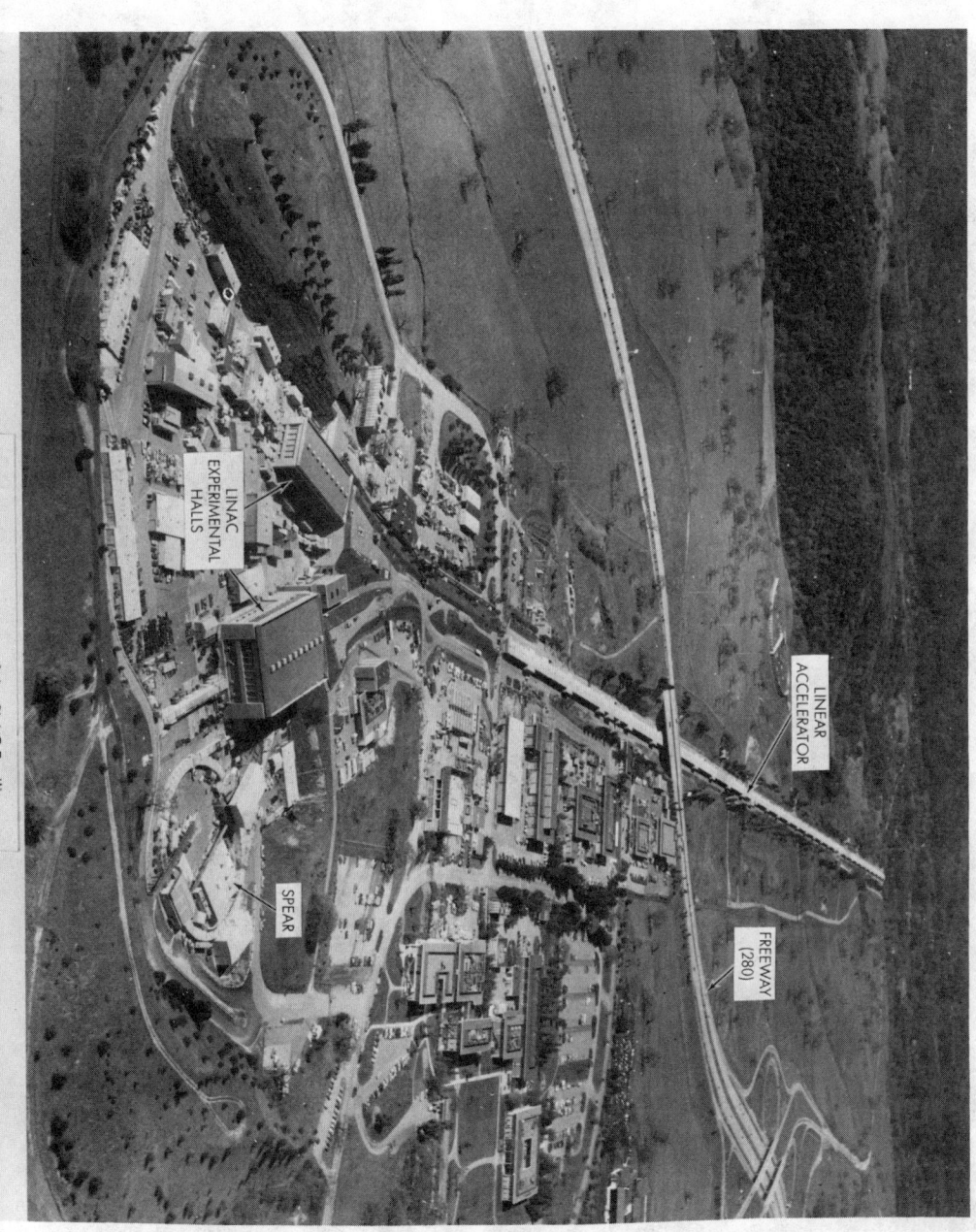

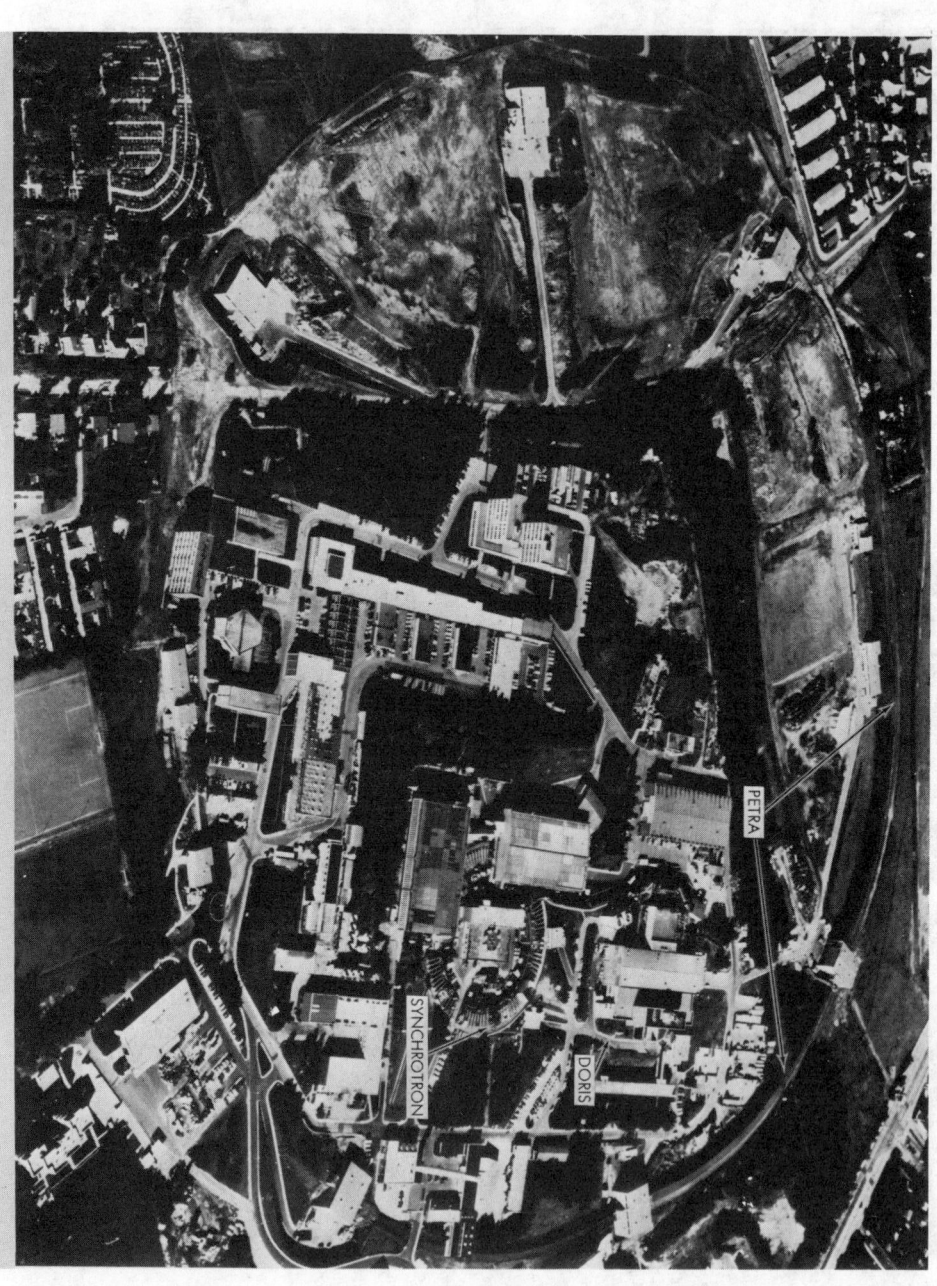

Figure 6. Aerial View of the DESY Facility.

injector for both DORIS and PETRA. A photograph of the magnets in the PETRA tunnel is shown in Fig. 7. Here one can see the long dipoles (6 m), which produce the bending field for the circulating beams, and the shorter quadrupole magnets (2.5 m), which keep the beams focused in the transverse plane as they follow their long paths. The particles circulate on these paths for a few hours, a time corresponding to more than 10^{10} revolutions. It is to be noted that there is a single ring in the tunnel. This is an advantage for a machine in which oppositely charged particles collide: both beams can circulate in opposite directions in the same magnetic field. PEP, a project similar to PETRA, is scheduled for operation in early 1980 at SLAC (see figs. 5 and 8).

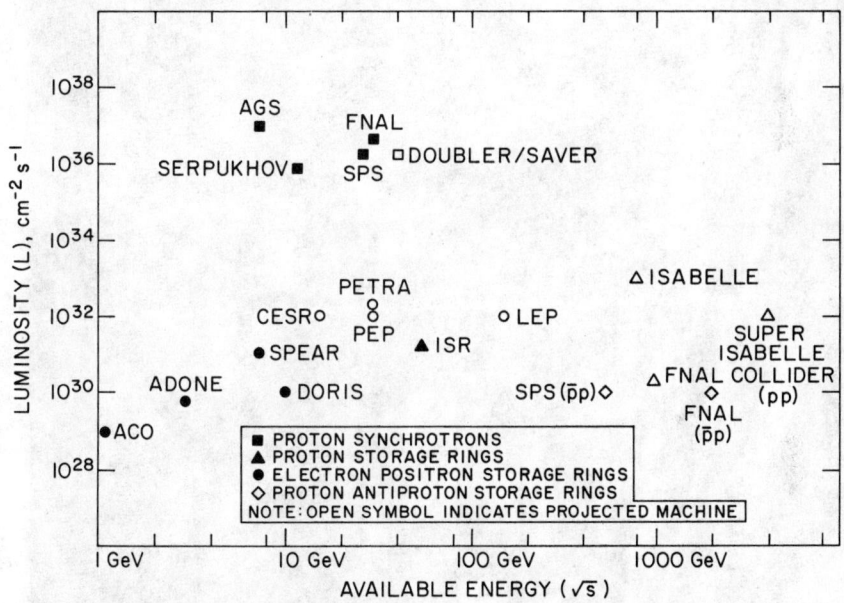

Fig. 2. Luminosity vs available energy for accelerator and storage rings.

Anticipation of possible experimental breakthroughs in very high energy electron-positron collisions has led to suggestions for the construction of even larger rings to accommodate electrons and positrons with energies of 100 GeV and even higher per beam. A specific proposal to build such a facility at CERN, to be called LEP (Large Electron Positron), has been made. Much discussion has taken place as to exactly what ring size is appropriate for the project. Figure 9 is a map of the Geneva area. Drawn on this map is the 20 kilometer long circumference of a LEP design. Such a machine would yield 70 GeV per beam and require a total power of 100 MW in order to supply the

8

Figure 7. Tunnel and Magnets of the PETRA Ring.

Figure 8. SLAC Site Showing PEP Facility.

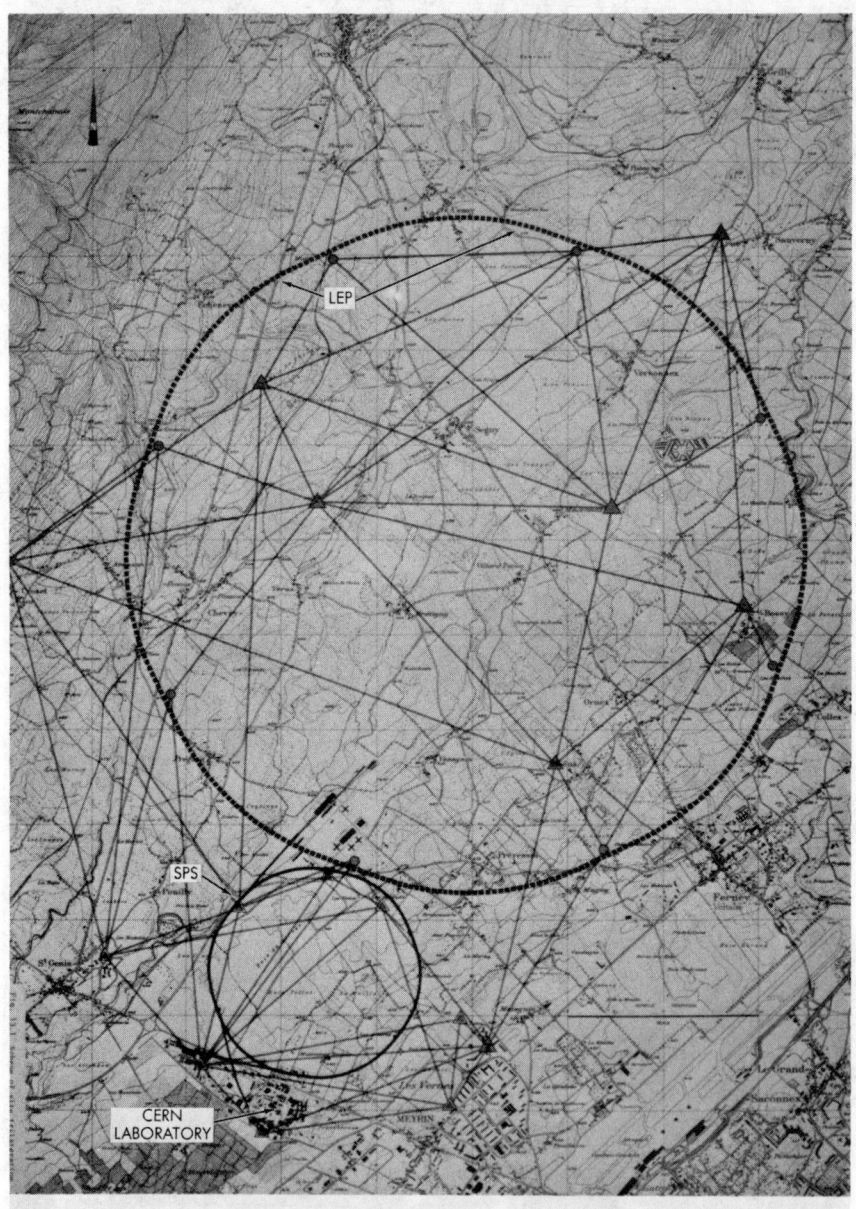

Figure 9. Geneva Area Showing Proposed LEP Facility.

TABLE I. Storage Rings: Energy and Cost

Machine	ADONE (Frascati)	SPEAR (Stanford)	PETRA (Hamburg)	LEP (Geneva)	SPS (Geneva)	ISR (Geneva)	ISABELLE (Brookhaven)	VBA
Type of beam Collisions	e^+e^-	e^+e^-	e^+e^-	e^+e^-	$p\bar{p}$	pp	pp	$pp, \bar{p}p$
Number of rings	1	1	1	1	1	2	2	2,1
Maximum energy (GeV)[a]	1 x 1	4 x 4	18 x 18	85 x 85	270 x 270	30 x 30	400 x 400	5000 x 5000
Maximum Luminosity (cm^{-2}sec^{-1})	10^{29}	10^{31}	10^{32}[b]	10^{32}[b]	10^{30}[b]	10^{31}	10^{33}[b]	10^{33}[b]
Time of Operation (typical year)	1970	1975	1979	1988	1982	1975	1986	1995
Size (km) (circumference)	0.1	0.25	2	30	7	1	4	25
Approximate Construction Cost	$5 M	$10 M	$100 M	$800 M	$250 M +$100 M	$100 M	$275 M	$1 B
Others in this Category	ACO (Orsay)	DORIS (Hamburg)	PEP (Stanford)		FNAL (Chicago)			Super ISABELLE

[a] 1 GeV = 1.6 x 10^{-10} joule
[b] Estimate

emitted synchrotron radiation. Klystrons are an efficient source of this rf power. An example of a klystron for the PEP storage ring is shown in Fig. 10. The energy is transferred to the beam via resonant cavities through which the beam passes. A straight section of the PEP ring with rf cavities is displayed in Fig. 11. To replace the large energy loss in LEP, more than 1.5 kilometers of active cavity length will be needed. It therefore appears that with the existing rf technology, LEP could well represent the limit in the construction of electron machines of the SPEAR type. This rf barrier is not a rigid one, and a promising technique on the horizon may be the introduction of superconducting cavities with the accompanying reduction in required rf power. Such an "improvement program" could raise the energy of LEP to approximately 110 GeV/beam. The hope at CERN is to have the first stage of LEP completed and operational by as early as 1988.

The recent advances in superconducting magnet technology have made possible the achievement of very high energies for proton beams. Proton-proton colliding beam facilities, and more recently proton-antiproton facilities, are under construction. After the successful operation of the ISR at CERN, a proposal was made by Brookhaven National Laboratory to construct a new facility for pp collisions with energies of up to 400 GeV per beam and maximum luminosity of 10^{33} cm^{-2} sec^{-1}. The project, called ISABELLE, (ISA for Intersecting Storage Accelerator and "BELLE" for beautiful), is now in the construction phase. An artist's conception of the overall ISABELLE facility on the Brookhaven site is presented in Fig. 12. Figure 13 is a view of the tunnel cross section showing the superconducting magnet dewars of the two rings which are required to guide the two counterrotating, interlaced proton beams. These rings will cross, or intersect, at six regions along the circumference and thereby provide six areas for high energy experiments. The ISABELLE schedule calls for completion by 1986.

At CERN, a project has been initiated to construct a facility that collides proton and antiprotons using one of the existing accelerators on the site, the SPS (Super Proton Synchrotron), (see Figs. 3 and 4). As one can imagine, such a project will be difficult to bring into operation. Not only are there the practical problems involved in converting an existing facility and, at the same time, minimizing the interference with an on going experimental program, but there is also the fundamental problem of accumulating enough antiprotons which are needed to carry out reasonable experiments. To obtain a source of antiprotons one bombards a heavy target with high energy protons and collects the antiprotons at some lower energy. In rough figures, to obtain 1 antiproton, 10^7 protons are needed. Thus, the antiproton accumulation must take a long time and the phase space of the beam must be continuously compressed. Difficult as this task appears, CERN has taken it on. A similar type of facility is under active consideration at the Fermi National Accelerator Laboratory (FNAL) near Chicago. Here, a new superconducting magnet ring is about to be installed in the 6.3 kilometer tunnel housing the 400 GeV accelerator. A panoramic view of the FNAL site is given in Fig. 14. Figure 15 is a view of the FNAL tunnel, showing some superconducting magnets installed underneath the

Figure 10. PEP Klystron Gallery.

Figure 11. PEP RF Cavities.

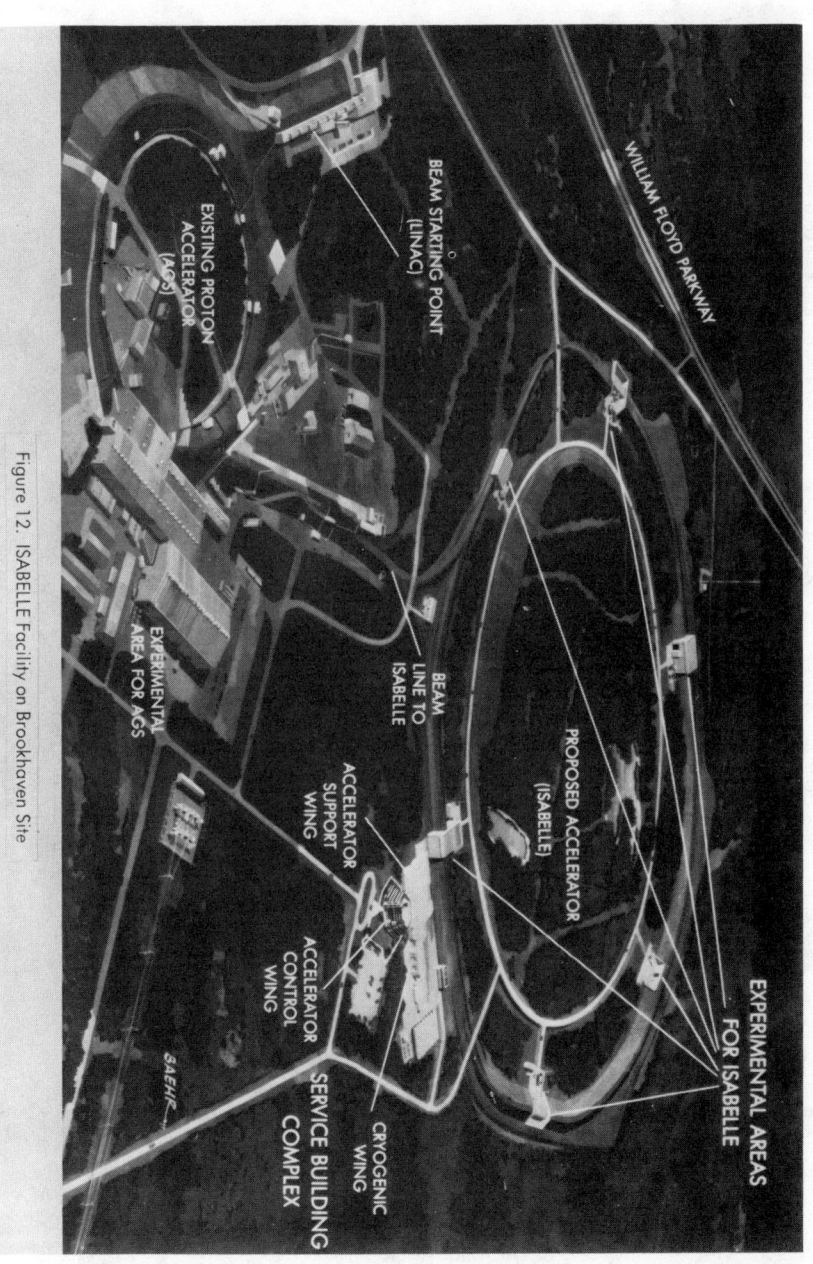

Figure 12. ISABELLE Facility on Brookhaven Site

Figure 13. Storage Ring Tunnel for ISABELLE.

Figure 14. Panoramic View of Fermilab.

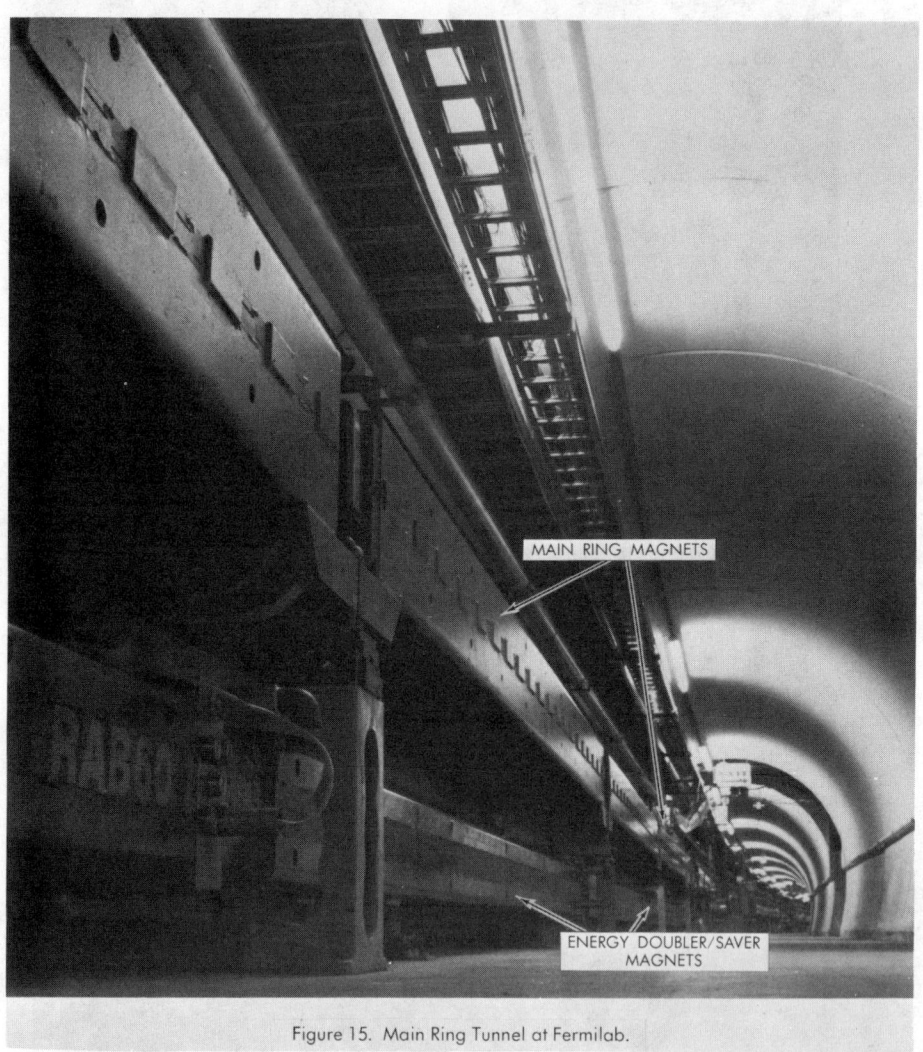

Figure 15. Main Ring Tunnel at Fermilab.

"Main Ring" magnets. Because of the higher magnetic field capability, and the lower overall power consumption of the superconducting magnets, the energy of the p$\bar{\text{p}}$ collisions at FNAL will be about four times higher than at the CERN p$\bar{\text{p}}$ project (2000 GeV in the center-of-mass compared to about 540 GeV).

III. COLLIDING BEAMS

There are two designs for colliding beam systems. The first is the single ring system which can be used in the case of short oppositely charged bunches circulating in opposite directions. As we have mentioned, because of the opposite charges, the same magnetic lattice will affect the counterrotating bunches identically. The bunches must be short in length so that they collide only in the specific regions where the experimental detector apparatus is located. Electron-positron colliding beam systems work in this way, and the p$\bar{\text{p}}$ facilities under development for the CERN SPS and for the FNAL superconducting ring also work in the same way. The second design is the one used at the ISR and also planned for ISABELLE. It consists of two counterrotating beams of the same charge contained in two separate, but interlaced, magnet systems. The beams move in parallel arcs for some distance, and at specified "interaction regions," they are bent toward each other and cross at an angle. This collision angle essentially determines the length of the "interaction diamond". After the collision, the beams diverge and again follow parallel paths, interchanged with respect to the center of the machine. Eight such crossings exist at the ISR, and six are planned for ISABELLE.

In Fig. 16 the electron-proton single ring collision design is schematically shown, with some magnitudes for the collision length ($\sim \frac{1}{2}$ the bunch length, since the bunches are moving at equal speeds in opposite directions) and the interaction cross section. Figure 17 is a photograph of the MK-II Solenoid detector at SPEAR. One can visualize the tiny bunches of particles entering the center of the detector from both sides, some electrons and positrons interacting, and producing secondary particles that spiral radially toward the outer cylinders of detectors. Superimposed on the photographs of the SPEAR facility in Fig. 18 are arrows which indicate how the electrons and positrons enter the ring and subsequently collide in specific areas.

The proton-antiproton systems that are being developed are similar in concept to the electron-positron setup. However, the magnitudes of the interaction length and the collision cross section are different. Figure 19 is a schematism for the collision geometry of this system.

A sketch of the geometry of p-p collision is given in Fig. 20. The significant difference between these collisions and the p$\bar{\text{p}}$ ones, for example, is that the collision region has a diamond shape and its overall length is determined by the crossing angle. In addition, the ISR beams are not bunched, but rather they are continuous around the two rings. This is also the mode of operation presently planned for ISABELLE. It might be mentioned, though, that bunching in this latter case is an optional mode of operation which has no influence on the

Figure 17. Mark II Solenoid Detector at SPEAR.

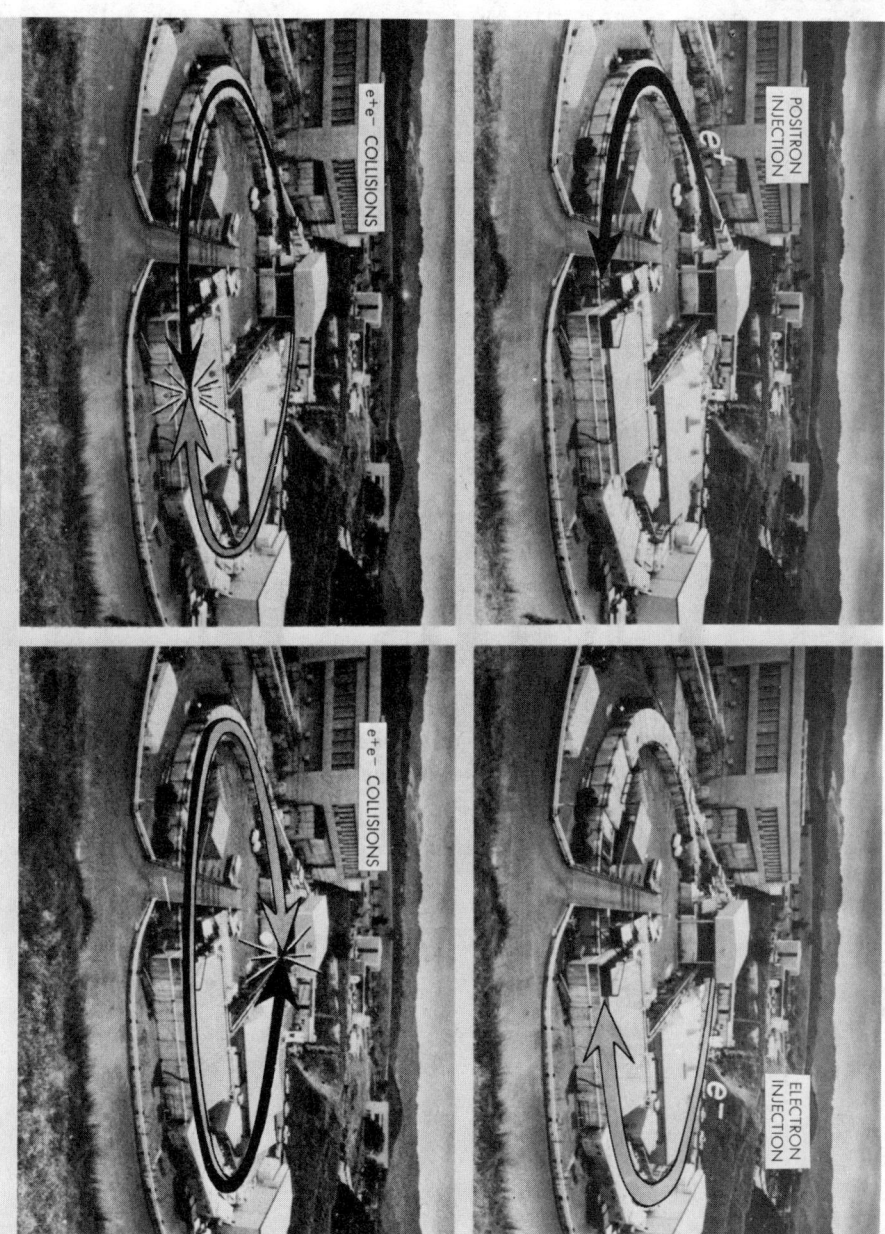

Figure 18. Colliding e+e− Beams at SPEAR.

collision geometry. The point is that, for long bunches, the collision length does not depend on the bunch length. A view of the crossing region at the ISR (the split field magnet interaction region) is shown in Fig. 21, while Fig. 22 shows an artist's conceptual view of the lepton detector region proposed for ISABELLE. To be noted are some implications of the great difference in crossing angles in the two cases (at the ISR, the crossing angle is twenty-five times larger than at ISABELLE, ~ 250 mrad compared to 10 mrad): (1) the beam pipes remain close together in ISABELLE, requiring that the region close to the collision region remain free of magnets; and (2) the final ring separation is much larger at the ISR necessitating a much wider tunnel (~ 15 m tunnel diameter at the ISR compared to ~ 5 m at ISABELLE).

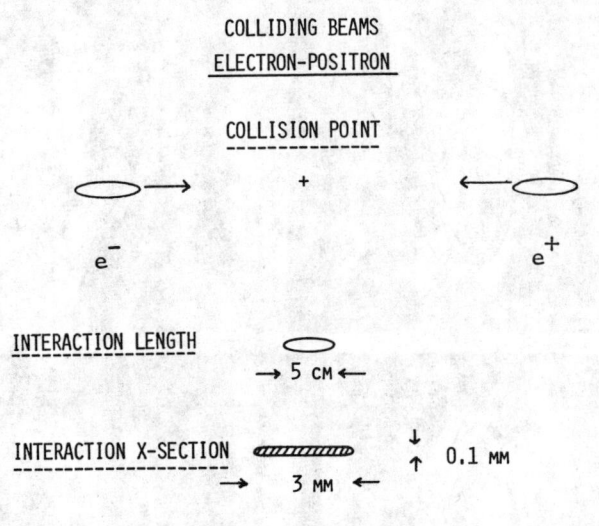

Fig. 16. Collision geometry for e^+e^- beams.

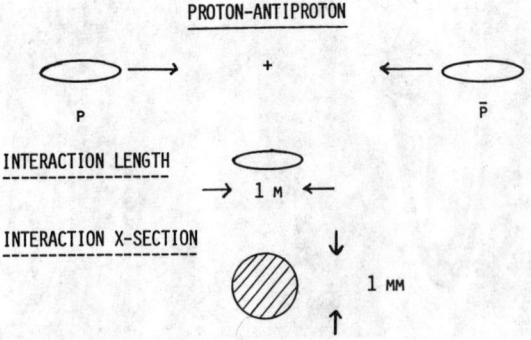

Fig. 19. Collision geometry for $\bar{p}p$ beams.

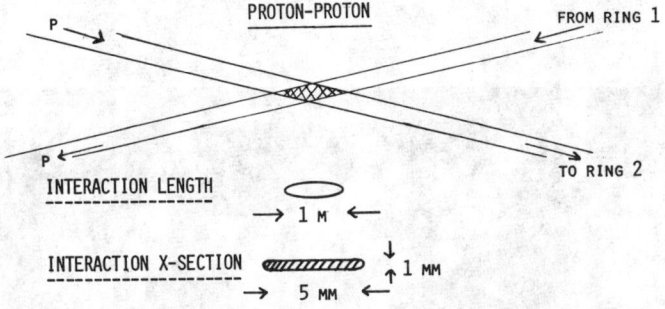

Fig. 20. Collision geometry for pp beams.

IV. SUMMARY

In summary, we give in Table II a complete list of the major high energy physics facilities, those that are operating, those under construction and those being proposed formally as national projects. Given are the different accelerators and storage rings that make up these facilities and the performance either obtained or anticipated. They are divided into world regions: facilities in the U.S., in Western Europe, in the Soviet Union, and in East Asia. It is apparent that the world physics community has embarked on an adventure of major proportions. This paper has attempted to give a glimpse into the construction and technologies that are needed to transform this vision into reality.

TABLE II.
LIST OF HEP FACILITIES

(A) Fixed Target (Protons)

Laboratory	Date of First Beam	Maximum Beam Energy (GeV)	Beam Intensity (Protons per Second)	Status
U.S.				
BNL, AGS	1960	28	8×10^{12}	Operating
ARGONNE, ZGS[a]	1963	12.5	2×10^{12}	Operating
FERMILAB, MAIN Ring	1972	400	3×10^{12}	Operating
FERMILAB, ENERGY SAVER	1981	500	1×10^{12}	Construction
FERMILAB, TEVATRON	1984	1000	1×10^{12}	Planned

Figure 21. Split-Field Intersection Region at the ISR.

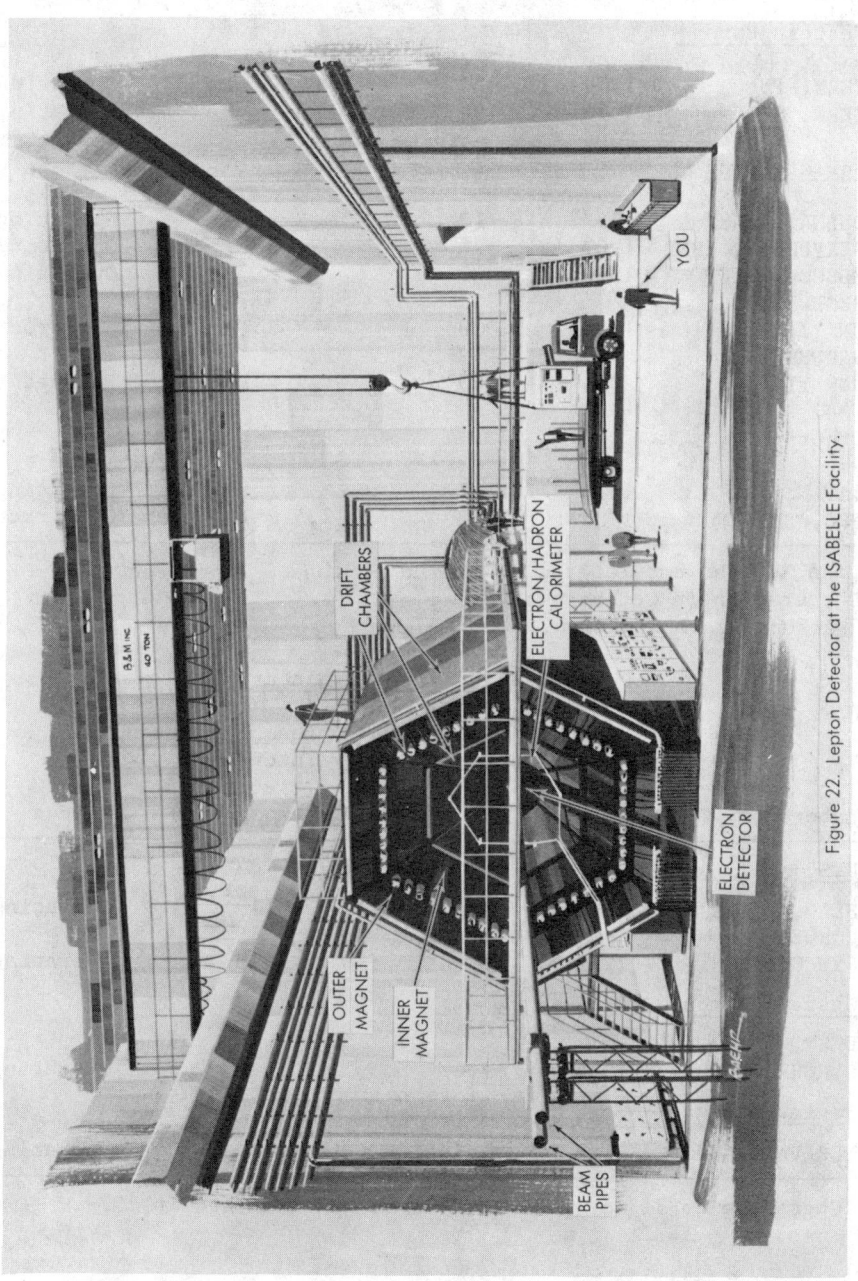

Figure 22. Lepton Detector at the ISABELLE Facility.

TABLE II. (Continued)

Western Europe

CERN, PS	1959	26	8×10^{12}	Operating
CERN, SPS	1976	400	3×10^{12}	Operating

Soviet Union

DUBNA, JINR	1957	10	2×10^{10}	Operating
SERPUKHOV, IHEP	1967	76	4×10^{11}	Operating
MOSCOW, ITEP	1973	10		Operating
SERPUKHOV, UNK I	~1985	400	7×10^{12}	Construction[b]
SERPUKHOV, UNK II	?	3000	7×10^{12}	Planned

Others

JAPAN, KEK	1976	12	10^{12}	Operating
PRC, BEIJING	1985	50	2.5×10^{12}	Construction

[a] The ZGS has accelerated polarized protons. It ceased operation in October, 1979.
[b] Personal communication.

(B) Fixed Target (Electrons)

Laboratory	Date of First Beam	Maximum Beam Energy (GeV)	Beam Intensity (Electrons per second)	Status
U.S.				
SLAC, LINAC[a]	1966	31	2×10^{14}	Operating
CORNELL, SYNCHROTRON	1967	12	2×10^{11}	Operating
Western Europe				
DESY, SYNCHROTRON	1964	7.4	3×10^{13}	Operating
Soviet Union				
YEREVAN, ARUS	1967	6.1	5×10^{12}	Operating

[a] Upgrading to 31 GeV from 22 GeV was accomplished in 1979.

TABLE II. (Continued)

(C) Colliding Beams (Electrons X Positrons)

Laboratory	Date of First Beam	Maximum Beam Energy (10^9 eV)	Luminosity (Events per cm^2 per second)	Status
U.S.				
SLAC, SPEAR	1972	4 x 4	10^{31}	Operating
CORNELL, CESR	1979	8 x 8	10^{32a}	Operating
SLAC, PEP	1979	15 x 15	10^{32}	Construction
Western Europe				
FRASCATI, ADONE	1969	1.5 x 1.5	10^{30}	Operating
DESY, DORIS	1974	5 x 5	10^{30}	Operating
ORSAY, DCI	1976	1.8 x 1.8	10^{31}	Operating
DESY, PETRA	1978	15 x 15	10^{32b}	Operating
DESY	1986	33 x 33	10^{32}	Planned
CERN, LEP	1988	85 x 85	10^{32}	Planned
Soviet Union				
NOVOSIBIRSK, VEPP3	1969	3.5 x 3.5		Operating
NOVOSIBIRSK, VEPP4	1979	7 x 7		Construction

[a] Circulating beam just obtained--expected performance not yet achieved.
[b] PETRA has achieved 17 GeV per beam. Its luminosity is presently about 1% of its projected value.

(D) Colliding Beams (Protons/Antiprotons)

Laboratory	Date of First Beam	Maximum Beam Energy (GeV)	Luminosity (Events per cm^2 per second)	Status
U.S.				
FERMILAB, TEVATRON($p\bar{p}$)	1983	1000 x 1000	10^{30}	Planned
BNL, ISABELLE (pp)	1986	400 x 400	10^{33}	Construction
Western Europe				
CERN, ISR pp	1971	31 x 31	2×10^{31}	Operating
CERN, SPS $p\bar{p}$	1982	270 x 270	10^{30}	Construction

TABLE II. (Continued)

<u>Soviet Union</u>

SERPUKHOV, UNK pp	?	3000 × 3000	10^{30}	Planned

<u>Other</u>

JAPAN, TRISTAN pp	?	100 × 100	10^{30}	Planned

(E) Colliding Beams (Electrons × Protons)

<u>Western Europe</u>

DESY	?	33(e^-) × 1000(p)	10^{30}	Planned

<u>Other</u>

JAPAN, TRISTAN	?	20(e^-) × 300(p)	10^{30}	Planned

ELECTROMAGNETIC INTERACTION OF

COLLIDING BEAMS IN STORAGE RINGS

J. C. HERRERA

BROOKHAVEN NATIONAL LABORATORY

I. INTRODUCTION

When the beams cross each other in the intersection region of a storage ring, two distinct effects occur. A particle of one beam may undergo a fundamental interaction with a particle of the other beam. The average number of such events is equal to the product of the machine luminosity and the particle-particle total cross section. It is, of course, these particle interactions which the experimental physicist studies. In contrast to what happens for these fundamental interactions, a particle of one beam passing through the interaction region will always be influenced by the electric and magnetic fields produced by the entire other beam. Clearly, this classical electromagnetic interaction between the two beams will perturb the motion of the particles circulating in the machine. It is our purpose in this paper to present the basic aspects of this beam-particle force. We shall indicate how the effect of this electrical force is quite different in the case of continuous (unbunched) proton beams which cross each other at an angle (as in the ISR at CERN and in ISABELLE at Brookhaven), as compared to the case of electron-positron bunches which collide head on (as in the SPEAR II Ring and in PETRA and PEP).

II. UNBUNCHED INTERSECTING BEAMS

In Fig. (1) we show a particle trajectory crossing a continuous beam at an angle α_o. We assume that the transverse (x,y) distribution of current (I) in the beam is Gaussian, that is,

$$\rho(x,y) = \frac{I}{2\pi \sigma_V \sigma_H} \exp\left[-\frac{x^2}{2\sigma_H^2} - \frac{y^2}{2\sigma_V^2}\right]. \quad (1)$$

It is clear from the figure that, during the passage of the particle from one side of the beam to the other side, it will sample a succession of gradient fields, with the result that its horizontal and vertical betatron tunes will be changed. For a central trajectory (y=0), these changes can be calculated by perturbation theory and are

$$\Delta \nu_V = \frac{\beta_V^* r_p I}{\sqrt{2\pi} \, ec \, \gamma \beta^2 \, \sigma_V^* \, \tan(\alpha_o/2)}, \quad (2)$$

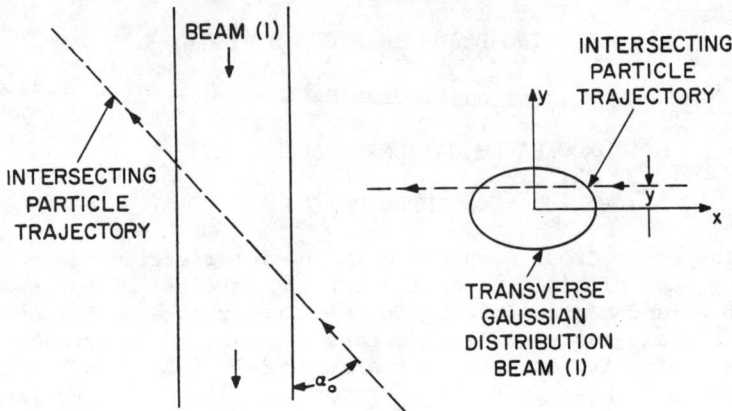

Fig. 1. Particle Crossing a Continuous Beam.

and

$$\Delta \nu_H = 0, \qquad (3)$$

where

- α_o = the horizontal total crossing angle (rad),
- σ_V^* = Gaussian rms (vertical) size at crossing (m),
- β_V^* = vertical amplitude function at crossing (m),
- r_p = classical radius of proton (1.535×10^{-18} m),
- (γ, β) = relativistic particle factors of colliding protons,
- I = average current per beam (A),
- $(\Delta \nu)_{V,H}$ = vertical or horizontal tune shift per crossing,
- e = charge of proton (1.602×10^{-19} C), and
- c = velocity of light (2.9979×10^8 ms^{-1}).

It is important to point out that in deriving these equations for the tune shift we have assumed that the cross-sectional size of the beam does not vary over the intersection region, and therefore that the betatron amplitude functions are constant.

A pertinent question to ask is: "What is the deflection of the particle trajectory as it crosses the continuous beam?". Because of the symmetrical way in which the particle approaches and leaves the center of the other beam, the net angular deflection in the horizontal plane is zero, similar to the tune shift, Eq. (3). Instead for the case of the vertical plane we have for the net change in slope,

$$\Delta\left(\frac{dy}{ds}\right) = \frac{2\pi r_p I}{ec\gamma\beta^2 \tan(\alpha_0/2)} \Phi(y\sqrt{2}\,\sigma_V^*) \quad (4)$$

where Φ is the error function defined as

$$\Phi(x) = \frac{2}{\sqrt{\pi}} \int_0^x dt\, e^{-t^2}. \quad (5)$$

In Fig. (2) we have sketched the variation of this angular deflection as a function of the displacement (y) of the particle path from the horizontal symmetry plane of the colliding beams.

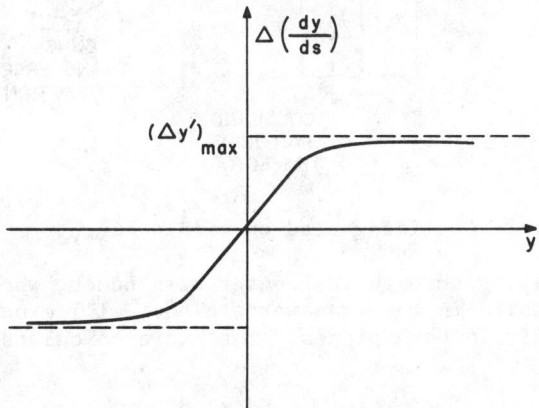

Fig. 2. Angular deflection for a particle crossing an unbunched beam at an angle.

We note that the magnitude of the maximum deflection tends toward

$$(\Delta y')_{max} = \frac{2\pi r_p I}{ec\gamma\beta^2 \tan(\alpha_0/2)} \quad (6)$$

for large vertical amplitudes of the incident particle. It is important to point out that since the linear term obtainable from Eq. (4) is proportional to the change in betatron tune expressed by Eq. (2), the magnitude of this value of $\Delta\nu$ has been conventionally designated as the beam-beam strength parameter. Thus, the change in tune per crossing, calculated from first order perturbation theory, serves as a measure of the strength of the nonlinear beam-particle interaction specified by the net angular deflection of an intersecting particle trajectory. This can best be seen by writing the net deflection in the form

$$\Delta\left(\frac{dy}{ds}\right) = (2\pi)^{3/2} |\Delta\nu_V| \frac{\sigma_V^*}{\beta_V^*} \Phi\left(\frac{y}{\sqrt{2}\sigma_V^*}\right). \quad (7)$$

III. HEAD-ON COLLISIONS OF BUNCHED BEAMS

For an electron traversing a bunch of positrons, such as occurs in e^+e^- storage rings, the situation is that depicted in Fig. (3).

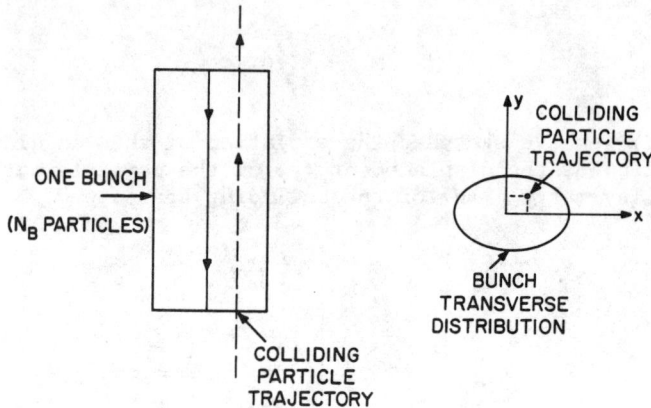

Fig. 3. Particle Colliding head-on with a bunch.

A particle passing through the center of a bunch, whose density of charge is Gaussian in the transverse plane, will experience a betatron tune shift in both planes. These are calculated to be

$$\Delta \nu_V = \frac{\beta_V^* r_e N_B}{2\pi \gamma \beta^2 \sigma_V^*(\sigma_H^* + \sigma_V^*)} \; , \tag{8}$$

and

$$\Delta \nu_H = \frac{\beta_H^* r_e N_B}{2\pi \gamma \beta^2 \sigma_H^*(\sigma_H^* + \sigma_V^*)} \; . \tag{9}$$

We have used a notation such that $\sigma_{H,V}^*$ = rms horizontal or vertical size at the collision point (m), $\beta_{H,V}^*$ = betatron amplitude function over the collision region (m), N_B = number of particles per bunch, r_e = classical radius of an electron (2.818×10^{-15} m), and $\Delta \nu_{V,H}$ = vertical or horizontal tune shift per collision.

For head-on collisions the angular deflection of the particle path is not confined to the vertical plane as is the case for continuous beams crossing at an angle. When the bunch has a symmetrical Gaussian distribution of charge in the transverse plane, the net deflections of a colliding particle in the two planes are

$$\Delta\left(\frac{dx}{ds}\right) = \frac{-2N_B r_e x}{\gamma \beta^2 r^2}(1 - e^{-r^2/2\sigma^2}), \tag{10}$$

and

$$\Delta\left(\frac{dy}{ds}\right) = \frac{-2N_B r_e}{\gamma \beta^2} \frac{y}{r^2}(1 - e^{-r^2/2\sigma^2}), \tag{11}$$

where x and y are the horizontal and vertical offsets of the particle trajectory and r is the associated radial distance from the center line of the bunch. In Fig. (4) we show the variation of this net angular deflection in one of the planes.

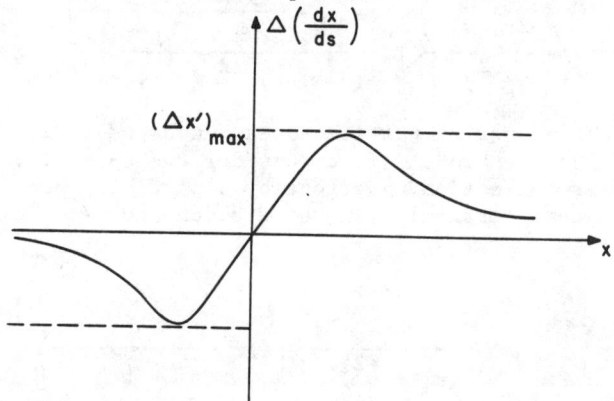

Fig. 4. Angular deflection for a particle colliding head-on with a bunched beam.

In contrast to the unbunched case, the deflection after rising to a maximum value of about

$$(\Delta x')_{max} \simeq \frac{0.9 \, N_B r_e}{\gamma \beta^2 \, \sigma} \tag{12}$$

decreases to zero for particles passing through the outer fringes of a bunch.

Though the strength parameters, Eqs. (8) and (9), are applicable to unsymmetrical Gaussian bunches, the net deflections, Eqs. (10) and (11), only apply to cylindrically symmetrical beams, with an rms half-width of σ. It is therefore of interest to obtain the expression for the angular deflections for a Gaussian distribution with arbitrary aspect ratio, particularly since e^+e^- rings operate with a ratio of horizontal to vertical size at the collision point of about 20 to 1. To do this we start with the potential[1] produced by a two dimensional Gaussian distribution of charge having a total charge per unit length equal to λ_L. Thus we have

$$V(x,y) = \frac{\lambda_L}{4\pi \epsilon_0} \int_0^\infty \frac{1 - \exp\left[-\frac{x^2}{a^2 + t} - \frac{y^2}{b^2 + t}\right]}{(a^2 + t)^{1/2}(b^2 + t)^{1/2}} \, dt. \tag{13}$$

Differentiating this expression to obtain the electric field and using the fact that the particles are moving in opposite directions with essentially the speed of light, we find that the net angular deflection in the y and x planes are

$$\Delta\left(\frac{dy}{ds}\right) = \frac{-2N_B \, r_e \, y}{\gamma \beta^2} \int_0^\infty \frac{\exp\left[-\frac{x^2}{a^2+t} - \frac{y^2}{b^2+t}\right]}{(a^2+t)^{1/2}(b^2+t)^{3/2}} \, dt, \quad (14)$$

and

$$\Delta\left(\frac{dy}{ds}\right) = \frac{-2N_B \, r_e \, x}{\gamma \beta^2} \int_0^\infty \frac{\exp\left[-\frac{x^2}{a^2+t} - \frac{y^2}{b^2+t}\right]}{(a^2+t)^{3/2}(b^2+t)^{1/2}} \, dt \quad (15)$$

Again, N_B is the number of particles per bunch and r_e is the classical radius of the electron. If we consider the special case of the vertical angular deflection undergone by a particle when it is incident in the central vertical plane of a bunch [Eq. (14) with x = 0], then we derive

$$\Delta\left(\frac{dy}{ds}\right)_{x=0} = \frac{-2\sqrt{\pi} N_B \, r_e}{\gamma \beta^2 (a^2-b^2)^{1/2}} \left[\Phi\left(\frac{a}{b} \frac{y}{(a^2-b^2)^{1/2}}\right)\right.$$

$$\left. - \Phi\left(\frac{y}{(a^2-b^2)^{1/2}}\right)\right] \exp\left(\frac{y^2}{a^2-b^2}\right), \quad (16)$$

with Φ equal to the error function. When, in addition, the horizontal width of the bunch is considerably greater than its height (a >> b), this result simplifies to

$$\Delta\left(\frac{dy}{ds}\right)_{x=0} = \frac{-2\sqrt{\pi} N_e \, r_e}{\gamma \beta^2 \, a} \left[\Phi\left(\frac{y}{b}\right) - \Phi\left(\frac{y}{a}\right)\right] \exp\left(\frac{y^2}{a^2}\right) \quad (17)$$

Under these conditions it is then possible to find the approximate value of the maximum deflection experienced by a particle passing through an unsymmetrical Gaussian bunch. Equation (17) yields for this value:

$$\Delta(y')_{max} \simeq \frac{\sqrt{2\pi} \, N_B \, r_e}{\gamma \beta^2 \, \sigma_H^*}, \quad (18)$$

into which we have inserted the rms half width at the collision point, $\sigma_H^* = a/\sqrt{2}$.

In general the integrals appearing in Eqs. (14) and (15) cannot be written in closed form. However, with the stipulation that $b/a = \rho < 1$, they can be written in the following more convenient way:

$$\Delta\left(\frac{dy}{ds}\right) = \frac{-4N_B r_e y}{\gamma\beta^2 a^2 (1-\rho^2)} \exp\left[-\frac{x^2 - y^2}{a^2(1-\rho^2)}\right] \times$$

$$\int_\rho^1 \frac{dw}{w^2} \exp\left[\frac{1}{a^2(1-\rho^2)}\left(x^2 w^2 - \frac{y^2}{w^2}\right)\right], \quad (19)$$

and

$$\Delta\left(\frac{dx}{ds}\right) = \frac{-4N_B r_e x}{\gamma\beta^2 a^2 (a-\rho^2)} \exp\left[-\frac{x^2 - y^2}{a^2(1-\rho^2)}\right] \times$$

$$\int_\rho^1 dw \exp\left[\frac{1}{a^2(1-\rho^2)}\left(x^2 w^2 - \frac{y^2}{w^2}\right)\right]. \quad (20)$$

Imposing the condition $\rho \ll 1$ for wide beams, we now obtain

$$\Delta\left(\frac{dy}{ds}\right) = -4\pi |\Delta\nu_V| \frac{y}{\beta_V^*} \frac{\sigma_V^*}{\sigma_H^*} \exp\left[-\frac{x^2 - y^2}{a^2}\right] \int_0^1 \frac{dw}{w^2} \exp\left[\frac{1}{a^2}\left(x^2 w^2 - \frac{y^2}{w^2}\right)\right], \quad (21)$$

and

$$\Delta\left(\frac{dx}{ds}\right) = -4\pi |\Delta\nu_H| \frac{x}{\beta_H^*} \exp\left[-\frac{x^2 - y^2}{a^2}\right] \int_0^1 dw \exp\left[\frac{1}{a^2}\left(x^2 w^2 - \frac{y^2}{w^2}\right)\right]. \quad (22)$$

IV. BEAM-BEAM PARAMETERS OF (pp) AND (e^+e^-) MACHINES

Table I is a presentation of the parameters determining the luminosity and the beam-beam strength for the CERN ISR and the Brookhaven pp rings, ISABELLE. An attempt has been made to give a set of consistent values for these parameters which can be calculated from the listed equations. It may be of interest to mention that, for ISABELLE at 30 GeV, the beam-beam strength parameter and the maximum angular deflection are significantly higher than those at the ISR.

In Table II, we give an analogous list of parameters for the SPEAR II ring as well as for the PETRA and PEP machines. It is evident that e^+e^- rings operate with a beam-beam strength parameter which is larger by more than an order of magnitude than that for pp unbunched machines. In addition, these strengths are approximately equal in magnitude in the two transverse planes. The tabulation also indicates that the maximum deflection undergone by a particle in traversing the other beam is three orders of magnitude greater for e^+e^- rings than the corresponding deflection in pp rings.

REFERENCES

1. B. W. Montague, CERN Rpt. 68-38 (1968).

Table IA. Beam characteristics of pp storage rings.

pp Machines		ISR CERN $(p,p) \equiv (1,2)$	ISABELLE 30 × 30 GeV [Standard Ins.]	ISABELLE 400 × 400 GeV [Standard Ins.]
Energy	$E_1 = E_2 = E$	26.78 GeV	29.4 GeV	400. GeV
Particle Gamma	$\gamma = E/M_p$	28.49	31.3	426.3
Circumference	$2\pi R$	942.48 m	3833.8 m	3833.8 m
Current	$I_1 = I_2 = I$	28.6 A	8 A	8 A
Crossing Angle	α	14.77°	11.88 mrad	11.88 mrad
Number of Intersections		8	6	6
Betatron Tunes	$\nu_H \simeq \nu_H$	8.9	22.6	22.6
Normalized Betatron Emittance (95% Phase Space)	$E_V = \dfrac{6\pi\sigma_V^2(\gamma\beta)}{\beta_V}$	$19.6\pi \times 10^{-6}$ radm	$15\pi \times 10^{-6}$ radm	$15\pi \times 10^{-6}$ radm
Beta Functions (Intersection)	β_H^*	22 m	41.24	41.24
	β_V^*	14.25 m	7.5 m	7.5 m
Revolution Frequency	$f_o = c/2\pi R$	3.181×10^5 Hz	78.197×10^3 Hz	78.197×10^3 Hz
Number of Particles	$N = I/ef_o$	5.61×10^{14}	6.38×10^{14}	6.38×10^{14}
Classical Radius of Proton	r_p	1.535×10^{-18} m		

Table IB.

p-p Machines		ISR CERN	ISABELLE 30 × 30 GeV	ISABELLE 400 × 400 GeV
rms Vertical Size (Intersection)	$\sigma_V^* = \sqrt{\dfrac{E_V \beta_V^*}{6\pi(\gamma\beta)}}$	1.278 mm	0.776 mm	0.21 mm
Effective Height	$h_e = 2\sqrt{\pi}\, \sigma_V^*$	4.53 mm	2.74 mm	0.743 mm
Angular Divergence (Intersection)	$y'^* = \sqrt{6}\, \dfrac{\sigma_V^*}{\beta_V^*}$	2.2×10^{-4} rad	2.5×10^{-4} rad	0.68×10^{-4} rad
Luminosity	$L = \dfrac{I_1 I_2}{2\sqrt{\pi}\, e^2 c \sigma_V^* \tan\frac{\alpha}{2}}$	1.8×10^{31} cm^{-2} sec^{-1}	5.4×10^{31} cm^{-2} sec^{-1}	2.0×10^{32} cm^{-2} sec^{-1}
Vertical Beam-Beam Strength Parameter Per intersection	$\Delta\nu_V = \dfrac{-\beta_V^* r_p I}{\sqrt{2\pi}\, ec\gamma\beta^2 \sigma_V^* \tan\frac{\alpha}{2}}$	-1.1×10^{-3}	-5.6×10^{-3}	-1.5×10^{-3}
Horizontal Beam-Beam Tune Shift	$\Delta\nu_H$	0.	0.	0.
Beam-Beam Maximum Vertical Deflection	$(\Delta y')_{max} = \dfrac{2\pi r_p I}{ec\gamma\beta^2 \tan\frac{\alpha}{2}}$	1.56×10^{-6} rad	9.2×10^{-6} rad	0.67×10^{-6} rad

Table IIA. Beam characteristics of e^+e^- storage rings.

e^+e^- Machines		SPEAR II (1,2) ≡ (e^-,e^+)	PEP (e^-,e^+)	PETRA (e^-,e^+)
Energy	$E_1 = E_2 = E$	3.71 GeV	15 GeV	15 GeV
Particle Gamma	$\gamma = E/m_e$	7260.2	29354	29354
Circumference	$2\pi R$	234.13 m	2200 m	2304 m
Magnetic Radius of Curvature	ρ	12.7 m	165.52 m	197.14 m
Number of Particles	$N_1 = N_2 = N$	1.75×10^{11}	2.52×10^{12}	3.84×10^{12}
Number of Bunches	n_B	1	3	4
Particles Per Bunch	$\frac{N}{n_B} = \frac{q}{e}$	1.75×10^{11}	8.4×10^{11}	9.6×10^{11}
Crossing Angle	α	0.	0.	0.
Betatron Tunes	ν_H	5.27	21.25	22.2
	ν_V	5.19	18.75	22.2
Beta Function at Intersection	β_H^*	1.197 m	2.8 m	3.0 m
	β_V^*	0.0998 m	0.11 m	0.15 m
ETA Function at Intersection	η^*	0.	0.49 m	
Revolution Frequency	$f_o = c/2\pi R$	1.28×10^6 Hz	1.36×10^5 Hz	1.3×10^5 Hz
Average Current	$I = e\left(\frac{q}{e}\right) n_B f_o$	35.9 mA	54.9 mA	80.0 mA
Classical Radius of Electron	r_e	2.818×10^{-15} m		

Table IIB.

e^+e^- Machines			SPEAR (e^-,e^+)	PEP (e^-,e^+)	PETRA (e^-,e^+)
Betatron Emittance	$\epsilon = \sigma^2/\beta$	ϵ_H	0.665×10^{-6} radm	0.138×10^{-6} radm	0.15×10^{-6} radm
		ϵ_V	0.902×10^{-8} radm	0.933×10^{-8} radm	0.75×10^{-8} radm
Coupling Coefficient	$K = \sqrt{\epsilon_V/\epsilon_H}$		0.12	0.26	0.22
Betatron rms Size (Intersection)	$\sigma^2 = \beta\epsilon$	$(\sigma_H^*)_\beta$	0.892 mm	0.662 mm	0.67 mm
		σ_V^*	0.030 mm	0.032 mm	0.034 mm
Energy Spread	$\frac{\sigma_\epsilon}{E} = \left[\frac{3.84 \times 10^{-13} \gamma^2}{2\rho}\right]^{1/2}$		0.89×10^{-3}	1.0×10^{-3}	0.91×10^{-3}
Energy rms Size (Intersection)	$\sigma_\epsilon^* = \eta^* \frac{\sigma_\epsilon}{E}$		0.	0.49 mm	
Total Horz. rms Size	$\sigma_H^2 = \sigma_\epsilon^{*2} + (\sigma_H^*)_\beta^2$		0.892 mm	0.79 mm	0.67 mm
(Energy Loss/Rest Mass) Per Turn Due To Syn. Radiation	$\left(\frac{\Delta W}{mc^2}\right)_{turn} = \frac{4\pi}{3} \frac{r_e}{\rho} \gamma^4 \beta^3$		2.582	52.94	44.45
Energy Loss (SR) Per Turn	ΔW		1.32 MeV	27.1 MeV	22.7 MeV
Total Power Loss Per Beam	$(\Delta W)I$		47.4 kW	1.5 MW	1.8 MW
Average Radiated Power Per Particle	$P_\gamma = (\Delta W) f_o$		1.69×10^{12} eV/sec	3.68×10^{12} eV/sec	2.95×10^{12} eV/sec
Transverse Damping Time	$\tau_T = 2E/P_\gamma$		4.4×10^{-3} sec	8.1×10^{-3} sec	10×10^{-3} sec

Table IIC.

e^+e^- Machines		SPEAR II	PEP	PETRA
Angular Divergence (Intersection)	$y'^* = \sqrt{6}\, \dfrac{\sigma_V^*}{\beta_V^*}$	7.6×10^{-4} rad	7.1×10^{-4} rad	5.5×10^{-4} rad
Luminosity Per Bunch Collision	$\mathcal{L}_{BC} = \dfrac{q_1 q_2}{e^2 4\pi \sigma_V^* \sigma_H^*}$	9.1×10^{24} cm^{-2}	2.2×10^{26} cm^{-2}	3.2×10^{26} cm^{-2}
Average Luminosity	$L = n_B f_o \mathcal{L}_{BC}$	1.16×10^{31} cm^{-2} sec^{-1}	9.1×10^{31} cm^{-2} sec^{-1}	1.7×10^{32} cm^{-2} sec^{-1}
Vertical Beam-Beam Strength Parameter	$\Delta\nu_V = \dfrac{\beta_V^* r_e (N/n_B)}{2\pi\gamma\beta^2 \sigma_V^* (\sigma_H^* + \sigma_V^*)}$	0.039	0.054	0.092
Horizontal Beam-Beam Strength Parameter	$\Delta\nu_H = \dfrac{\beta_H^* r_e (N/n_B)}{2\pi\gamma\beta^2 \sigma_H^* (\sigma_H^* + \sigma_V^*)}$	0.016	0.055	0.093
Approx. Max Particle Deflection Per Collision	$\sim \dfrac{\sqrt{2\pi}\, (N/n_B)\, r_e}{\sigma_H^* \gamma}$	2×10^{-4} rad	2.6×10^{-4} rad	3.5×10^{-4} rad

A Summary of Some Beam-Beam Models

A. W. Chao

Stanford Linear Accelerator Center

ABSTRACT

Two categories of theoretical models for the beam-beam interaction are reviewed: the linear-lens models and the single-resonance models. In a linear-lens model, the beam-beam force is linearized and represented by a localized linear lens. Analyses of incoherent single particle effects can be performed exactly in these models by using matrix techniques. Although the results do not agree with the experimental observations in many respects, the linear-lens models constitute a starting point of our understanding of the beam-beam interaction. In the single-resonance models, one is concerned with the possible incoherent instabilities as the betatron tune of some of the particles is close to a certain rational number. It is assumed in these models that one and only one such rational number dominates the single-particle beam-beam effects. It is found that static single resonances cannot explain many of the experimental results. Some attempts have been made to modify the static single-resonance theory by including some mechanisms for diffusive tune fluctuations or periodic tune modulations. These modified single-resonance models have met only with some limited qualitative success.

TABLE OF CONTENTS

1. Introduction
2. Linear-Lens Models
   2.1 Incoherent Betatron Effects
   2.2 Incoherent Synchrotron Effects
   2.3 Coherent Effects
3. Single-Resonance Models
   3.1 The Static Model
   3.2 The Dynamic Models

1. INTRODUCTION

The subject of the beam-beam interaction has been studied ever since the colliding beam devices were first considered.[1-4] Today, with many colliding beam storage rings in operation, this subject is not yet fully understood. This is evidenced by the fact that many of the existing storage rings have not reached the beam-beam limited design luminosities. The main difficulty is, in fact, a familar one, i.e., to solve a system with a "nonintegrable" Hamiltonian. Similar difficulty has appeared, or eventually will appear, in other branches of physics.

Figure 1 shows the number of papers (that I found in the SLAC library) the storage ring physicists have written on the subject of the beam-beam interaction versus the calendar years. One notices a peak around 1973-1974 and a drop after 1975. This might explain why we urgently need to have a symposium on the subject of beam-beam interaction, since writing review articles (like this one) prevents the curve in Fig. 1 from dropping to zero, at least for 1978-79.

The references considered in Fig. 1 contain experimental studies, numerical simulations, theoretical models and review articles. The theoretical models can roughly be divided into three catagories, the linear-lens models, the single-resonance models and the multiple-resonance models. In the following, we will review some of the linear-lens and the single-resonance models. Other topics are covered by other speakers of this symposium. For simplicity of discussion, we will mainly concern ourselves with collisions of bunched beams.

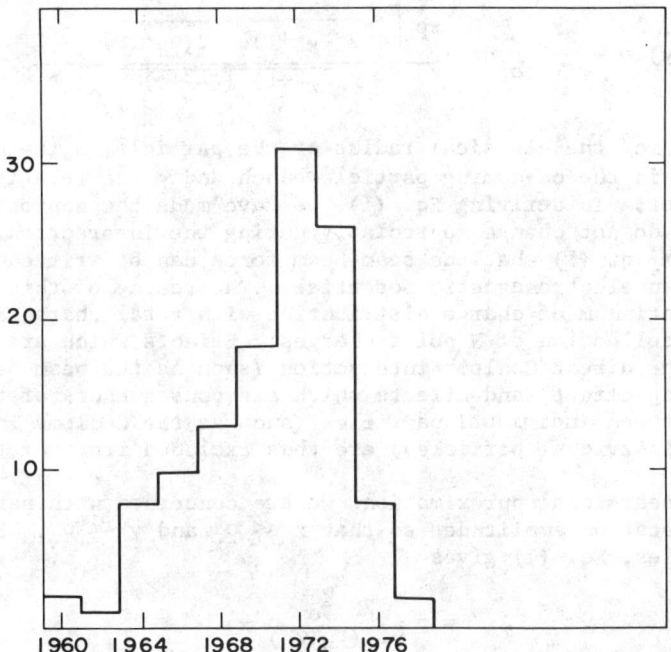

Fig. 1. Number of papers written by storage ring physicists on the subject of beam-beam interaction versus the calendar years.

2. LINEAR-LENS MODELS

When two bunched beams collide head-on, a particle in one beam passes through the electromagnetic fields produced by particles in the other beam and receives a transverse impulse. This transverse impulse can be decomposed into a horizontal x-component and a vertical y-component to yield two kicking angles $\Delta x'$ and $\Delta y'$. If the bunch length is short compared with the horizontal and vertical betatron wavelengths at the interaction point, the transverse particle distribution in the x-y plane does not change appreciably during the interaction and the impulse can be considered to be a localized delta function perturbation. For a given transverse particle distribution, the kicking angles $\Delta x'$ and $\Delta y'$ depend on the horizontal displacement x and the vertical displacement y of the particle as it passes through the interaction region. For an upright biGaussian distribution with beam dimensions σ_x and σ_y, for example, we have[5]

$$\Delta x' = - \frac{\partial U(x,y)}{\partial x} \text{ and } \Delta y' = - \frac{\partial U(x,y)}{\partial y}, \tag{1}$$

where $U(x,y)$ is the electromagnetic potential well found to be

$$U(x,y) = - \frac{Nr_o}{\gamma} \int_0^\infty dt \frac{\exp\left[-\frac{x^2}{2(\sigma_x^2+t)} - \frac{y^2}{2(\sigma_y^2+t)}\right] - 1}{\sqrt{(\sigma_x^2+t)(\sigma_y^2+t)}} \tag{2}$$

with $r_o = e^2/mc^2$ the classical radius of the particle, N the number of particles in the on-coming particle bunch and γ the relativistic Lorentz factor. In deriving Eq. (1), we have made the approximation that x and y do not change appreciably during the interaction. It is clear from Eq. (1) that the beam-beam force can be written as the gradient of an electromagnetic potential. The source of this potential is a continuum of charge distribution with total charge Ne rather than a collection of N point charges. Effects which are not results of the direct Coulomb interaction (such as the beam-beam Bremsstrahlung effect) and effects which are consequences of the interaction between individual particles (such as the Coulomb interaction of two individual particles) are thus excluded from our treatment.

In a linear-lens approximation, we are concerned with particles with small betatron amplitudes so that $x \ll \sigma_x$ and $y \ll \sigma_y$. For those particles, Eq. (1) gives

$$\Delta x' = \pm \frac{2N\, r_o}{\gamma \sigma_x (\sigma_x + \sigma_y)} x$$

$$\Delta y' = \pm \frac{2N\, r_o}{\gamma \sigma_y (\sigma_x + \sigma_y)} y, \tag{3}$$

where the - signs are used if the electric charges of the two beams have opposite signs and the + signs are used otherwise. In the following we will assume - signs. As seen from Eq. (3), the motion becomes linear and decoupled. In the linear-lens model, it is therefore possible to consider only one degree of freedom, which we choose to be the vertical dimension.

2.1 Incoherent Betatron Effects

The motion of small amplitude particles is conveniently analyzed by using matrix techniques. The transformation matrix for the vector

$$\begin{bmatrix} y \\ y' \end{bmatrix}$$

across the interaction region is, from Eq. (3),

$$\begin{bmatrix} 1 & 0 \\ -1/f & 1 \end{bmatrix}, \quad \frac{1}{f} = \frac{2N\, r_o}{\gamma \sigma_y (\sigma_x + \sigma_y)} \tag{4}$$

which, we recognize, is the same as the transfer matrix of a thin-lens quadrupole magnet with focal length f. If we let there be one interaction region per superperiod of the storage ring, ν be the unperturbed betatron tune per superperiod and β_o^* be the unperturbed beta-function at the interaction point, the transformation from the middle of one interaction region to the middle of a neighboring interaction region can be written as[6]

$$\begin{bmatrix} \cos 2\pi(\nu+\Delta\nu) & \beta^* \sin 2\pi(\nu+\Delta\nu) \\ -\frac{1}{\beta^*} \sin 2\pi(\nu+\Delta\nu) & \cos 2\pi(\nu+\Delta\nu) \end{bmatrix}$$

$$= \begin{bmatrix} 1 & 0 \\ -\frac{1}{2f} & 1 \end{bmatrix} \begin{bmatrix} \cos 2\pi\nu & \beta_o^* \sin 2\pi\nu \\ -\frac{1}{\beta_o^*} \sin 2\pi\nu & \cos 2\pi\nu \end{bmatrix} \begin{bmatrix} 1 & 0 \\ -\frac{1}{2f} & 1 \end{bmatrix} \tag{5}$$

where $\Delta\nu$ is the tune shift caused by the beam-beam perturbation, β^* is the perturbed beta-function at the middle of the interaction region. In the thin-lens model, the transverse beam size scales with the square root of the perturbed beta-function. Solving Eq. (5) yields

$$\cos 2\pi(\nu+\Delta\nu) = \cos 2\pi\nu - \frac{\beta_o^*}{2f} \sin 2\pi\nu$$

$$\beta^*/\beta_o^* = \sin 2\pi\nu / \sin 2\pi(\nu+\Delta\nu). \tag{6}$$

It had been hoped, and seriously studied experimentally and theoretically, that one or two dimensionless scaling parameters might describe the beam-beam interaction completely or almost completely. In the linear-lens model, it is clear from Eq. (6) that two such parameters are the unperturbed betatron tune ν and a beam-beam strength parameter defined as[1-4]

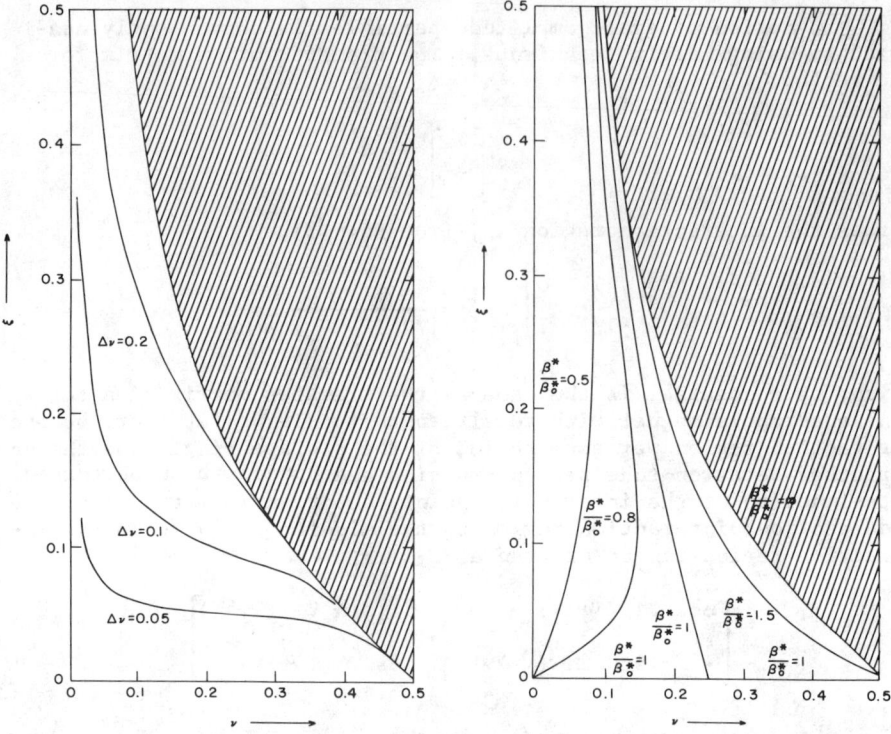

Fig. 2. Contours of constant tune shift $\Delta\nu$ in the plane of the two scaling variables (ν,ξ).

Fig. 3. Contours of constant β^*/β_o^* in the (ν,ξ) plane.

spread, an unmistakable sign of nonlinearities. On the other hand, the linear-lens model does describe physically the motion of small amplitude particles. It also has the advantage of being easy to analyze and can serve as a starting point of further adventures.

2.2 Incoherent Synchrotron Effects

Although the linearized beam-beam perturbation does not couple the x- and y-motion as is indicated by Eq. (3), it can still couple the x- or y-motion to the longitudinal synchrotron degree of freedom under some conditions.[7,8] In this section, we will show this by an example, using the linear-lens approximation and matrix techniques.

Consider two beams, one strong and one weak, crossing each other vertically at an angle 2α. We can look for the betatron and synchrotron tune shifts as well as changes in particle distribution

$$\xi = \frac{\beta_o^*}{4\pi f} \tag{7}$$

A negative sign must be added to Eq. (7), yielding $\xi < 0$, if the electric charges of the two beams are of the same sign. The factor 4π is included so that, when ξ is small, the beam-beam tune shift $\Delta \nu$ is equal to ξ. Note that it is the unperturbed β_o^* that goes into the definition of ξ. The factor β_o^* can be thought of as a magnification factor which, when applied to a perturbation, gives the effectiveness of the perturbation on the particle's betatron motion. To reduce the effect of the beam-beam interaction, it is advantageous to reduce the value of β_o^*. This leads to the idea of low-β^* insertion that has been successfully used to increase the luminosity of many storage rings.

Using the two parameters ν and ξ as coordinates, we have plotted in Figs. 2 and 3 countours of constant $\Delta \nu$ and β^*/β_o^*. The value of ν is restricted in 0 and 1/2 since it is periodic with period 1/2. The shaded area indicates a region with no stable solution. The width, $\delta \nu$, of this unstable region for a given ξ is given by

$$\delta \nu = \frac{1}{\pi} \tan^{-1} 2\pi \xi \tag{8}$$

for small ξ, $\delta \nu = 2\xi$. From Fig. 2, we notice that $\Delta \nu$ is approximately equal to ξ provided that ξ is small and ν is not too close to 0 or $\frac{1}{2}$.

For a given storage ring configuration, one would like to know, at least qualitatively, the behavior of luminosity $\mathcal{L}$ as a function of the number of particles N in each beam. To do this, one first notices that as N is changed, the values of ξ and β^* change according to Eq. (6) and the expression $\xi \propto N/\beta^*$ in a self-consistent manner, and that it is this self-consistent solution of β^* that must be substituted in the expression $\mathcal{L} \propto N^2/\beta^*$ to find the dependence of $\mathcal{L}$ on the number of particles N. This is done using a computer and the results are illustrated in Fig. 4 on a full logarithmic graph paper. The units are arbitrary. The dotted line represents the value of $\mathcal{L}$ without taking into account the changes in β^*. For large ν (i.e., close to 1/2), $\mathcal{L}$ is consistently below the dotted line, while for small ν, $\mathcal{L}$ can be above the dotted line; but in both cases, $\mathcal{L}$ tends to level off as N is increased indefinitely. To optimize the luminosity, this result suggests that the betatron tune per superperiod be slightly above (below if $\xi < 0$) a half-integer or integer.

Many predictions of the linear-lens model do not agree with the experimental observations. For example, we seen to observe that the beam-beam perturbation does couple the x- and y-motions; the transverse beam sizes do not seem to scale with the square root of the perturbed beta-functions; the beam-beam instability seems to appear with a beam intensity well below the stability limit indicated in Figs. 2 and 3; and the beam has been observed to have a finite tune

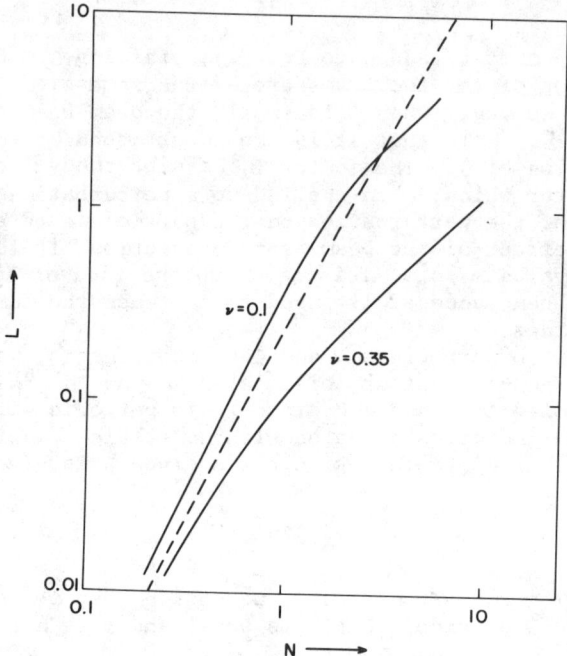

Fig. 4. Expected behavior of luminosity $\mathcal{L}$ versus the number of particles N for two values of the unperturbed tune ν.

of the weak beam under the influence of the collisions with the strong beam. To do this, we consider a particle in the weak beam with no horizontal displacement, and with vertical displacement y and longitudinal displacement z relative to the beam center. The impulse that this particle receives from each crossing is given by, in the relativistic limit,

$$\frac{\Delta \vec{p}}{p_o} = - \frac{G \hat{n}}{\cos \alpha} \int_0^{y \cos \alpha - z \sin \alpha} dy \exp\left(- \frac{t^2}{2\Sigma^2}\right) \tag{9}$$

which, after linearization, becomes[7]

$$\frac{\Delta \vec{p}}{p_o} = - G \hat{n} (y - z \tan \alpha) \tag{10}$$

where

$$G = \frac{2 N r_o}{\gamma \sigma_x \Sigma}$$

$$\Sigma^2 = \sigma_z^2 \sin^2 \alpha + \sigma_y^2 \cos^2 \alpha$$

and $\hat{n}$ is the unit vector shown in Fig. 5; σ_x, σ_y and σ_z are the rms beam width, beam height and bunch length of the strong beam. This impulse, which contains a vertical component and a longitudinal component, provides a coupling mechanism between the y- and z-motions. The longitudinal component of $\Delta \vec{p}$ causes the energy of the particle to change by an amount

$$\Delta \delta = \frac{\Delta p_z}{p_o} = G \sin \alpha \ (y - z \tan \alpha), \tag{11}$$

δ = relative energy error $\Delta E/E$,

and the vertical component gives a kick to the particle

$$\Delta y' = \frac{\Delta p_y}{p_o} = - G \cos \alpha (y - z \tan \alpha). \tag{12}$$

If we define a vector

$$\begin{bmatrix} y \\ y' \\ z \\ \delta \end{bmatrix}$$

we obtain from Eqs. (11) and (12) that the transfer matrix of this vector for the beam-beam crossing can be written as

$$T_1 = \begin{bmatrix} 1 & 0 & 0 & 0 \\ -G \cos \alpha & 1 & G \sin \alpha & 0 \\ 0 & 0 & 1 & 0 \\ G \sin \alpha & 0 & -G \tan \alpha \sin \alpha & 1 \end{bmatrix} \tag{13}$$

The transfer matrix between two neighboring interaction points, assuming no y-z coupling aside from the beam-beam perturbation, is

$$T_2 = \begin{bmatrix} \cos 2\pi \nu_y & \beta_o^* \sin 2\pi \nu_y & 0 & 0 \\ -\frac{1}{\beta_o^*} \sin 2\pi \nu_y & \cos 2\pi \nu_y & 0 & 0 \\ 0 & 0 & a & b \\ 0 & 0 & c & d \end{bmatrix} \tag{14}$$

where $ad - bc = 1$, $a + d = 2 \cos 2\pi \nu_s$ and $b = -\eta L$. The momentum compaction factor η, the total path length L and the tunes $\nu_{y,s}$ refer to the values per superperiod. The total transformation per superperiod is given by $T_o = T_2 T_1$.

The eigenvalues, λ, of T_o satisfy

$$\det (T_o - \lambda) = 0$$

or

$$\left(\frac{\lambda^2+1}{2\lambda} - \cos 2\pi\nu_y + \frac{\beta_o^* G}{2} \sin 2\pi\nu_y\right)\left(\frac{\lambda^2+1}{2\lambda} - \cos 2\pi\nu_s - \frac{\alpha^2}{2} G\eta L\right) = \quad (15)$$

$$-\frac{1}{4} \alpha^2 \beta_o^* G^2 \eta L \sin 2\pi\nu_y.$$

The four solutions of Eq. (15) for λ are related to the perturbed betatron and synchrotron tunes by $\lambda = \exp(\pm i2\pi\nu)$. For stability, the perturbed tunes must be real, or equivalently, all four λ's must satisfy $|\lambda| = 1$. It can be shown after some algebra that this stability condition is satisfied outside a stopband around the synchro-betatron resonances $\nu_y - \nu_s$ = integer, with a stopband width[8]

$$\delta\nu \approx \frac{1}{\pi}\left(\frac{\beta_o^* G^2 \eta L \alpha^2}{\sin 2\pi\nu_y}\right)^{\frac{1}{2}} \quad (16)$$

The fact that particle motion is stable around the sum resonances and unstable around the difference resonances, contrary to the case of coupling between the transverse dimensions, is due to the negative-mass effect of the longitudinal dimension. If we set $\beta_o^* G = 0.6$, $\alpha = 1$ mrad, $\nu_y = 0.1$, $\eta L = 2$ m and $\beta_o^* = 0.2$ m, the stopband width is found to be 8×10^{-4}.

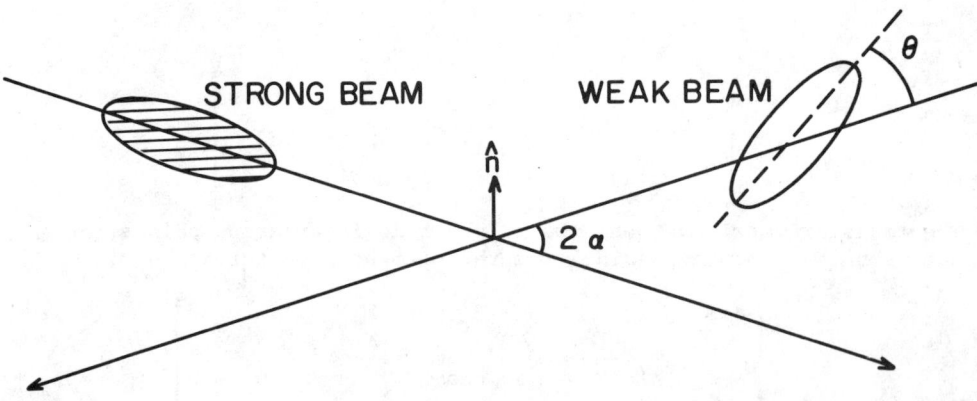

Fig. 5. Collision of a strong and a weak beam at an angle. The weak beam is tilted relative to its direction of motion.

It is also possible to find from the transfer matrix T_o the changes in particle distribution. The matrix T_o, which describes the motion of a single particle in the weak beam, has two eigenmodes. The eigenmode axes lie in the y-z plane but are tilted relative to the y-z axes. The weak beam is oriented in such a way that the principle axes of the beam distribution coincide with these eigenmode axes. Knowing T_0, we can thus obtain the tilting angle θ (see Fig. 5) by looking for the eigenmode axes of T_0. The result, assuming $|\alpha| \ll 1$ and $|\beta_0^* G| \ll 1$, is

$$\theta \approx \frac{\beta_0^* G \alpha \sin 2\pi \nu_y}{2(\cos 2\pi \nu_y - \cos 2\pi \nu_s)} \qquad (17)$$

To see the effect of this beam tilt on luminosity, we note that in order to avoid any appreciable loss of luminosity, one must have

$$\theta \lesssim \sigma_y/\sigma_z$$

From Eq. (17) we find that this condition is fulfilled if we stay enough distance away from the synchro-betatron resonances, i.e.,

$$|\nu_y \pm \nu_s - K| \geq \delta\nu$$

where K is some integer and $\delta\nu$ is given by

$$\delta\nu = \frac{\sigma_z}{\sigma_y} \frac{\beta_0^* G \alpha}{2\pi} \qquad (18)$$

If $\sigma_z/\sigma_y = 600$ $\beta_0^* G = 0.6$ and $\alpha = 1$ mrad, we find $\delta\nu = 0.06$. Comparing this value with the stopband width found following Eq. (16), we find that the luminosity may suffer from the beam-beam synchro-betatron coupling even well outside of the unstable stopband region.

It should be mentioned that another possible example of beam-beam longitudinal effects is obtained by considering a head-on collision of two beams at a location with finite energy dispersion.[8] The x- and z-motions are coupled in this case. The linear-lens model and the matrix techniques can be used here as well.

Still another synchro-betatron coupling effect caused by the beam-beam perturbation occurs when the bunch length σ_z is not small compared with the beta-function β_0^* at the interaction point. Particles with different longitudinal positions, z, within a bunch arrive at the interaction point at different times, with a time spread of σ_z/c. If σ_z is not small compared with β_0^*, the transverse particle distribution will change appreciably in the time duration of σ_z/c. The transverse beam-beam kick received by a particle then depends on the longitudinal position of the particle, causing a synchro-betatron coupling effect. This effect, whose leading term is quadratic in z, cannot be treated by the linear-lens model.

2.3 Coherent Effects

So far we have been considering the incoherent motion of a single particle under the influence of the beam-beam collisions. It turns out that the beam-beam interaction can also excite coherent bunch oscillations, in which each bunch behaves like a rigid distribution of particles, while the center-of-mass of the bunches oscillate.[9] In the linear-lens approximation, coherent effects are again analyzed by using matrix methods.

Let us consider a storage ring with two counter-rotating colliding bunches of equal intensity, one bunch in each beam. There are two interaction points and two superperiods. The kicking angles experienced by the two rigid bunches passing through each other are

$$\Delta y'_+ = -\frac{1}{f}(y_+ - y_-) = -\Delta y'_- \tag{19}$$

where $y_{+,-}$ are the vertical displacements of the bunch centers relative to the unperturbed trajectory and the focal length f has been defined in Eq. (4). If we describe the coherent motion by the vector

$$\begin{bmatrix} y_+ \\ y'_+ \\ y_- \\ y'_- \end{bmatrix},$$

the transfer matrix for the beam-beam collision is

$$T_1 = \begin{bmatrix} 1 & 0 & 0 & 0 \\ -\frac{1}{f} & 1 & \frac{1}{f} & 0 \\ 0 & 0 & 1 & 0 \\ \frac{1}{f} & 0 & -\frac{1}{f} & 1 \end{bmatrix} \tag{20}$$

The total transformation per superperiod is $T_o = T_2 T_1$, where T_2 is the transfer matrix between two neighboring collision points:

$$T_2 = \begin{bmatrix} \cos 2\pi\nu & \beta_o^* \sin 2\pi\nu & 0 & 0 \\ -\frac{1}{\beta_o^*} \sin 2\pi\nu & \cos 2\pi\nu & 0 & 0 \\ 0 & 0 & \cos 2\pi\nu & \beta_o^* \sin 2\pi\nu \\ 0 & 0 & -\frac{1}{\beta_o^*} \sin 2\pi\nu & \cos 2\pi\nu \end{bmatrix} \tag{21}$$

The eigenvalues of T_o, determined from $\det(T_o - \lambda) = 0$, are found to be two complex conjugate pairs

$$e^{\pm 2\pi i \nu} \quad \text{and} \quad e^{\pm 2\pi i (\nu + \Delta\nu)}$$

with

$$\cos 2\pi(\nu + \Delta\nu) = \cos 2\pi\nu - 4\pi\xi \sin 2\pi\nu \tag{22}$$

The first pair corresponds to the "0-mode" in which the two bunches move up and down together in phase at the interaction points. The bunches do not feel the beam-beam forces and the mode frequency is equal to the unperturbed betatron frequency. This mode is always stable. The second pair of eigenvalues corresponds to the "π-mode" in which the two bunches move out of phase. The beam-beam force seen by a bunch center with displacement y is the same as that for a single particle with displacement 2y in the incoherent motion. Equation (22) is thus the same as Eq. (6) with $\xi \rightarrow 2\xi$. In order for the π-mode to be stable, the absolute value of the right hand side of Eq. (22) must be less than 1. For a given ξ, the value of ξ must be less than a certain stability limit ξ_{limit}. Figure 6 shows the behavior of ξ_{limit} as a function of ν. It repeats with a period of $\nu = 1/2$. The stability limit for the single particle motion is twice of ξ_{limit}. For a given ξ, the coherent motion is unstable if ν is within a distance $(1/\pi)\tan^{-1}(4\pi\xi) \approx 4\xi$ below (above if $\xi < 0$) a half-integer or integer.

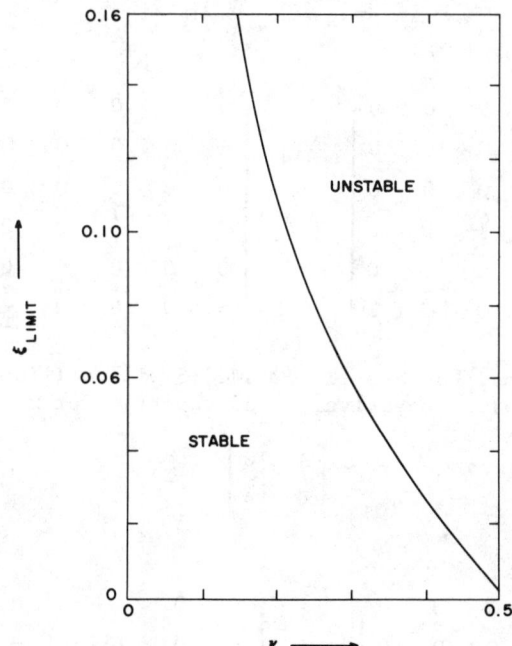

Fig. 6. Region of stability against dipole mode coherent oscillations in the (ν,ξ) plane for two bunches.

It should be pointed out that operating colliding beams inside the stopband does not necessarily mean losing the bunch because as soon as the oscillation amplitude grows beyond a sizeable fraction of the beam size, the beam-beam force becomes much weaker than the linear-lens model predicts. However, even a stable π-mode oscillation can be harmful to the luminosity since the beams simply miss each other most of the time.

We now consider another example with six colliding bunches, three bunches in each beam. The storage ring has six interaction points. The analysis of this example follows closely that of the previous example of two bunches; but the vector, shown below in a horizontal row, is now 12-dimensional:

$$[y_{+1}\ y'_{+1}\ y_{+2}\ y'_{+2}\ y_{+3}\ y'_{+3}\ y_{-1}\ y'_{-1}\ y_{-2}\ y'_{-2}\ y_{-3}\ y'_{-3}],$$

where the first six coordinates refer to the + beam, the other six refer to the - beam. Indices 1, 2, 3 refer to different bunches within one beam. The transfer matrix for bunch +1 to collide with bunch -1, and bunch +2 to collide with bunch -2, and bunch +3 to collide with bunch -3 is a 12 × 12 matrix

$$T_1 = \begin{bmatrix} A & B \\ B & A \end{bmatrix}$$

where

$$A = \begin{bmatrix} 1 & 0 & 0 & 0 & 0 & 0 \\ -\frac{1}{f} & 1 & 0 & 0 & 0 & 0 \\ 0 & 0 & 1 & 0 & 0 & 0 \\ 0 & 0 & -\frac{1}{f} & 1 & 0 & 0 \\ 0 & 0 & 0 & 0 & 1 & 0 \\ 0 & 0 & 0 & 0 & -\frac{1}{f} & 1 \end{bmatrix} \qquad B = \begin{bmatrix} 0 & 0 & 0 & 0 & 0 & 0 \\ \frac{1}{f} & 0 & 0 & 0 & 0 & 0 \\ 0 & 0 & 0 & 0 & 0 & 0 \\ 0 & 0 & \frac{1}{f} & 0 & 0 & 0 \\ 0 & 0 & 0 & 0 & 0 & 0 \\ 0 & 0 & 0 & 0 & \frac{1}{f} & 0 \end{bmatrix} \quad (23)$$

At the next collision, bunches +1, +2 and +3 will collide with bunches -2, -3 and -1, respectively, and the transfer matrix is

$$T_2 = \begin{bmatrix} A & C \\ D & A \end{bmatrix}$$

where

$$C = \begin{bmatrix} 0 & 0 & 0 & 0 & 0 & 0 \\ 0 & 0 & \frac{1}{f} & 0 & 0 & 0 \\ 0 & 0 & 0 & 0 & 0 & 0 \\ 0 & 0 & 0 & 0 & \frac{1}{f} & 0 \\ 0 & 0 & 0 & 0 & 0 & 0 \\ \frac{1}{f} & 0 & 0 & 0 & 0 & 0 \end{bmatrix} \qquad D = \begin{bmatrix} 0 & 0 & 0 & 0 & 0 & 0 \\ 0 & 0 & 0 & 0 & \frac{1}{f} & 0 \\ 0 & 0 & 0 & 0 & 0 & 0 \\ \frac{1}{f} & 0 & 0 & 0 & 0 & 0 \\ 0 & 0 & 0 & 0 & 0 & 0 \\ 0 & 0 & \frac{1}{f} & 0 & 0 & 0 \end{bmatrix} \quad (24)$$

The next collision is +1, +2 and +3 against -3, -1 and -2, with the transfer matrix

$$T_3 = \begin{bmatrix} A & D \\ C & A \end{bmatrix} \quad (25)$$

In between collisions, the transfer matrix is

$$R = \begin{bmatrix} r & 0 & 0 & 0 & 0 & 0 \\ 0 & r & 0 & 0 & 0 & 0 \\ 0 & 0 & r & 0 & 0 & 0 \\ 0 & 0 & 0 & r & 0 & 0 \\ 0 & 0 & 0 & 0 & r & 0 \\ 0 & 0 & 0 & 0 & 0 & r \end{bmatrix} ; \quad r = \begin{bmatrix} \cos 2\pi\nu & \beta_o^* \sin 2\pi\nu \\ -\frac{1}{\beta_o^*} \sin 2\pi\nu & \cos 2\pi\nu \end{bmatrix} \quad (26)$$

where $2\pi\nu$ is the betatron phase advance between two neighboring interaction points. The total transformation for one superperiod (3 collisions, half a revolution) is

$$T_o = RT_3 \, RT_2 \, RT_1 \quad (27)$$

The six eigenmodes of T_o describe the six coherent beam-beam modes. The coherent motion is stable if and only if all eigenvalues of T_o have absolute values of unity. In Fig. 7, the stability limit of the beam-beam parameter, ξ_{limit}, is plotted vs ν (which is 1/6 of the unperturbed betatron tune). For a given $\xi \ll 1$, the coherent motion is unstable if ξ is within a distance 4ξ below (above if $\xi < 0$) 1/2, or within a distance 2ξ below 1/3, or within a distance 2ξ below 1/6. In a range of ν of 1/2, this amounts to a total stopband width of 8ξ. It is clear that the unstable region occupies most of the available ν-space if ξ is something like 0.06.

3. SINGLE-RESONANCE MODELS

In a linear-lens model, the highly nonlinear beam-beam force is linearized. The particle motion can be analyzed exactly by using matrix methods. We found that the single particle motion is completely described by two parameters: the betatron tune per superperiod ν and the strength parameter ξ. We also found that if ν is sufficiently close to n/2 with n an integer, the particle motion becomes unstable.

It turns out[10-15] that the particle motion is affected by the beam-beam perturbation when ν is close to a rational number n/p, and the linear-lens model has only treated the case for p = 2. Unfortunately, it is not possible to perform an exact analysis for $p \neq 2$,

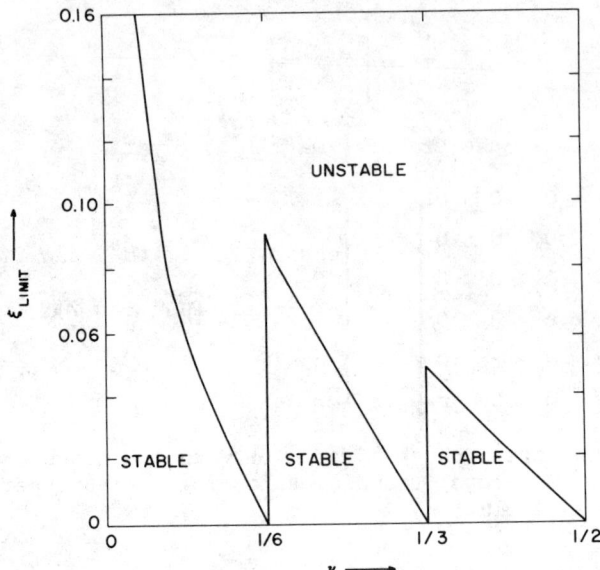

Fig. 7. Region of stability against dipole mode coherent oscillations in the (ν, ξ) plane for six bunches.

since nonlinear terms in the beam-beam force must be included. To proceed, we make two drastic assumptions:
- there is one and only one rational number n/p which dominates the single particle motion; and (28)
- the "smooth approximation" (see later) holds.

These assumptions are drastic because, first of all, there must be infinitely more rational numbers which are closer to ν than n/p as long as ν is not exactly equal to n/p. One then argues that only smaller values of p, say p < 10, are worth considering because resonances of order higher than ~ 10 do not affect the particle motion too much. This will be partially justified later. Secondly, the "smoothing" procedure has excluded some physical phenomena from being studied, one example of which is the existence of "stochastic layers" in the phase space.[16]

In the following, we will insist on the assumptions (28). A static single-resonance model, for which parameters such as ν and ξ are static in time, is first discussed. The method of analysis is well-defined in this model and will be included in some detail. Unfortunately, according to this model, the particle motion is always stable against the beam-beam perturbation, in disagreement with the experimentally observed decrease of beam lifetime when beams collide.

To explain this discrepancy, we seem to have two alternatives. One alternative is to keep the two assumptions (28) but modify the static model to include time-dependent parameters. This alternative is discussed in Section 3.2, but only very briefly for the reason that the conclusions of some of these models seem more convincing that the analyses which lead to these conclusions. Another alternative is to relax the assumptions (28) by considering multiple resonances or by numerical studies of the stochastic layers. This alternative will not be discussed here since it is better covered by other speakers of this symposium.

3.1 The Static Model

The motion of a particle under the influence of the beam-beam perturbation is described by the Hamiltonian

$$H = \frac{1}{2}(p_x^2 + K_x x^2) + \frac{1}{2}(p_y^2 + K_y y^2) + U(x,y)\delta(s), \qquad (29)$$

where $U(x,y)$ is defined in Eq. (2), and the aximuthal coordinate s has been defined so that the beam-beam collision occurs at $s = 0$. The equations of motion are

$$d^2z/d^2s + K_z(s)z = -\partial U/\partial z\, \delta(s), \quad z = x,y. \qquad (30)$$

Before we apply the single-resonance approximation, we need to make three successive canonical transformations on this Hamiltonian: a Courant-Snyder transformation,[6] an action-angle transformation and a slow-variable transformation.[10] It is not necessary to carry out each transformation individually. The three transformations can be combined into a single one whose generating function is[17]

$$G(x,y,\psi_x,\psi_y) = -\frac{x^2}{2\beta_x(s)}\left[\tan\phi_x - \frac{\beta_x'(s)}{2}\right] - \frac{y^2}{2\beta_y(s)}\left[\tan\phi_y - \frac{\beta_y'(s)}{2}\right]$$

with (31)

$$\phi_z = \psi_z + \nu_{zo}\frac{s}{R} + \int_0^s ds'\left(\frac{1}{\beta_z(s')} - \frac{\nu_z}{R}\right), \quad z = x,y$$

where $\beta_{x,y}$ are the beta-functions, $2\pi R$ is the circumference of the storage ring, $\phi_{x,y}$ are the betatron phases, ν_{xo} and ν_{yo} are the resonant tune values satisfying $\nu_x \approx \nu_{xo}$, $\nu_y \approx \nu_{yo}$ and $q\nu_{xo} + p\nu_{yo} = n$ (q, p and n are integers). Since $\phi_{x,y} \approx \nu_{x,yo}s/R$, the variables $\psi_{x,y}$ changes slowly in time. The canonical variables before transformation are x, p_x, y, p_y. After transformation the canonical variables are ψ_x, J_x, ψ_y, J_y. The two sets of coordinates are related by

$$z = \sqrt{2J_z\beta_z}\cos\phi_z$$

$$p_z = -\sqrt{\frac{2J_z}{\beta_z}}\left[\sin\phi_z - \frac{\beta_z'}{2}\cos\phi_z\right], \quad z = x,y. \qquad (32)$$

The Hamiltonian after the canonical transformation is

$$K = (\nu_x - \nu_{xo}) J_x + (\nu_y - \nu_{yo}) J_y + K_1$$

with (33)

$$K_1 = \delta(\theta) U\left[\sqrt{2J_x \beta_x} \cos(\psi_x + \nu_{xo}\theta), \sqrt{2J_y \beta_y} \cos(\psi_y + \nu_{yo}\theta)\right]$$

We have changed the time variable from s to $\theta = s/R$. The value of $\beta_{x,y}$ in K_1 are taken at the collision point.

So far no approximation has been made. We now make the two assumptions mentioned previously, i.e., (1) we consider one and only one set of integers q, p and n so that $q\nu_x + p\nu_y \approx n$ and (2) we will separate the beam-beam perturbation Hamiltonian K_1 into a fast oscillating term and a slowly changing term and make the smooth approximation that the fast oscillating term can be ignored.

The perturbation Hamiltonian K_1 can be decomposed into Fourier series:

$$K_1 = \sum_{\alpha,\beta=-\infty}^{\infty} f_{\alpha\beta}(J_x, J_y) e^{i\alpha(\psi_x + \nu_{xo}\theta) + i\beta(\psi_y + \nu_{yo}\theta)} \cdot \frac{1}{2\pi} \sum_{\gamma} e^{i\gamma\theta} \quad (34)$$

The slowly changing terms in K_1 are those terms in Eq. (34) whose exponent satisfies $\alpha = kp$, $\beta = kp$ and $\gamma = -kn$. Keeping only those terms in K_1 yields

$$K_1 \approx \frac{1}{2\pi} \sum_{k=-\infty}^{\infty} f_{(kq)(kp)}(J_x, J_y) e^{ik(q\psi_x + p\psi_x)} \quad (35)$$

Note that we have included all the slow terms corresponding to the resonant conditions $kq\nu_x + kp\nu_y = kn$ (k = any integer). Substituting the explicit expression of U from Eq. (2) and the Fourier coefficients

$$f_{(kq)(kp)} = \frac{1}{4\pi^2} \int_0^{2\pi} d\theta_x \int_0^{2\pi} d\theta_y \, e^{-ikq\theta_x - ikp\theta_y}$$

$$\times U(\sqrt{2J_x\beta_x} \cos\theta_x, \sqrt{2J_y\beta_y} \cos\theta_y)$$

into Eq. (35) and summing over k, one gets

$$K_1 \approx -\frac{1}{4\pi^2} \frac{Nr_0}{\gamma} \int_0^{2\pi} d\theta_x \int_0^{2\pi} d\theta_y \sum_{s=-\infty}^{\infty} \delta(q\theta_x + p\theta_y - q\psi_x - p\psi_y + 2\pi s)$$

$$\times \int_0^{\infty} \frac{dt}{\sqrt{(\sigma_x^2 + t)(\sigma_y^2 + t)}} \left\{ \exp\left[-\frac{J_x\beta_x \cos^2\theta_x}{\sigma_x^2 + t} - \frac{J_y\beta_y \cos^2\theta_y}{\sigma_y^2 + t}\right] - 1 \right\}$$

(36)

Expression (36) can be written as Nr_0/γ times a dimensionless factor which depends on the dimensionless quantities

$$q, \; p, \; q\psi_x + p\psi_y, \; J_x\beta_x/\sigma_x^2, \; J_y\beta_y/\sigma_y^2 \text{ and } \sigma_y/\sigma_x$$

The quantities $q\psi_x + p\psi_y$ and $J_z\beta_z/\sigma_z^2 \equiv \alpha_z$, $z = x,y$ are the dynamical quantities describing the particle's coordinates. The equations of motion, derived from the Hamiltonian K and written in terms of $q\psi_x + p\psi_y$, α_x and α_y, contains these dimensionless parameters:

$$\frac{Nr_0\beta_x}{\gamma\sigma_x^2}, \; \frac{Nr_0\beta_y}{\gamma\sigma_y^2}, \; \frac{\sigma_y}{\sigma_x}, \; p, \; q, \; q\nu_x + p\nu_y - n.$$

The first three are replaceable by

$$\xi_x \equiv \frac{Nr_0\beta_x}{2\pi\gamma\sigma_x(\sigma_x+\sigma_y)}, \; \xi_y \equiv \frac{Nr_0\beta_y}{2\pi\gamma\sigma_y(\sigma_x+\sigma_y)}, \; \frac{\sigma_y}{\sigma_x}$$

The other three quantities are determined from the values of ν_x and ν_y. We thus conclude that the single-resonance model is completely described by the scaling parameters ξ_x, ξ_y, ν_x, ν_y and σ_y/σ_x.

In the following, we will ignore the x-y coupling by the beam-beam interaction and consider only the vertical motion of a particle. The Hamiltonian of this case is obtained from Eq. (36) by setting $\sigma_x \gg \sigma_y$ and $q = 0$. The resonant terms included in the single-resonance model are

$$\nu \approx \frac{\pm n}{\pm p}, \; \frac{\pm 2n}{\pm 2p}, \; \frac{\pm 3n}{\pm 3p}, \text{ etc.}$$

The Hamiltonian is

$$K = (\nu - \frac{n}{p}) J + K_1(J, \psi)$$

$$K_1 = -\frac{\xi}{\beta} \sum_{s=0}^{p-1} \frac{1}{p} \int_0^\infty \frac{dt}{\sqrt{1+t/\sigma_y^2}} \left\{ \exp\left[-\frac{\alpha \cos^2(\psi - \frac{2\pi s}{p})}{1 + t/\sigma_y^2}\right] - 1 \right\} \quad (37)$$

$$\alpha = J\beta/\sigma_y^2$$

We have dropped some subscripts y. The perturbation Hamiltonian K_1 is independent of n. This is due to the fact that the beam-beam perturbation is a δ-function kick.

It is customary to expand K_1 into Fourier series in ψ. The result is

$$K_1 = -\frac{Nr_0\sigma_y}{\pi\gamma\sigma_x} \left\{ 1 + {\sum_s}' G_{sp}(\alpha) \cos sp\psi \right\}$$

where (38)

$$G_{sp}(\alpha) = (-1)^{\frac{sp}{2}} \frac{2}{(1+\delta_{so})(s^2p^2-1)} e^{-\frac{\alpha}{2}} \left[(1+\alpha) I_{\frac{sp}{2}}\left(\frac{\alpha}{2}\right) + \alpha I'_{\frac{sp}{2}}\left(\frac{\alpha}{2}\right) \right]$$

and Σ' means summing over s from 0 to ∞, but keeping only terms with sp = even, δ_{so} is the Kronecker delta function and $I_{sp/2}$ is the Bessel function. Note that if p is odd, the perturbation Hamiltonian for a resonance of order p is identical to that of order 2p. Note also that sometimes we keep only the first two terms in the Fourier expansion (38). This is based on the approximation that ν is close to $\pm n/\pm p$ but away from $\pm 2n/\pm 2p$, $\pm 3n/\pm 3p$, etc. We decided not to do that since we discovered that $\pm n/\pm p$ is equal to $\pm 2n/\pm 2p$, etc. The reason that only even terms appear in Eq. (38) is due to the fact that we have assumed head-on collision of symmetric bunches.

The equations of motion, obtained from the Hamiltonian K, are

$$\psi' = \nu - \frac{n}{p} - 2\xi \sum_s{}' G'_{sp}(\alpha) \cos sp\psi$$

$$\alpha' = 2\xi \sum_s{}' sp\, G_{sp}(\alpha) \sin sp\psi.$$
(39)

The particle motion is completely described by the two scaling parameters ν (and hence p and $\nu - n/p$) and ξ.

The third term on the right hand side of the ψ' equation gives the effective tune shift of a particle whose slow coordinates are α and ψ. For small amplitude particles with $\alpha \to 0$, it is equal to $\xi(1 + \cos 2\psi)$ if p = 1 or 2 and ξ otherwise. The s = 0 term in the Σ' summation defines a "detuning" contribution:

$$\Delta\nu(\alpha) = -2\xi\, G'_{op}(\alpha) = \xi e^{-\frac{\alpha}{2}} \left[I_0\left(\frac{\alpha}{2}\right) + I'_0\left(\frac{\alpha}{2}\right) \right] \quad (40)$$

Note that the detuning $\Delta\nu$ is independent of p, the order of the resonance under consideration. A particle with $\alpha = 0$ has a perturbed tune $\nu + \xi$, a particle with $\alpha \to \infty$ has a tune ν and a particle with α between 0 and ∞ has a tune between ν and $\nu + \xi$. The rest of the terms in the summation define a "resonance width" (in tune unit):

$$\delta\nu_p(\alpha) = 4\xi \sum_{s\neq 0}{}' |G'_{sp}(\alpha)| \quad (41)$$

The effective tunes of a group of particles with a given amplitude α and arbitrary phase ψ occupy a spread within $\pm\delta\nu_p/2$ around the detuned value $\nu + \Delta\nu(\alpha)$. The influence of resonances on particle motion is qualitative sketched in Fig. 8. Particles with amplitude α are strongly influenced by resonances if $\nu + \Delta\nu(\alpha)$ lies inside a shaded region.

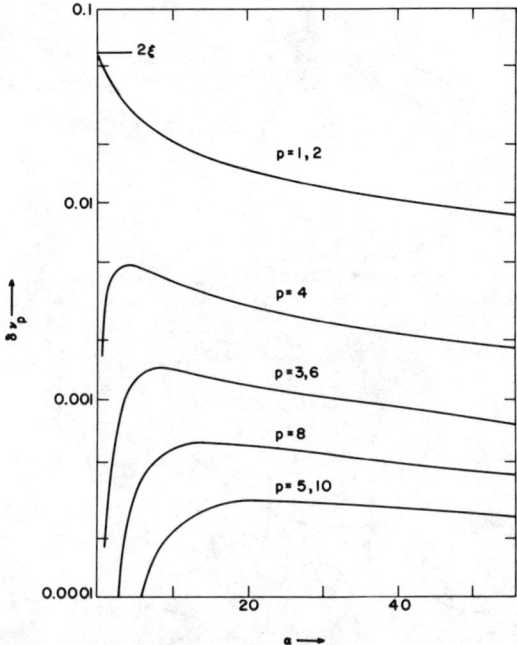

Fig. 9. Resonance widths $\delta_p \nu$ versus particle amplitude α for $\xi = 0.03$ and several values of p.

seriously since they are not rigorously defined quantities and serve only as order of magnitude concepts. More accurate description of particle motion must be obtained from the equations of motion (39).

The particle trajectories in phase space follow constant Hamiltonian contours. Topologically, there are two types of constant Hamiltonian contours, as illustrated in Figs. 10a and 10b. Particle motion is stable if the contours look like Fig. 10a because the amplitudes are always bounded. The same thing is not true if the contours look like that of Fig. 10b.

For the case of the beam-beam interaction, consider the Hamiltonian K, Eq. (37), at large values of J. As $J \to \infty$, we find that $K_1 \propto \sqrt{J}$ is dominated by the $(\nu - n/p)J$ term, indicating that the constant Hamiltonian contours at large amplitudes are circles. The Hamiltonian describing the beam-beam interaction, therefore, belongs to the type shown in Fig. 10a, indicating stable single particle motion. Physically, this follows from the fact that a particle crossing the collision point with large amplitude experiences little perturbation from the on-coming beam. One explicit example of beam-beam phase space topology with $p = 6$, $\nu - n/p = -0.01$ and $\xi = 0.06$ is shown in Fig. 11.

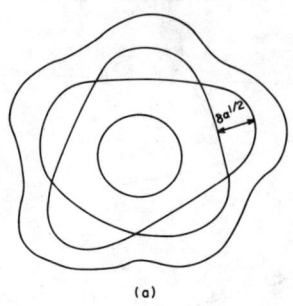

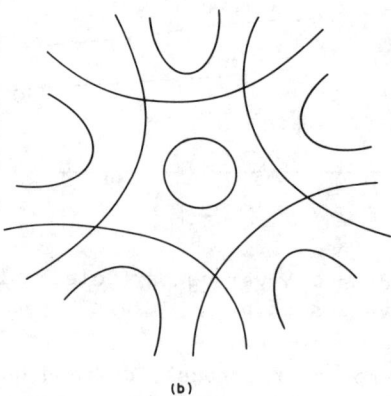

Fig. 10. Two types of constant Hamiltonian contours. 10(a) describes stable particle motion. 10(b) describes unstable motion.

3.2 Dynamic Models

In the previous section, we found that a static single-resonance model, in which all beam-beam parameters such as ν and ξ stay constant in time, does not provide an instability mechanism. This somewhat unexpected result disagrees with the experimental observations. In this section, we will briefly discuss the possibility of modifying the static model to include time-varying parameters. It is hoped that such dynamic single-resonance models may explain away the discrepancy.

There seem to be two types of time-dependences that have been studied. In the first type (the trapping model)[18], a beam-beam parameter modulates in time more or less sinusoidally with a certain frequency, while in the second type (the diffusion model),[19,20] a diffusion behavior of the parameter is believed to be the source of the beam-beam instability. Each model suggests a mechanism which continuously brings particles from small betatron amplitudes to larger amplitudes. A physical aperture limitation on the amplitude then explains the observed particle loss in colliding beams. For simplicity of discussion, we will consider the case of electron-positron collisions and assume that the time-varying parameter is

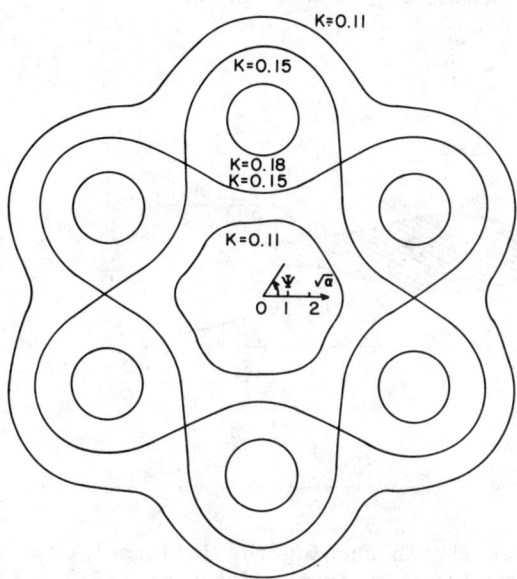

Fig. 11. Constant Hamiltonian contours for the beam-beam interaction. The particle motion is stable.

the tune ν.

We first consider the trapping model. In the static model, a particle streams along a constant Hamiltonian contour in the phase space. The topology of these contours depends on the Hamiltonian and, from Eq. (37), depends in turn on the distance from resonance $\nu - n/p$. In the trapping model, the tune of a particle oscillates slowly around or close to the resonant value n/p. As a consequence, resonance islands like those shown in Fig. 11 move in and out quasi-statically and, when doing so, distort and relocate phase space area elements. An analysis of the particle motion under these conditions is extremely difficult. As a rough qualitative description, however, we note that a particle is either trapped or not trapped by the resonance islands which are moving from small to large amplitudes. For particles that are trapped, the amplitude grows with the islands. For particles that are not trapped, the amplitude is essentially unperturbed by the passing islands. Particle loss is then a consequence of having the islands trap particles and move them to the physical aperture limit.

To determine the probability that a particle of amplitude α will be trapped by the passing islands, one needs to know the total area A_0 of the islands and the instantaneous growth rate of the amplitude of the islands α_0'. It is suggested that the trapping probability per

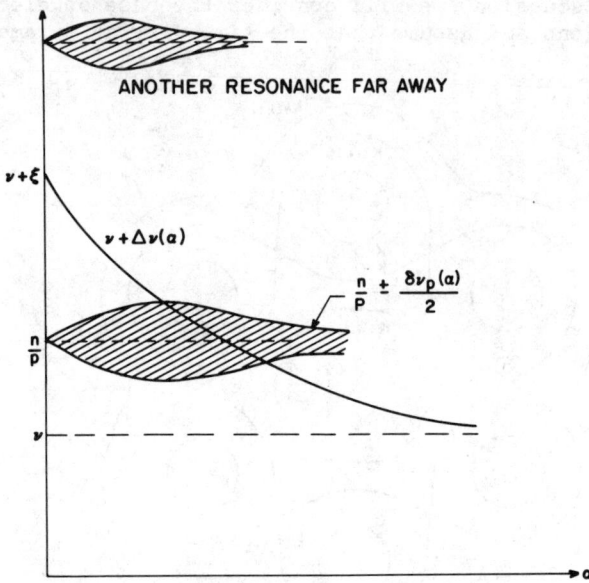

Fig. 8. Qualitative sketch showing the influence of a single resonance on particle motion. $\Delta\nu$ is the detuning and $\delta_p\nu$ is the resonance width.

If the tune is far away from all significant resonances, the above results are still applicable if we keep only the $s = 0$ term in the summation Σ'. In this special case, the detuning is the same as Eq. (40) but there are no resonance widths. It follows from Eq. (39) that α is a constant of the motion and, since the perturbed tune $\nu + \Delta\nu$ depends only on α, the tune distribution of the beam can be obtained once the distribution in α is known.[11]

In Fig. 9, we have plotted $\delta\nu_p$ versus particle amplitude α for $\xi = 0.03$ and for several values of p. For odd p, $\delta\nu_p$ is identical to $\delta\nu_{2p}$. For even p, $\delta\alpha_p$ is generally smaller for larger p. This partially justifies the assumptions made in Eq. (28).

Sometimes it is more convenient to define a resonance width $\delta\alpha^{\frac{1}{2}}$ (in unit of $\alpha^{\frac{1}{2}}$) to be the difference between the maximum and the minimum values of $\alpha^{\frac{1}{2}}$ along the separatrix (see Fig. 10a below). Let $\alpha_o^{\frac{1}{2}}$ be the average value of $\alpha^{\frac{1}{2}}$ around the separatrix, $\delta\alpha^{\frac{1}{2}}$ is roughly given by

$$\delta\alpha^{\frac{1}{2}} \approx \left(\frac{\sum'_{s \neq 0} |G_{sp}(\alpha_o)|}{2\alpha_o |G''_{op}(\alpha_o)|} \right)^{\frac{1}{2}} \qquad (42)$$

Note that under this definition, we are no longer interested in the dependence of resonance width on the particle amplitude α. One should not, however, take the definitions of resonance width too island passing is proportional to A_0 and decreases more or less exponentially with increasing α_0'.

Figure 12 sketches the expected behavior of trapping probability P_T as a function of particle amplitude α as the islands sweep through this amplitude with a certain speed. One notices that P_T increases with α for small α and decreases with α for large α. A peak occurs somewhere around $\alpha \sim 1$-3, depending on the order of resonance p under consideration. A typical particle trapping event looks perhaps like this: As the islands move outward in phase space, their trapping efficiency increases and trap a particle at a relatively small amplitude (say, $\alpha \sim 0.5$). The particle then moves outward together with the islands. As the islands move out further, their trapping efficiency starts to decrease and the islands become leaky. The particle will then be dropped from the islands at a larger amplitude (say, $\alpha \sim 5$). We have sketched this process in Fig. 13. The shaded area indicates the region where trapping is efficient. Outside this region, the islands are leaky. The aperture limit can be either inside or outside the shaded region. It is believed in the trapping model that this mechanism may explain the observed decrease of colliding beam lifetime. This trapping mechanism also suggests a possible halo structure in the beam distribution.

The time modulation assumed in the trapping model may come from several sources. For example, the tune may be modulated with the synchrotron frequency f_s in accordance with the particle energy through chromaticity effects; or, if the bunch length is not small compared with the beta-function at the collision point, the parameter ξ may be modulated by the longitudinal position of the particle with the frequency $2f_s$.

A second type of dynamic single-resonance model is the diffusion model. In the presence of the usual radiation damping and quantum diffusion effects of an electron beam,[21] the particle distribution reaches an equilibrium state ρ. Without perturbations, ρ is a Gaussian distribution in the particle amplitude. A quantum lifetime of the beam is then determined by the particle diffusion rate of this distribution across the physical aperture limit. It is conceivable that the distortion of phase space by the presence of a single resonance will also distort the equilibrium distribution, hence reducing the associated quantum lifetime for a given aperture limit. It is clear that one needs to solve for ρ from a Fokker-Planck type of diffusion equation[21], taking simultaneously into account the damping, diffusion and the single resonance contributions. Some attempts have been made in this direction but unfortunately hindered by the very difficult mathematics.[19,20] Nevertheless, as an illustration, let us perhaps write

$$\rho \sim \exp(-gK) \qquad (43)$$

where K is the Hamiltonian found in Eq. (37). This expression reduces to the correct Gaussian distribution in the limit $\xi \to 0$ if we choose $g = \beta/|\nu - n/p|\sigma_0^2$ where β is the beta-function and σ_0 is the natural rms beam size in the absence of beam-beam perturbation. Equation (43) is approximately valid if the beam-beam perturbation is not

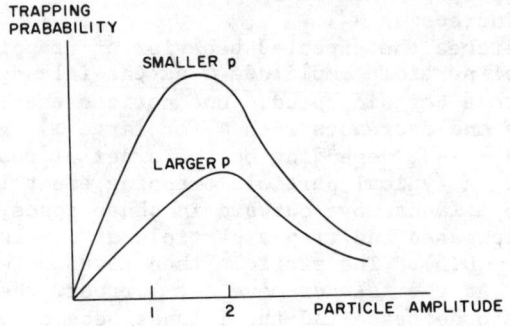

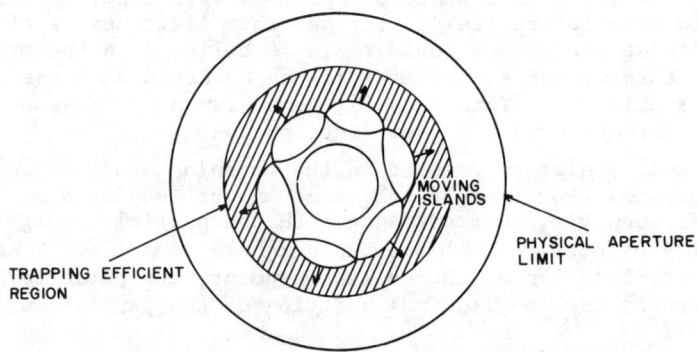

Fig. 12. Expected behavior of the trapping probability of a particle by a moving phase space island.

Fig. 13. A possible mechanism of transporting particles to larger amplitudes by a slowly changing phase space topology.

too strong so that $p|\nu - n/p| \gg \xi$. It is also correct in the limit of no damping, no diffusion and arbitrary ξ. Under these conditions, we expect the effective physical aperture limit to be reduced by an amount roughly of the order of the resonance width $\delta\alpha^{\frac{1}{2}}$ and the beam lifetime to be shortened accordingly.

Note that the equilibrium particle distribution, Eq. (43), contains a structure of islands just like the Hamiltonian does. Note also that if the tune is away from all resonances, the beam-beam perturbation contributes only a phasewise streaming term in the ψ' equation while keeping $\alpha' = 0$; the beam lifetime will not be affected in that case.

The diffusion mechanism mentioned above does not have to be the quantum diffusion. The momentum diffusion caused by intrabeam scattering for example, can cause a tune diffusion through the chromaticity effect.[19,20]

In addition to those parameters such as $\nu - n/p$, p and ξ, typical for a single-resonance theory, the dynamic models suggest that a few more parameters might be important candidates for determining the beam-beam stability limit: the synchrotron frequency f_s, the chromaticity C and the physical aperture limit A.

REFERENCES

Out of the 119 references considered in Fig. 1, I have read about 25% and understood about 10%. Most of the references listed below belong to those I happened to have read and not necessarily those I have understood.

1. F. Amman and D. Ritson, "Space-Charge Effects in Electron-Electron and Positron-Electron Colliding or Crossing Beam Rings", Intern. Conf. on High Energy Accel., Brookhaven, 471 (1961).
2. E. D. Courant, "Beam Instabilities in Circular Accelerators", IEEE Trans. on Nucl. Science, NS-12, 550 (1965).
3. D. Ritson and J. Rees, "Limitations on Storage Ring Reaction Rates", SLAC-TN-65-39 (1965).
4. M. Bassetti, "Numerical Computations of Space Charge Effects in a Positron and Electron Storage Ring", Vth Intern. Conf. on High Energy Accel., Frascati, 708 (1965).
5. B. W. Montague, "Calculation of Luminosity and Beam-Beam Detuning in Coasting Beam Interaction Regions", CERN/ISR-GS/75-36 (1975).
6. E. D. Courant and H. S. Snyder, Ann. of Phys. 3, 1 (1958).
7. L. E. Augustin, "Space Charge Effects in e+e- Storage Ring with Beams Crossing at an Angle", Orsay Note Interne 35-69 (1969).
8. A. Piwinski, "Limitation of the Luminosity by Satellite Resonances", DESY 77/18 (1977).
9. A. Piwinski, "Coherent Beam Break-up due to Space Charge", VIIIth Intern. Conf. on High Energy Accel., Geneva, 357. A. W. Chao and E. Keil, CERN/ISR note, in preparation (1979).
10. A. Schoch, "Theory of Linear and Nonlinear Perturbations of Betatron Oscillations in Alternating Gradient Synchrotrons", CERN Report 57-23 (1957).
11. A. Jejcic and J. LeDuff, "Non-Linear Beam-Beam Effect", VIIIth Intern. Conf. on High Energy Accel., Geneva 354 (1971).
12. E. Keil, "Nonlinear Space Charge Effects, I", CERN/ISR-TH/72-7 (1972).
13. E. Keil, "Nonlinear Space Charge Effects, II", CERN/ISR-TH/72-25 (1972).
14. A. G. Ruggiero and L. Smith, "Calculation of Resonance Effects Due to a Localized Gaussian Charge Distribution". PEP-52 (1973).
15. G. H. Rees, W. T. Toner and J. V. Trotman, "Effects of Beam-Beam Forces in Large Electron-Positron Storage Rings", IEEE Trans. on Nucl. Science, NS-22, 1447 (1975).
16. L. J. Laslett, IXth Intern. Conf. on High Energy Accel., Stanford, 394 (1974).
17. L. Smith, private communications (1975).
18. M. Month, "Nature of the Beam-Beam Limit in Storage Rings", IEEE Trans. on Nucl. Science, NS-22, 1376 (1975).

19. H. G. Hereward, "Diffusion in the Presence of Resonances", CERN/ISR-DI/72-26 (1972).
20. J. LeDuff, "About the Diffusion Process in the Nonlinear Beam-Beam Effect", CERN/ISR-AS/74-53 (1974).
21. H. Bruck, Acceleratuers Circulaire de Particules (Press Universitaires de France, Paris, 1966).

REVIEW OF THE INVESTIGATIONS ON THE BEAM-BEAM INTERACTIONS AT THE ISR

G. Guignard

I. INTRODUCTION

At the ISR, an extensive theoretical and experimental program on the beam-beam interactions has been carried out. For coasting beams, analytical theories for beam-beam tune shifts and nonlinear resonances have been developed and checked experimentally. Other theories, such as for Arnold diffusion, stochasticity and migration due to adiabatic change of the parameters have also been studied experimentally. The resonance theory has been generalized to the case of bunched beams, and the resonances due to the so-called beam-beam overlap knock-out have been observed in some detail. The present paper tries to summarize this work and to give the most interesting experimental results.

II. ELECTROMAGNETIC FIELDS DUE TO A PROTON BEAM

When two beams are crossing, a particle of beam 1 sees the electric field due to the charges in beam 2 and the magnetic field associated with the motion of these charges. The electric and magnetic forces are proportional and this may be shown from Maxwell's equations for a cylindrical beam of any section.

$$\oint |\vec{E} \times d\vec{l}| = \iint \frac{\rho_e}{\epsilon_o} d\sigma = \mu_o c^2 \iint \rho_e d\sigma$$

$$\oint \vec{B} \cdot d\vec{l} = \mu_o \oint \vec{H} \cdot d\vec{l} = \mu_o v \iint \rho_e d\sigma$$

(1)

where ρ_e is the charge density of beam 2, l is distance around the perimeter of cylinder and $d\sigma$ an element of area of the cylinder cross section.

Equation (1) shows that the fields are in a ratio v/c^2. Since the magnetic force is proportional to the longitudinal speed v, the ratio between the forces is $v^2/c^2 = \beta^2$. Hence, if Φ is the electric potential, the effects of the electromagnetic fields can be described by a total potential equal to

$$\Phi_{tot} = (1 \pm \beta^2)\Phi$$

(2)

The positive and negative signs are associated with antiparallel and parallel motion of the two beams, respectively. Using the Hamiltonian formalism, the Hamiltonian function for the perturbation can be calculated for an electric potential[1]

$$H = \frac{\Phi_{tot}}{\beta c |B\rho|} + \frac{1}{2}\left(\frac{\Phi_{tot}}{c|B\rho|}\right)^2 \simeq \frac{\Phi_{tot}}{\beta c |B\rho|} = \frac{(1 \pm \beta^2)\Phi}{\beta c |B\rho|} \qquad (3)$$

where $|B\rho|$ is the magnetic rigidity for beam 1 ($= p_1/e$).

Equations (2) and (3) are basic expressions which enable the beam-beam interaction problems to be solved, provided that Φ can be calculated. For colliding beams, the positive sign is valid.

III. BEAM-BEAM TUNE SHIFT

The first effect to be considered is the focusing change associated with the electromagnetic fields of a beam. There is a linear effect due to a field gradient and commonly characterized by the linear tune shift, which can be calculated from[2]

$$\Delta Q_y = \frac{1}{4\pi} \int_0^C \beta_y K_y \, ds \qquad (4)$$

where y stands for x or z, s is the longitudinal coordinate and K_y is the focusing force. In the case of interest, this force has to be deduced from Eq. (3),

$$K_y = \frac{1+\beta^2}{\beta c |B\rho|} \frac{\partial^2 \Phi}{\partial y^2} \qquad (5)$$

Φ being given by $\overrightarrow{\text{div grad}}\ \Phi = \rho_e/\varepsilon_o$.

Starting from Eqs. (4) and (5), different analytical expressions of ΔQ_y have been deduced, depending on the complexity of the beam model and on the geometry of the interaction region. For instance, when two beams are crossing at zero angle, the electric field can be calculated for an elliptic beam of Gaussian distribution, as in the space charge theory.[3] It is even possible to integrate the contribution of elliptic pulses centered at different radial positions, in order to obtain the field of a wide stack.[3] More relevant to the ISR is the case where the two beams cross horizontally at a small but nonzero angle. In this case, the geometry is more complicated but the beam model may be simplified. Analytical expressions of ΔQ_y have been worked out for a cylindrically symmetric Gaussian beam[4] and are given in an integral form,

$$\Delta Q_y = \frac{r_p E_o}{4\pi e c^2} \frac{1+\beta^2}{\beta^2 p_1} \frac{I_2 \beta_{y1}^* d}{\sigma_{y2}^{*2}} J_y(\xi,\omega) \qquad (6)$$

where d is the interaction length, I_2 the beam 2 current, σ_{y2}^* the rms radius of beam 2, p_1 the momentum of beam 1, β_{y1}^* the betatron amplitude of beam 1 at the crossing point (taken as a local minimum) and

$$\zeta = \frac{d}{2\beta^*_{y_1}}$$

$$\omega = \frac{\beta^*_{y_1}}{\sigma^*_{y_2}} \sin \psi$$

ψ being the crossing angle. For the ISR conditions where $\omega \gg 1$ and $\omega \gg \zeta$, the integral J_x vanishes while J_z is equal to $\sqrt{2\pi}/\zeta\omega$. Hence, the horizontal tune shift is zero but the vertical one is given by

$$\Delta Q_z = \frac{r_p}{\sqrt{2\pi}} \frac{E_o}{ec^2} \frac{1+\beta^2}{\beta^2 P_1} \frac{I_2 \beta^*_{z1}}{\sigma^*_{z2} \sin\psi} \tag{7}$$

where the physical constants have their standard meaning.

So far, only the linear tune shift was considered, but nonlinear effects on off-centered particles do exist. If the two beams are vertically steered, the first nonlinear field shifting the tune is the octupole component and this tune shift has been shown to be opposed to the linear one.[5] Hence the total shift seen by the particles of beam 1 is distributed between zero and ΔQ_z [Eq. (7)] around an average value calculated to be [5]

$$\Delta Q_{z,tot} \cong 0.7 \Delta Q_z \tag{8}$$

Combining (7) and (8) gives for the ISR

$$\Delta Q_{z,tot} = 2.48 \times 10^{-7} \frac{I_2 \beta^*_{z1}}{\gamma_1 h_{eff_2}} \tag{9}$$

where γ_1 is the relativistic parameter of beam 1 and h_{eff_2} is the effective height of beam 2 (= $2\sqrt{\pi}\sigma^*_{z2}$).

Some measurements of the total tune shift have been done at 26 GeV/c for different current levels, as shown in Table I.

Table I. Total tune shifts measured in the ISR.

Line	I_1 (A)	I_2 (A)	h_{eff_2} (mm)	Constant	Comments
ELSA	6	23	5.0	2.18×10^{-7}	Not optimum steering
	Pulse	32	5.5	2.78×10^{-7}	Not optimum steering
FP	1.6	18.5	4.4	2.22×10^{-7}	Optimum steering
	5.8	11.2	7.0	2.67×10^{-7}	Optimum steering

The agreement between measurements and Eq. (9) is very good.

IV. NONLINEAR RESONANCES

Having the Hamiltonian (3) and following the perturbation treatment presented in Refs. 1 and 6, it is possible to calculate the bandwidths Δe associated with isolated resonances. For an isolated resonance given by

$$e = n_x Q_x + n_z Q_z - p = 0, \quad N = n_x + n_z \tag{10}$$

the bandwidth can be written[1,6]

$$\Delta e = \frac{(1+\beta^2) E_x^{n_x/2} E_z^{n_z/2}}{2^{(N-1)} \pi^{N/2} \beta c |B\rho| \, n_x! n_z!} \left(\frac{n_x^2}{E_x} + \frac{n_z^2}{E_z} \right) \times$$

$$\int_0^C \beta_x^{n_x/2} \beta_z^{n_z/2} \frac{\partial^N \Phi}{\partial x^{n_x} \partial z^{n_z}} \exp[i(n_x \mu_x + n_z \mu_z - e\frac{S}{R})] ds \tag{11}$$

where E_x and E_z are the transverse emittances of a particle in beam 1, μ_x and μ_z the phase advances, and R the radius of the machine. Here again, the final expressions for Δe depend on the beam model and on the geometry of the interaction region. In the case mentioned in Section III where the two beams cross horizontally at a small angle, the bandwidths become[7]

$$\Delta e = K_{n_x n_z} \Delta Q_z \, J_{n_x n_z}(\zeta, \omega) \tag{12}$$

where $K_{n_z n_z}$ is a numerical factor, ΔQ_z is the linear tune shift given by Eq. (6), $J_{n_x n_z}$ is a correction factor of integral form and ζ and ω are the parameters defined in Section III. In the ISR, where $\omega \gg 1$ and $\omega \gg \zeta$, the correction factor $J_{n_x n_z}$ vanishes when $n_x \neq 0$ and is close to one and to z/σ_{z2}^* for the vertical resonances of even and odd order, respectively. Hence, vertical beam-beam resonances are dominant in the ISR and their bandwidths are proportional to the linear tune shift [Eq. (7)].

$$\Delta e = \frac{N}{2^{(N/2-2)}(N-1)(N/2-1)!} \Delta Q_z \quad \text{for N even}$$

$$\Delta e = \frac{2N}{2^{(N-1)/2}\left(\frac{N-1}{2}\right)!} \frac{z}{\sigma_{z2}^*} \Delta Q_z \quad \text{for N odd,} \quad \frac{z}{\sigma_{z2}^*} \ll 1 \tag{13}$$

where z is the vertical distance between particle 1 and beam 2. From the bandwidth (13), it is possible to calculate the effects either of a rapid, nonadiabatic crossing of the resonance[1,6] or of an adiabatic diffusion of particles into the resonance.[8] The vertical size increase due to nonadiabatic crossing[1,6] is approximated by

$$\frac{\Delta \sigma_z}{\sigma_z} = \sqrt{\frac{\pi c}{2R}} \frac{\Delta e}{N^{3/2}\sqrt{dQ_z/dt}} \qquad (14)$$

while the loss rate associated with a tune diffusion can be estimated from[8]

$$\left(\frac{dN_p}{N_p dt}\right)_{max} = - \frac{P_T}{\Delta Q} \frac{D_Q}{\delta Q} \qquad (15)$$

where P_T is the trapping probability, D_Q the tune diffusion constant, ΔQ the total tune spread in the beam and δQ the tune change which causes the trapped particles to reach the walls.

It is interesting to calculate the orders of magnitude of the resonances and of their effects, for an intense beam (30 A) at 26 GeV/c with a size of $\sigma_{z2}^* = 1$ mm and a vertical distance of $z = 0.2$ mm. The results are summarized in Table II, assuming either a rapid change of Q of 0.16 s^{-1} or a tune diffusion associated with an intrabeam momentum diffusion of 6.7 × 10^{-11} s^{-1}.

Table II. Calculated bandwidths, blow-ups and loss rates for beam-beam resonances.

Order N	4	5	6	7	8
Bandwidth Δe	1.9×10^{-3}	3.6×10^{-4}	4.0×10^{-4}	8.2×10^{-5}	6.1×10^{-5}
Blow-up $\Delta\sigma_z/\sigma_z$	1.052	0.143	0.121	0.020	0.012
Loss rate $dN_p/N_p dt$ (ppm/min)	50	35	22	9	5
Working line	4C	5C	ELSA	ELSA	8C

In comparison with these beam-beam bandwidths, the measurement of the vertical 4th order resonance, made by octupole compensation, gives a bandwidth due to the machine hardware equal to 4.5 × 10^{-6}. In other words, the resonance excitation produced by machine hardware is not really harmful in the ISR, compared with the beam-beam resonance excitation. Another experiment shows this very well. The Schottky signal is monitored during a noise excitation of the beam and the transfer function is calculated using the Fast Fourier Transform.[9] The amplitude of this function is locally modified by the presence of resonances in the stack. Figure 1 shows that hardware resonances are not visible, while beam-beam resonances appear with a 15 A stack in the other ring. This technique should even give the possibility of measuring the resonance bandwidth, provided that the signal is not too small, and this will be tried in the near future.

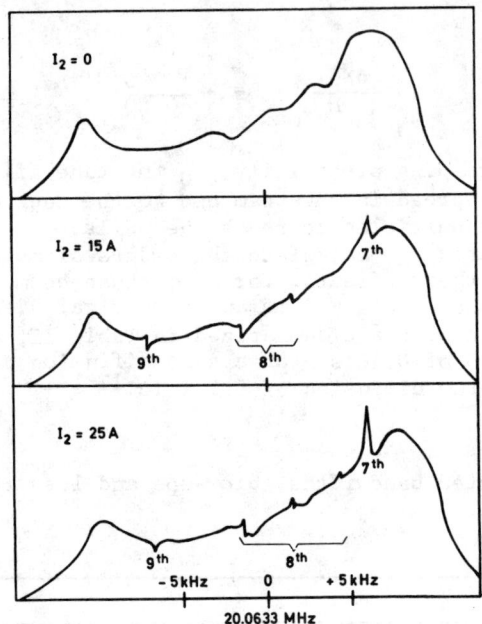

Fig. 1. Amplitude of the vertical transfer function of beam 1 as a function of the beam 2 current, showing beam-beam resonances of order 7 to 9.

It is also very interesting to compare the estimated loss rates of Table II with the loss rates observed in the ISR on different working lines, which cross resonances of different orders. The results, which do not depend on the actual beam current, but on the momentum p_1 and on $z/\sigma_{z_2}^*$, are given in Figs. 2 to 4. The measured loss rates and their dependence with the order of the resonances present in the stack are in fairly good agreement with the calculated values.

V. STRONG NONLINEARITIES

When the phase space energy or the nonlinearity is very large, the motion becomes unstable and the trajectories behave in a random fashion. In this case, stochastic regions start to appear in the phase space and the trajectories there have a strong dependence on

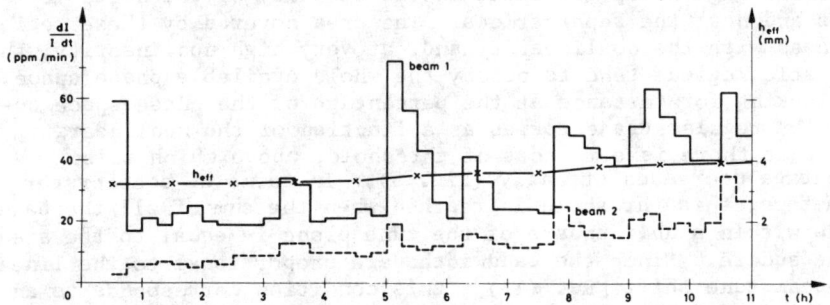

Fig. 2. Loss rates observed on the 5C line in the ISR run 353 with 8 A and 9 A stacks at 26 GeV/c. The initial effective height was 3.2 mm and the vacuum pressure was between 2.5 and 3×10^{-11} Torr.

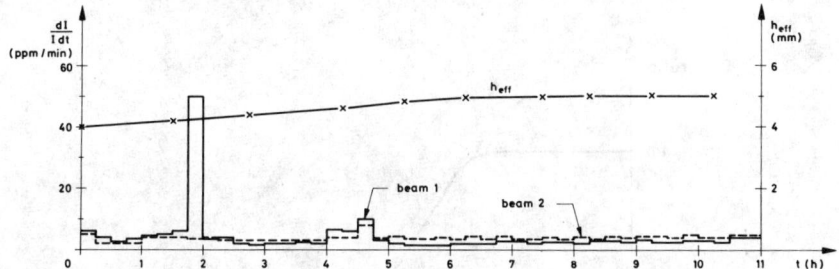

Fig. 3. Loss rates observed on the ELSA line in the ISR run 643 with 25 A and 24 A stacks at 26 GeV/c. The bottom of the stack crossed 7th order resonances, the initial effective height was 4 mm and the vacuum pressure was between 1.5 and 2×10^{-11} Torr.

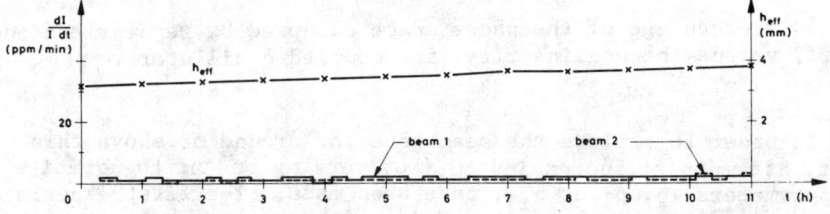

Fig. 4. Loss rates observed on the 8C line in the ISR run 418 with 10 A stacks at 26 GeV/c. The initial effective height was 3.2 mm and the vacuum pressure was between 2 and 3.5×10^{-11} Torr.

the initial conditions. For the transverse motions in a synchrotron, stochastic regions appear first in the vicinity of the unstable fixed points and near the separatrices. The area covered by these regions increases with the nonlinearity and, at very high nonlinearity, the stochastic regions tend to occupy the whole available phase space. When looking for instance at the percentage of the phase space occupied by regular trajectories as a function of the nonlinearity, it seems that there is some sort of threshold, above which this percentage decreases strongly (Fig. 5). In terms of accelerator parameters, this threshold is reached when the sum of all the bandwidths within a unit square of the tune plane is equal to the area of the square. Since the bandwidths are proportional to the linear beam-beam tune shift [Eq. (13)], this condition corresponds to an upper limit for ΔQ_z. Because of intrabeam scattering and Arnold diffusion, it has been suggested that the practical limit is even lower[10,11] and the limit adopted for proton machine design is

$$\Delta Q_z < 0.005 . \qquad (16)$$

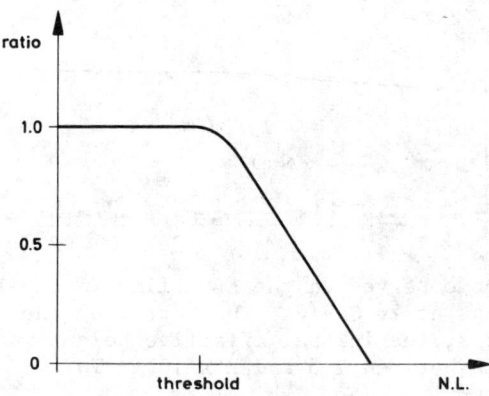

Fig. 5. Percentage of the phase space occupied by regular trajectories, versus the nonlinearity, for coupled oscillators.

In order to explore the beam behavior around or above this limit, attempts at increasing ΔQ_z[7] by varying one of the critical parameters I_2, p_1 or β^*_{z1}, have been made. The first experiment was to collide a weak beam at 2 GeV/c with a strong beam at 26 GeV. This was however not really conclusive because of the strong effect of intrabeam scattering on the lifetime of the weak beam, but it showed that there is no noticeable effect on the lifetime for a tune shift of ~ 0.005. The second experiment[13] simulated very strong beam-beam effects using a nonlinear lens, i.e. a pair of bars inside the vacuum chamber. This lens had the symmetry of the space charge corresponding to head-on collision and excites all resonances (10) for which n_x and n_z are even. When ΔQ_z changes between 0 and 0.09, the decay rate changes by about 3 orders of magnitude and depends strongly on the tunes. Figure 6 shows that the decay rate saturates

These experiments do not show the existence of a well-defined threshold, but indicate that below the chosen limit (16) for nonradiating particles the beam lifetime is long enough (> 100 hours) for storage rings.

VI. OVERLAP KNOCK-OUT RESONANCES

In a synchrotron, the Hamiltonian (3) is always periodic since the machine is circular, and Fourier development of this function gives rise to isolated resonances of the type (10). In the case where the perturbation, e.g. the electric potential Φ, varies with time, this variation can also be analyzed in Fourier series and a new frequency spectrum is thus superposed upon the spectrum due to the machine periodicity. This superposition corresponds to the appearance of a fine structure in the resonance lines and the exact treatment[1] shows that the new resonance condition is

$$e = n_x Q_x + n_z Q_z - p - b \frac{f_{per}}{f_{rev}} = 0 \quad (17)$$

where f_{per} is the basic perturbation frequency and f_{rev} is the revolution frequency of the particle of beam 1.

The bandwidth of such a resonance is given by an expression similar to Eq. (11), but multiplied by the harmonic coefficient c_b of the time varying part of Φ.

In the case of a bunched beam interacting with a coasting beam, the associated electric potential Φ varies with time at a given position of the circumference, since Φ is proportional to the bunched beam current I_2. The revolution frequency of the bunched beam is then f_{per} and the coefficients c_b are the harmonic amplitudes of the bunch spectrum. According to whether the two beams are in the same ring or in two different rings, the corresponding perturbation is called single-beam or two beam overlap knock-out respectively. For single-beam overlap knock-out, the electric potential Φ_{tot} is essentially due to the coupling forces induced by the bunches via the surrounding material and analytical predictions are difficult. For two-beam overlap knock-out, on the contrary, there is a finite number of rather short interaction regions and Φ_{tot} is essentially due to direct space charge forces of Eq. (2). Algebraic treatment becomes possible, in particular when the crossing angle is not vanishing. This has been done in the ISR and the expressions obtained for the bandwidths are[1]

$$\Delta e \left(Q_z = p + b \frac{f_{per}}{f_{rev}} \right) = \sum_{c.p.} K \, \text{erf} \left(\frac{z}{\sqrt{2} \, \sigma^*_{z_2}} \right) \exp\left[i(\mu_z - e \frac{S}{R}) \right]$$

$$\Delta e \left(N Q_z = p + b \frac{f_{per}}{f_{rev}} \right) = \sum_{c.p.} \frac{K \, N^2}{\sqrt{\pi} \, \sqrt{2}^{(N-3)} N!} \exp\left(-\frac{z^2}{2 \sigma^{*2}_{z_2}} \right) \quad (18)$$

$$H_{N-2} \left(\frac{z}{\sqrt{2} \, \sigma^*_{z_2}} \right) \exp\left[i(N \mu_z - e \frac{S}{R}) \right]$$

at tune shifts close to the stochasticity limit (reached at $\Delta Q_z = 0.05$ as shown by computer simulation). The third experiment[14] became possible with the installation of a low-β insertion in the ISR. The idea was to use this installation to increase β_{z1}^* up to 200, in order to avoid the difficulties of the two earlier experiments, i.e. the intrabeam scattering problems at too low energies and the difficult question of whether the nonlinear forces of the lens bear any resemblance to those of a beam. The measurements of the decay rate or of the beam lifetime have shown that both vary exponentially with the tune shift (Fig. 7). At the limit of $\Delta Q_z = 0.005$ the lifetime is of the order of 100 hours and at the ISR values (< 0.001) the lifetime is around 1000 hours. Extrapolating these data to the approximate limit for electrons (0.06), gives a lifetime of the order of milliseconds, which is typically the order of the damping time.

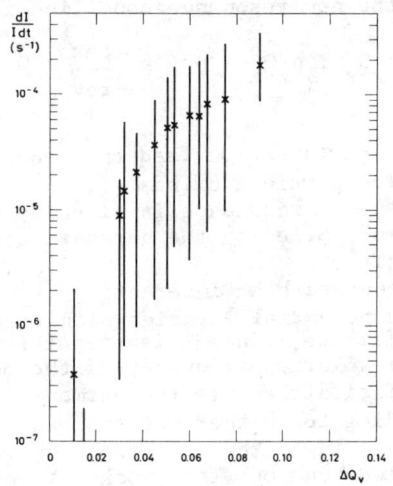

Fig. 6. Beam decay rate as a function of the excitation of a nonlinear lens.

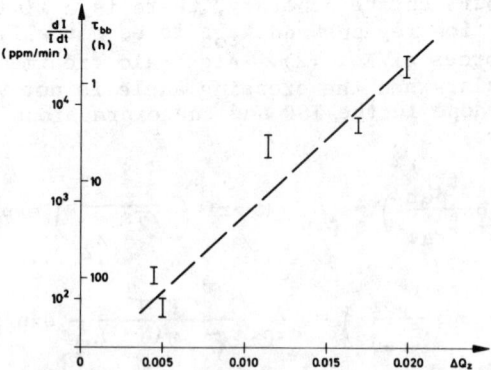

Fig. 7. Decay rate and beam lifetime, as functions of the beam-beam tune shift achieved with a high-β insertion.

where

$$K = \frac{\mu_o(1+\beta^2)}{4\pi\beta^2} \frac{I_2 \beta^*_{z1} c_b}{|Bp| \sigma^*_{z2} \sin\psi} = \frac{e^2 c^2 \mu_o c_b}{\sqrt{2}\,\pi E_o r_p} \Delta Q_z$$

and H_n are the Hermite polynominals, which satisfy the recursion formula

$$H_{n+1}(x) = 2xH_n(x) - 2nH_{n-1}(x)$$

The positions in the tune diagram of the overlap knock-out resonances (18) can be calculated from Eq. (17), using the fact that $n_x = 0$ and that b is a multiple of the rf harmonic number h, i.e. $b = nh$ (the side bands due to the fact that only 20 bunches occupy the 30 buckets can be neglected in a first approximation). If the two beams are at the same energy, the resonance positions are given in Fig. 8 for the ISR working line ELSA.

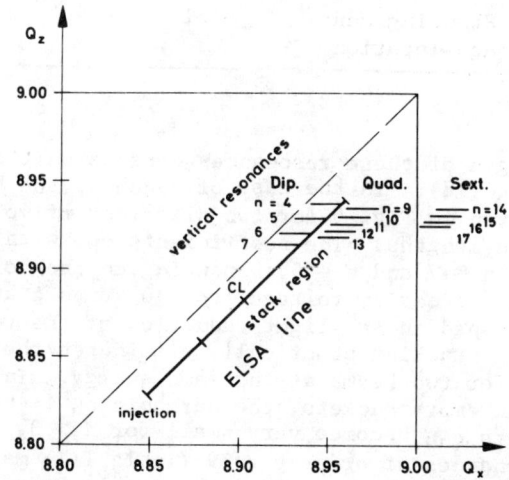

Fig. 8. Positions of the overlap knock-out resonances in a stack on the ELSA line, with a pulse at the injection in the other ring.

Other conditions of overlap knock-out can exist in the ISR, when the two beams have different energies and when the exciting beam consists of empty buckets inside an accelerated stack (the sole harmonies $b = nh$ are then present).

This is for instance the case when one beam is stacked at 26 GeV/c or accelerated between 26 and 31 GeV/c, while the other one is stable at 31 GeV/c. The resonance lines then appear at different positions in the stack and Table III summarizes very briefly which resonances may be harmful.

Table III. Overlap knock-out resonances in the ISR during stacking and acceleration against a 31 GeV/c beam.

Order N of the resonance	Beam 2 conditions (Beam 1 at 31 GeV on ELSA)	Values of n (b=nh) present in beam 1	Beam 1 regions where resonances appear
1	Stacking at 26 GeV/c	> 11	Top
	Acceleration to 31 GeV/c	> 7	Top and bottom
2 to 5	Stacking and acceleration	> 10	
6	Stacking and acceleration	> 8	Bottom
	Stacking and acceleration	> 3	Bottom, if beam 1 line is down to $Q_z = 8.840$
7 to 9	Stacking and acceleration	> 1	Bottom

The strength of these resonances depends on the values of the coefficients c_b (18). In the case of bunches, the bunch frequency spectra are given in Fig. 9 for two different rf voltages, i.e. two different bunch lengths. The coefficients c_b, with $b = nh$, become negligible for $n \geq 7$ and $n \geq 5$, depending on the voltage. Hence, a reduction of the cavity voltage from 16 kV to 4 kV as soon as the bunches are trapped and a slight reduction of the tunes by 0.015 (Fig. 8) during stacking practically eliminates the effects of overlap knock-out for two beams at the same energy. In the case of acceleration with empty buckets, the harmonic c_h is predominant and the coefficients c_{nh} becomes very small for $n > 3$. Hence, it seems that only resonances of order 6 to 9 (Table III) may be harmful during acceleration, at the bottom of the stable beam at 31 GeV/c.

Vertical blow-up and beam losses have effectively been observed at the top of ELSA, for the beams at 26 GeV/c and a cavity voltage of 16 kV, and at the bottom of a 31 GeV/c beam, when stacking and accelerating in the other ring, in agreement with these deductions. Specific measurements of current losses have also been made with an aperture limited stack on ELSA and a bunched pulse at injection, both at 26 GeV/c. The loss dependence on the rf voltage and on the tune value has been measured. Table IV gives the results obtained for single beam overlap knock-out resonances.

Similar measurements have been done for two-beam (26 GeV/c) overlap knock-out resonances. Since the two beams interact at the crossing points only, vertical resonances are excited and vertical

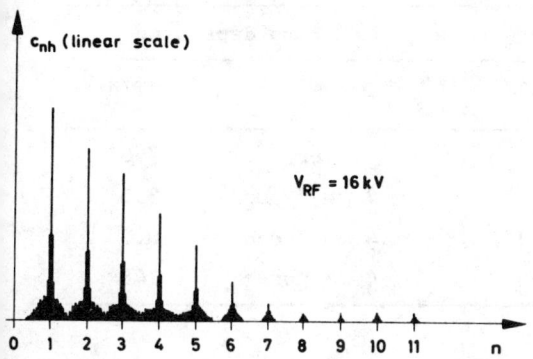 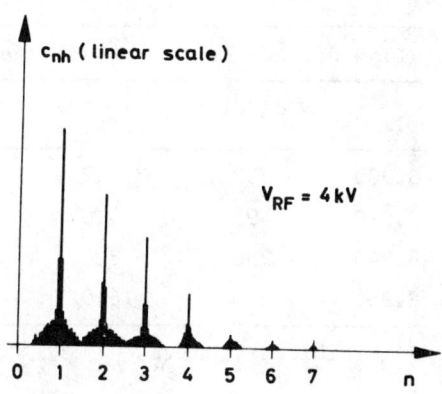

Fig. 9. Bunch frequency spectra for two different rf voltages.

beam separations become very important. Since Q_z is close to 9, large bumps in the 8 intersections cancel the effects (good bumps) while large bumps of opposite sign in diametrically opposed intersections are additive (bad bumps). Table V summarizes the effects on beam losses of the tune, of the rf voltage and of the bumps.

Table IV. Single beam overlap knock-out resonance effects.

	Tune dependence			rf voltage dependence	
Q_x	Q_z	Plane	dI/dt[a]	V_{rf} (kV)	dI/dt[a]
8.955	8.935	H	17	0.7	0.2
		V	12	4.0	8.2
8.975	8.955	H	25	8.0	17.5
		V	32	16.0	25.0

[a] in arbitrary units

In another experiment, a small and vertically limited beam of $\sim$ 80 mA was left circulating at the top of ELSA while moving bunches in the other ring. Enhancements of current loss when overlap knock-out resonances are crossed, have been observed (Fig. 10).

It is finally interesting to mention the measurement made in the ISR of two bunched beam overlap knock-out resonances.[15] The measured effects are indeed very strong under the conditions of the experiment, since the first harmonic c_h excited the resonances and

Table V. Two-beam overlap knock-out resonance effects.

Tune dependence		rf voltage dependence		Bump dependence	
Q_z	dI/dt [a]	V_{rf} (kV)	dI/dt [a]	Bumps	dI/dt [a]
8.900	30	0.7	6	zero	7.0
8.920	95	4.0	13	2 mm "good"	3.0
8.940	200	8.0	110	4 mm "good"	0.7
8.950	> 250	16.0	150	4 mm "bad"	145

[a] in arbitrary units

since the resonances were crossed many times because of the synchrotron oscillations. Figure 11 gives an example of the observation of a dipole resonance, the two beams having different energies, i.e. 11 and 26 GeV/c. This figure shows that a bunched beam of only ∼ 150 mA induced by overlap knock-out a loss of ∼ 30% in the other ring.

VII. CONCLUSIONS

The theories, briefly presented here, can satisfactorily account for all the beam-beam effects observed in the ISR. In the case of Keil, strong nonlinearities, experiments have shown that there is no

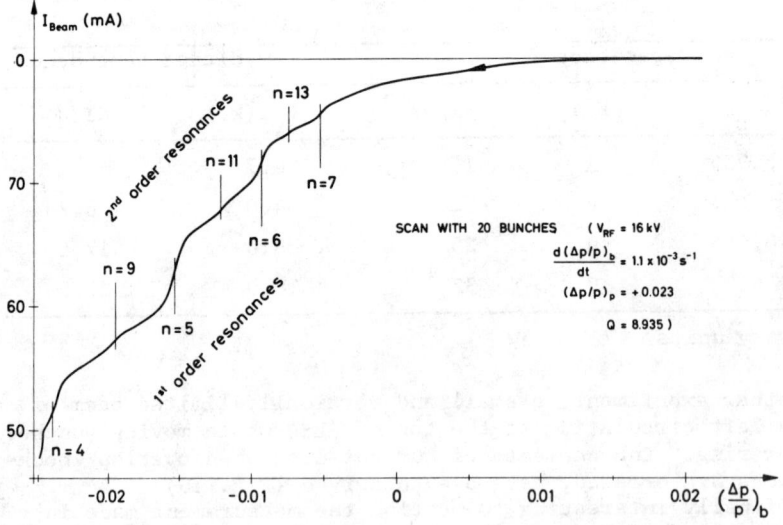

Fig. 10. Overlap knock-out resonances of order 1 and 2 for two beams at the same energy.

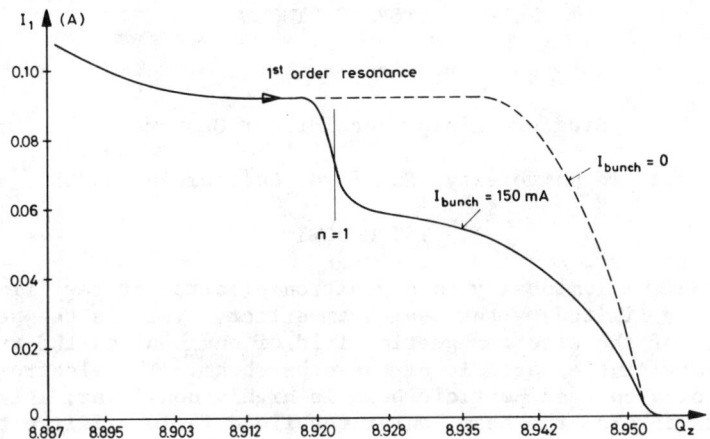

Fig. 11. Dipole overlap knock-out resonance with two bunched beams.

noticeable effect on the beam lifetime below the threshold deduced from the stochasticity theory and adopted for machine design. It is perhaps interesting to note that most of the theories presented can be applied to electron beams and that some experiments, such as overlap knock-out measurements, may give useful information for electron storage rings.

REFERENCES

1. G. Guignard, CERN 78-11 (1978).
2. E. D. Courant and H. S. Snyder, Ann. Phys. <u>3</u>, 1 (1958).
3. B. Zotter, CERN ISR-TH/75-5 (1975).
4. E. Keil, C. Pellegrini, A.M. Sessler, CERN ISR-TH/73-44 (1973).
5. E. Keil, CERN ISR-TH/72-7 (1972).
6. G. Guignard, CERN 76-06 (1976).
7. P. M. Hanney and E. Keil, CERN ISR-TH/73-55 (1973).
8. M. Month, <u>Proc. 9th Intern. Conf. on High Energy Accelerators,</u> Stanford, 1974 (CONF 740522, USAEC, Washington, 1974), p. 402.
9. J. Borer, G. Guignard, A. Hofmann, E. Peschardt, F. Sacherer, B. Zotter, IEEE Trans. Nucl. Sci. <u>NS-26</u>, 3405 (1979).
10. B. V. Chirikov, Nuclear Physics Institute, Novosibirsk, Report 267, 1969 (Translation, CERN 71-40, 1971).
11. E. Keil, <u>Proc. 8th Intern. Conf. on High Energy Accelerators,</u> CERN, 1971, p. 372.
12. K. Hübner, <u>Proc. 9th Intern. Conf. on High Energy Accelerators</u>, Stanford, 1974 (CONF 740522, USAEC, Washington, 1974), p. 63.
13. E. Keil and G. Leroy, IEEE Trans. Nucl. Sci. <u>NS-22</u>, 1370 (1975).
14. B. Zotter, <u>Proc. 10th Intern. Conf. on High Energy Accelerators</u>, Protvino, USSR, 1977, Vol. 2, p. 23.
15. S. Myers, IEEE Trans. Nucl. Sci. <u>NS-26</u>, 3574 (1979).

EXPERIMENTS ON THE BEAM-BEAM EFFECT IN e^+e^- STORAGE RINGS*

H. Wiedemann

Stanford Linear Accelerator Center

Stanford University, Stanford, California 94305

I. INTRODUCTION

The maximum luminosity in a positron electron storage ring is fundamentally limited by the beam-beam effect. This is the perturbing effect of the electromagnetic field of one beam on the trajectory of every single particle of the other beam. The electromagnetic field of a charged particle beam is highly nonlinear. At a large distance from the beam center the field falls off like the inverse of the distance. Right in the center of the beam where we have a more or less uniform density distribution over a limited region the field rises linear with the distance from the center. So the field rises from the center of the beam, saturates at some distance and going further out falls off again as 1/r. The amplitude of the field depends on the charge density as well as the aspect ratio of the beam. If the electromagnetic field of one beam gets too large, particles of the other beam get lost which leads to a reduced beam lifetime. This effect has been observed in all electron-positron or electron-electron storage rings built so far and is considered the fundamental limit on the performance of electron-positron storage rings.

Over the years many measurements have been performed at various laboratories to find the maximum permissible perturbation and the parametric dependences of that perturbation. In this paper new measurements performed at SPEAR are presented and compared with measurements from ACO,[1] ADONE,[2] and VEEP-2M.[3] Results from other storage rings have been ignored because they either show lower permissible perturbations or because of insufficient published data.

II. PHENOMENOLOGY OF THE BEAM-BEAM EFFECT

During the process of filling a storage ring with electrons or positrons both beams are customarily separated at the collision points by the use of electrostatic fields. When a current in both beams sufficiently large to exhibit a beam-beam effect but not too large to be destructive is stored, the electric fields are turned off within a few microseconds. In all experiments the intensity of both beams is equal. With both beams colliding we make the following general observation by looking at the beam cross section as transmitted via the synchrotron light:
- both beams are blown up vertically
- there is no significant horizontal blow up ($\lesssim 10\%$)
- one of the beams is blown up much more than the other one.

Figure 1(a) shows the two beam cross sections while the beams are still separated. In Fig. 1(b) the blown up cross sections of colliding beams are shown. The beam cross sections as shown in Fig. 1 are those at two different points in the ring both not at the interaction point. Therefore, the horizontal width of both beams is slightly different even where separated [Fig. 1(a)]. In Fig. 2 photometric measurements of the horizontal (x) and vertical (y) beam cross section for separated and colliding beams are shown for two different beam currents. Within the accuracy of the measurement there is no beam blow up for two colliding beams of 1 mA each, however, for a beam current of 4 mA we measure an increase of the beam height of about a factor two. The resolution of the system is about 0.25 mm.

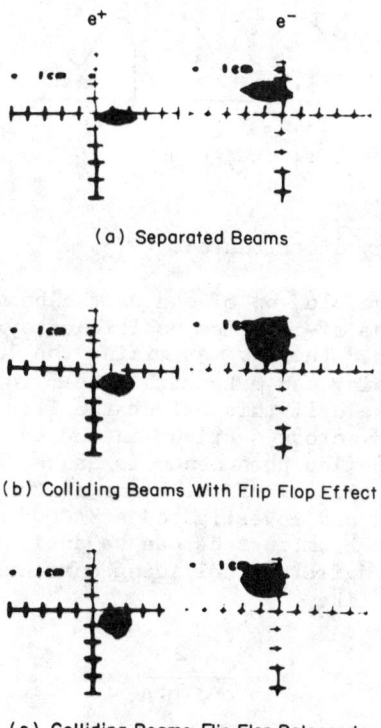

Fig. 1. Beam cross sections for separated and colliding beams.

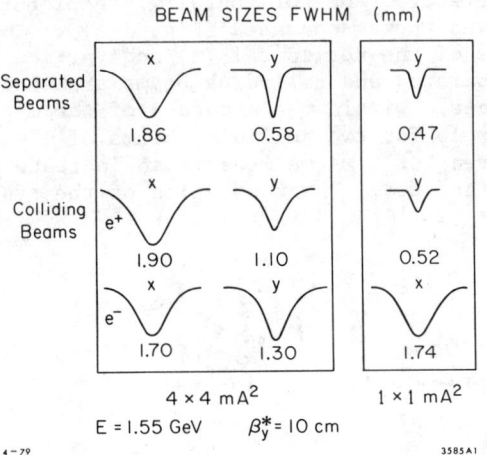

Fig. 2. Charge density distribution.

In SPEAR the large blow up of one of the beams can be avoided by properly phasing the rf-cavities positioned symmetrically about the interaction points,[4] thereby maximizing the luminosity. Because this cavity phasing can make either beam to be blown up with respect to the other we call this effect the flip-flop effect. Figure 1(c) shows the beam cross sections in the balanced state. It is believed that the flip-flop phenomenon is caused by a small horizontal separation of the centers of the beams at the interaction. This also has been observed and investigated at ACO.[5] The variation of the cross section with beam current can be derived from the luminosity measurements for different colliding currents. We expect the luminosity L to scale like

$$L = \frac{I^2}{e^2 \cdot f_0 \cdot A} \tag{1}$$

with $I = I^+ = I^-$ the beam current, f_0 the revolution frequency and A the beam cross section at the interaction point assumed to be the same for both beams. Figure 3 shows that up to a threshold current the luminosity scales like I^2. Above this threshold the luminosity is lower than expected. From Eq. (1) we expect that this is the regime where the beams get blown up. Since there is no horizontal beam blow up the ratio of the expected to actual luminosity is just equal to the vertical blow up of the core of the beams. As we increase the current beyond the threshold the luminosity increases[6] like $L \sim I^{1.3}$ and the beams get more and more blown up. The current limit and with it the maximum luminosity is reached when after bringing both beams into collision one of them exhibits a short lifetime. In practice, the limit is reached at somewhat lower currents because before any reduction in lifetime is observed the amount of background for the experimental detectors increases significantly.

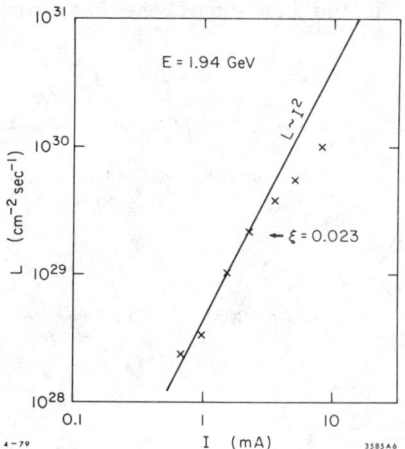

Fig. 3. Luminosity vs Beam Current.

We conclude from these observations that the beam-beam effect blows up the vertical beam size as a function of the intensity of the colliding beams. The limit is reached when the extreme tails of the density distribution reach the aperture limit which first causes increased background and then a reduction in beam lifetime. From the luminosity measurements or the beam size measurements using the synchrotron light we only can derive the dimensions of the core of the beam (up to 2σ for a Gaussian beam). Background and beam lifetime, however, are determined by the particles in the far out tails of the distribution. Since it is not obvious from luminosity measurements how the large amplitude particles are affected by the beam-beam effect the density distribution in the tails has to be measured separately. This we have done in SPEAR for single as well as colliding beams.

The density distribution in the tails is measured with the use of beam scrapers by observing the reduction of beam lifetime as a function of the position of the scrapers. In electron-positron storage rings the density distribution is expected to be Gaussian. For such a distribution the beam lifetime is given by[7]

$$\tau = \frac{1}{2} \tau_\beta \frac{e^\zeta}{\zeta} \qquad (2)$$

where τ_β is the transverse damping time and $\zeta = s^2/2\sigma_\beta^2$ with σ_β the transverse standard deviation of the beam size and s the position of the scrapers.

We measure the beam lifetime for a scraper position s_M, correct that lifetime for the residual gas lifetime and then calculate the ratio s/σ_β from Eq. (2) to give the same lifetime. For a truly

Gaussian beam the quantity s/σ_β should be a linear function of s_M. In Figs. 4 and 5 the results of these measurements together with the beam size measurement in the center of the beam are plotted for separated and colliding beam.

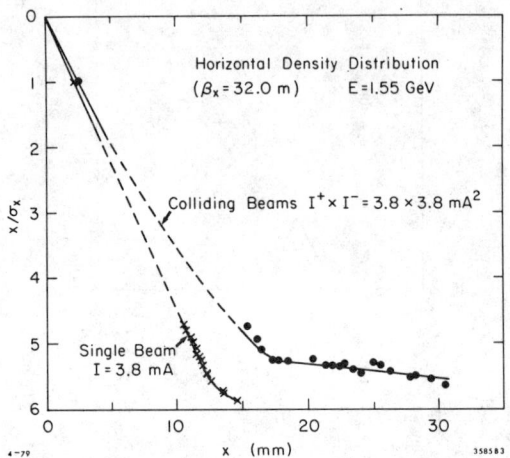

Fig. 4. Horizontal particle density distribution.

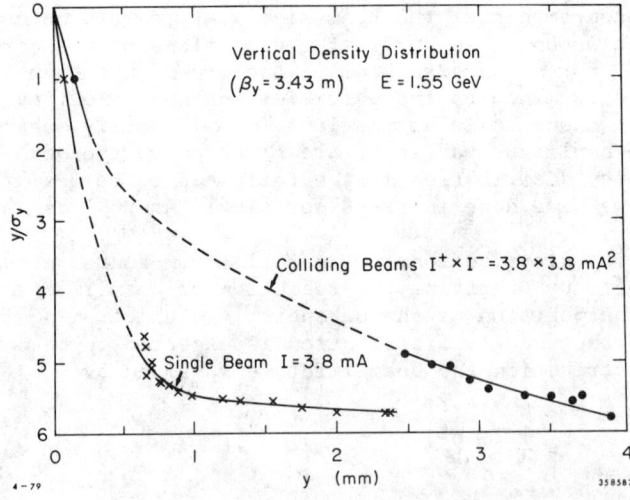

Fig. 5. Vertical particle density distribution.

It is seen clearly that in the horizontal plane the density distribution is Gaussian for a single beam up to about 5.5 $\sigma_{\beta x}$. At larger amplitudes there is a long tail which appears in all measurements.

These tails may be caused by nonlinearities. In the vertical plane the density distribution for a single beam is not quite Gaussian which might be due to nonlinear coupling between the horizontal and vertical plane.

For colliding beams we observe little blow up of the horizontal beam size in the center but significantly more in the tails. Much more dramatic, however, is the blow up of the vertical tails which is of the order of a factor 5 compared to a factor of 2 in the center of the beam. It is this large vertical blow up that requires a vertical acceptance of the storage ring that is larger than expected from beam size measurements of the core.

III. PARAMETRIC DEPENDENCIES OF THE BEAM-BEAM EFFECT

In order to understand more about the beam-beam effect, measurements of the limit as a function of many parameters have to be performed. To characterize the beam-beam effect it has become customary to measure or calculate the linear tune shift for small amplitude particles due to the space charge field of the other beam. As mentioned before, the electromagnetic field in the center of the beam increases linearly with the distance from the center. This is the characteristic field of a quadrupole which causes a shift in the betatron frequency, the tune of the storage ring, when both beams are brought into collision. Since all higher-order components of the space charge field are strictly proportional to the linear component, the linear tune shift is a characteristic quantity for the whole nonlinear field. The linear tune shift is given by

$$\xi_{x,y} = \frac{r_e}{2\pi e f_0 B \gamma} \frac{I \beta_{x,y}}{\sigma_{x,y}(\sigma_x + \sigma_y)} \qquad (3)$$

where r_e the classical electron radius, e the electric charge unit, f_0 the revolution frequency, B the number of bunches per beam, γ the beam energy in mc^2, I the beam current, $\beta_{x,y}$ the horizontal or vertical betatron function at the interaction point and σ_x σ_y the beam sizes at the interaction point. Since generally $\xi_x < \xi_y$, we use only $\xi_y = \xi$ in the rest of this paper. The Eq. (3) assumes a Gaussian distribution which is always sufficiently the case in the core of the beam. To calculate the linear tune shift from Eq. (3) requires the accurate knowledge of the beam sizes at the interaction point. This is best done by measuring the luminosity:

$$L = \frac{1}{4\pi e^2 f_0} \frac{I^2}{B \sigma_x \sigma_y} . \qquad (4)$$

If we combine Eqs. (3) and (4) and eliminate $\sigma_x \cdot \sigma_y$ we get

$$\xi = 2 r_e e \frac{L/I}{\gamma} \frac{\beta_y}{1 + \frac{\sigma_y}{\sigma_x}} . \qquad (4)$$

Here all quantities on the right hand side can be measured except for σ_x/σ_y. This ratio, however, is rather well known in cases where the beams are fully coupled like in ADONE or ACO or is very small like in SPEAR and, therefore, can be ignored. In a simple model used so far for the design of storage rings it is assumed that there is a maximum value for ξ (usually 0.06) independent of other parameters. With this assumption (ξ_{max} = constant) we would expect the luminosity to scale like

$$L \sim \xi_{max}^2 \, \gamma^2 \, \frac{B\epsilon_x}{\beta_y^{3/2}} \sim \xi_{max}^2 \, \gamma^4 \, \frac{B}{\beta_y^{3/2}} \tag{6}$$

with ϵ_x the horizontal beam emittance ($\epsilon_x \sim \gamma^2$). In the rest of this paragraph we will discuss the validity of this model.

In SPEAR and other storage rings, measurements have been performed to determine the parametric dependence of the maximum tune shift which we will discuss in the remainder of this section: $\xi_{max} = f(\beta_y)$: According to Eq. (6) the luminosity is expected to vary inversely proportional to $\beta_y^{3/2}$ with all other relevant parameters staying constant. In SPEAR the maximum luminosity was measured as a function of β_y and is shown in Fig. 6 together with the tune shift. It is evident that below a certain value of β_y - here about 10 cm - there is no gain in luminosity any more. In fact, the luminosity even drops to lower values. A possible explanation could be the fact that the bunch length $\sigma_z \approx 2$ to 3 cm becomes comparable with the values of the beta function. Especially particles in the tails (5 to 6 σ_z) of the longitudinal density distribution will collide with the core of the other beam at some distance from the interaction point and therefore at larger values of β_y. Further measurements on that point, however, are required.

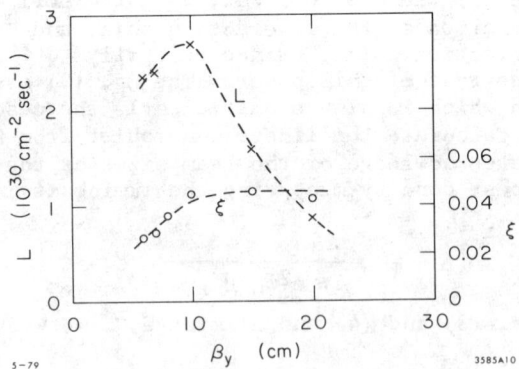

Fig. 6. Luminosity and tune shift parameter ξ vs the vertical betatron function.

$\xi_{max} = f(\nu_x, \nu_y, \nu_s)$: A strong dependence of the maximum achievable linear tune shift with the tune of the storage ring has been predicted.[8,9] Measurements at ADONE[2] and SPEAR[10] confirm this prediction which generally states that in electron-positron storage rings the beam-beam limit increases as the vertical tune approaches an integer from higher values. This can clearly be seen in Fig. 7 from Ref. 10.

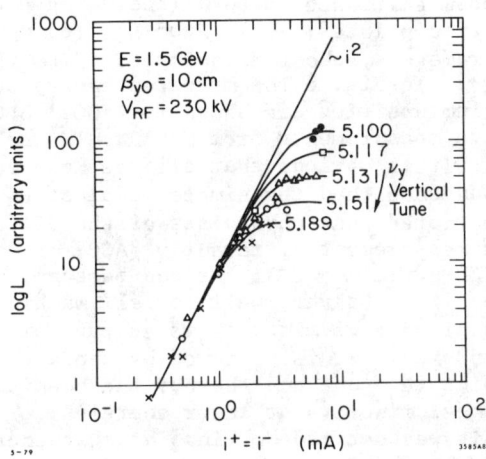

Fig. 7. Luminosity vs beam current and vertical tune.

During these measurements the horizontal tune was kept constant at $\nu_x = 5.20$ for $5.189 > \nu_y > 5.117$. For the case $\nu_y = 5.10$ the horizontal tune had to be lowered to $\nu_x = 5.15$ in order to avoid the resonance line $2\nu_y - \nu_x = 5$ which we would have had to cross for $\nu_x = 5.20$ on the way from the injection configuration to the collision configuration.

It is interesting to note that this resonance $2\nu_y - \nu_x = 5$ is a "forbidden" resonance in the 2-fold symmetry SPEAR. Indeed, for a single beam this resonance cannot be detected. In the beam operation, however, this line is strong enough to destroy one of the beams, an indication that due to the strong beam-beam nonlinearities "forbidden" ordinary betatron resonances appear very strong in higher-order terms. With the change in SPEAR of the rf-frequency from 50 MHz to 350 MHz in 1974 the strong dependence of the beam-beam limit with ν_y vanished almost completely. The only cause for that we can think of is that with the higher rf-frequency the synchrotron frequency increased from values $\nu_s < 0.01$ to values of $\nu_s \approx 0.03$. The synchrotron frequency causes synchrobetatron oscillations which appear as satellites of the integer tune. For small values of ν_s all relevant satellites are very close to the integer tune while for $\nu_s = 0.03$ they spread over the whole tune diagram. It is the appearance of these narrow spaced resonances which we believe makes the tune dependence of the beam-beam limit vanish in SPEAR. Obviously more measurements are in order to verify this.

$\xi_{max} = f(\delta_x, \delta_y)$: As discussed in Section II the luminosity and thereby the beam-beam limit depend very sensitively on the separation δx and δy of the beams at the interaction point. Every possible effort should be made to minimize beam center separation to avoid the before-mentioned flip-flop phenomenon.

$\xi_{max} = f(E, \epsilon_{ytot}, n_{IP})$: In this paragraph we want to investigate the dependence of the beam-beam limit on the energy (E), the storage ring acceptance or total beam emittance (ϵ_{ytot}) (including the tails) and the number of interaction points (n_{IP}). From Eq. (6) we expect, with otherwise constant parameters, the luminosity to scale like the fourth power of energy $L \sim E^4$. In Fig. 8 the measured energy dependence of the maximum achieved luminosities are shown for ACO,[1] ADONE,[2] and VEEP-2M,[3] together with measurements from SPEAR. The solid lines are fits with $L \sim E^6$. It is obvious that all the measurements show a stronger energy dependence than the simple model of Eq. (6) would predict. It should be noted that physicists of the different storage ring groups fit their measurement differently (ACO: $L \sim E^{5.4}$, ADONE, $L \sim E^7$, and VEEP-2M: $L \sim E^4$). As can be seen, however, from Fig. 8, a $L \sim E^6$ fit works very well for all machines. In the case of VEEP-2M the published[3] luminosity vs energy curve has a distinct step at about 450 MeV. It is believed by the author of this paper that at this point the tune and thereby the beam size has been changed to maximize the luminosity at lower energies. This was mentioned in Ref. 3 but it was not made obvious at what energy and by how much the tune was changed.

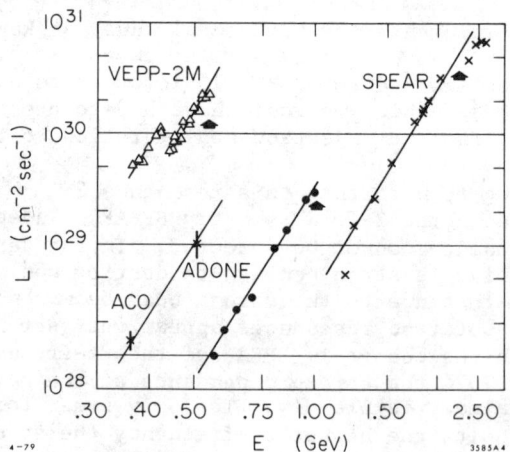

Fig. 8. Maximum luminosity vs energy.

For the rest of this paper we assume there is a universal beam-beam effect in all four storage rings that makes the maximum achievable luminosity scale like

$$L \sim E^6. \tag{7}$$

It should be mentioned, however, that above a certain energy indicated by an arrow ↑ in Fig. 8 the luminosity all of a sudden levels off or even drops again. For VEEP-2M this is so because of technical reasons which limit the current that can be stored. In SPEAR this "transition" energy coincides with the maximum injection energy. Operation at higher energies requires the beams to be brought from the injection energy up to higher energies. This prevents the operators in SPEAR from making little improvements over a longer period of time when all the ring magnets are in a steady state. In cases where energy ramping is required the end configuration never is exactly the same since the different magnets track differently. However, the difference between achieved and expected luminosity is large and one can suspect that some other effect as yet unexplored may come into play. This notation is also nourished by the ADONE data. The injection energy in ADONE is 350 MeV and therefore the energy always has to be ramped yet the luminosity levels off only above 1 GeV. For the rest of this paper we consider only that energy regime in which the luminosity scales like E^6. From the luminosity measurements of Fig. 8 we can calculate the linear tune shift if the beam current is known. In Fig. 9, the tune shift parameter ξ as a function of energy is plotted for ADONE[2] and SPEAR. All measurements from ADONE involve 6 interaction points while SPEAR only has 2 interaction points. For ACO and VEEP-2M no data for the beam currents were available to calculate the ξ-parameter. From Fig. 9, we conclude that the maximum linear tune shift is a linear function of energy and not as commonly assumed, a constant:

$$\xi_{max} \sim E \tag{8}$$

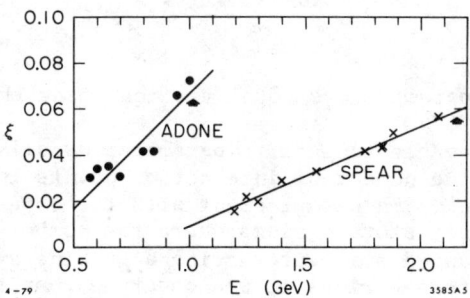

Fig. 9. Maximum tune shift vs energy.

In Section II it was described how the vertical beam blow up increases with increasing beam current until the aperture limit is

reached. We have investigated this in more detail in SPEAR. With
the help of a scraper the total vertical beam size was measured as
a function of the beam current. In Fig. 10, the result of such a
measurement is shown. We plot the tune shift parameter versus the
square root of the total beam emittance for reasons that will become apparent in the next section. The result of the measurement
looks very surprising. For a small beam-beam effect the beam size
is large then decreases with increasing beam-beam effect and finally
increases again. Since this variation of the beam size has been observed for different configurations it must be assumed to be real.
The large tails, and this is all we measure here, at low currents
are consistent with the tails observed in single beams (see Fig. 5)
and could be caused by the nonlinear magnetic field of the sextupoles. Addition of the beam-beam nonlinearity reduces at first the
amplitude of the tails for reasons as yet not investigated. Only
for ξ-parameters larger than 0.02 we observe the expected beam blow
up, which is consistent with the observation in Fig. 3.

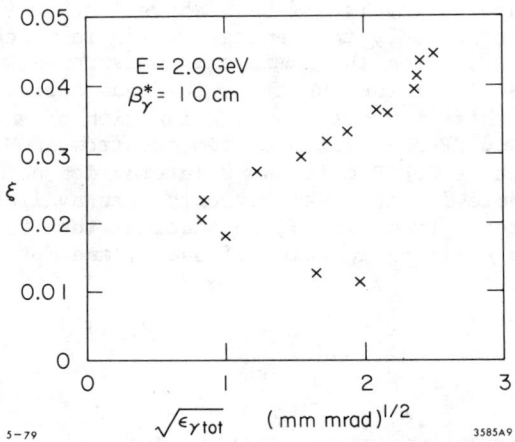

Fig. 10. Tune shift parameter vs total vertical beam emittance.

The maximum achievable tune shift parameter usually is assumed
to be independent of the number of interaction points in a storage
ring. This assumption is in disagreement with the observations from
ACO[1] and ADONE[2] the only storage rings where the number of interaction points can be changed still preserving a machine symmetry of at
least two. In both storage rings it was observed that the maximum
value of ξ is reduced as the number of interaction points is increased. The data are consistent with a scaling of

$$\xi_{max} \sim \frac{1}{\sqrt{n_{IP}}} \qquad (9)$$

which we will assume to be correct in the following section.

IV. TRY OF AN EMPIRICAL SCALING LAW FOR THE LINEAR TUNE SHIFT PARAMETER ξ_{max}

From experiments on the beam-beam effect we found the parameter ξ_{max} to vary like the energy of the beams. We also observe that, while the beam-beam effect blows up the vertical beam size, this blow up reaches a steady state amplitude depending on the beam current. This behavior rules out an instability or a strong resonance to be the cause for the beam-beam limit. Whatever the mechanism to blow up the beam, it seems in electron-positron storage rings to be successfully counteracted by the damping due to synchrotron radiation. In order to test the parametric dependence of the maximum tune shift parameter ξ_{max}, we try the following model:

The total vertical betatron emittance ϵ_{ytot} changes with time due to damping like[7]

$$\frac{d}{dt}\sqrt{\epsilon_{ytot}} = -\frac{\sqrt{\epsilon_{ytot}}}{2\tau_y} \qquad (10)$$

with τ_y the damping time. This damping is counteracted in the case of colliding beams by the beam-beam effect. We assume the blow up to be caused by a diffusion-like process as suggested, for example, by H. G. Hereward[11] and J. R. LeDuff.[12] Without using any theoretical derivation we assume the vertical beam size blow up to be proportional to the strength of the nonlinear field and the betatron amplitude at the interaction point or, in other words, the blow up is proportional to ξ. If one or more resonances are the driving source for the diffusion process not only the strength of the nonlinear field but also the amplitude of the particle plays an important role. It is well known that nonlinear resonances affect mostly large amplitude particles which are just the ones that limit the beam lifetime if they get scraped at the walls of the vacuum chamber. We try, therefore, the blow up rate to scale like $d/dt\sqrt{\epsilon_{ytot}} \sim \epsilon_x$ using ϵ_x rather than ϵ_y because the observation shows that the large amplitude particles are almost fully coupled in the horizontal and vertical phase plane and ϵ_x is a well known quantity in electron-positron storage rings. Since the beam-beam effect only changes the slope of the particle's trajectory but not the amplitude we assume the square root of the total emittance $\sqrt{\epsilon_{ytot}}$ to change linearly with time rather than ϵ_{ytot}. We further assume that the rate of growth of the vertical amplitude depends on the number of collisions per unit time. For a diffusion-like process the amplitude growth rate should scale like the square root of the number of collisions per second. With this model we get a growth rate which scales like:

$$\frac{d}{dt}\sqrt{\epsilon_{ytot}} \sim \xi_{max} \epsilon_x \sqrt{\frac{n_{IP}}{C}} \qquad (11)$$

where C is the circumference of the storage ring.

Both damping [Eq. (10)] and excitation [Eq. (11)] lead to an equilibrium with:

$$\xi_{max} = \mu \frac{\sqrt{\epsilon_{ytot}}}{\epsilon_x \tau_y} \sqrt{\frac{C}{n_{IP}}} \qquad (12)$$

where μ is a constant and ϵ_{ytot} the total vertical beam emittance.

From a dimensional point of view Eq. (11) is not very satisfactory, but since we do not know the exact mechanism of the beam-beam blow up we bury all dimensions in the proportionality constant μ and use Eq. (12) only as a guide for experiments and as a possible empirical scaling law.

In Eq. (12), we discover first the correct energy dependence. Since $\epsilon_x \sim E^2$ and $\tau_y \sim E^{-3}$ we find $\xi_{max} \sim E$ in agreement with measurements at ADONE and SPEAR (Fig. 9). With the maximum beam-beam parameter $\xi_{max} \sim E$ we arrive at a luminosity scaling like $L \sim E^6$. This is in agreement with all the observations as shown in Fig. 8. The dependence of ξ_{max} on the square root of the number of interaction points is in agreement with observations in ADONE and ACO and supports therefore, strongly a diffusion process.

Fitting the measurements of Fig. 8 we get

$$\xi_{ADONE} = -0.033 + 0.099 \, E(GeV)$$
$$\xi_{SPEAR} = -0.035 + 0.044 \, E(GeV) \qquad (13)$$

and with the parameters of Table I:

$$\xi_{ADONE} = -0.033 + 1.6 \times 10^{-8} \frac{\sqrt{\epsilon_{ytot}}}{\epsilon_x \tau_y} \sqrt{\frac{C}{n_{IP}}}$$

$$\xi_{SPEAR} = -0.035 + 1.33 \times 10^{-8} \frac{\sqrt{\epsilon_{ytot}}}{\epsilon_x \tau_y} \sqrt{\frac{C}{n_{IP}}} \qquad (14)$$

TABLE I

	ϵ_{ytot} (rad m)	ϵ_x/E^2 (rad m GeV^{-2})	τ_y/E^3 (sec GeV^{-3})	C (m)	n_{IP}
ADONE	45×10^{-6}	11.7×10^{-8}	0.038	100	6
SPEAR	12×10^{-6}	5.0×10^{-8}	0.226	234	2

In spite of the very different values of the parameters involved for both machines the proportionality factors are close enough to be considered equal.

As opposed to Eq. (12), however, in Eqs. (13) and (14) there appears a constant offset. There is no explanation available to date what the nature of this offset might be. Further measurements are required.

V. CONCLUSION

The maximum achieved beam-beam parameter ξ has been measured in SPEAR and compared with other storage rings. At lower energies we found a consistent behavior of the beam-beam effect leading to a maximum luminosity to scale like $L \sim E^6$ and a maximum beam-beam parameter scaling like $\xi_{max} \sim E$. This is in contradiction to the generally assumed constant value of ξ_{max} for the design of new storage rings. An empirical scaling law has been described which is consistent with the measurements available. The author is aware of the lack of detailed theoretical background for this model, but it has helped to perform specific measurements which might be useful to finally understand the beam-beam effect in electron-positron storage rings.

In general, we conclude that the damping time and the vertical acceptance of the storage ring are of prime importance in reaching large luminosities.

ACKNOWLEDGMENTS

The author wishes to sincerely thank Tom Taylor and the SPEAR Operations Group for their continuous help and support in performing the measurements. This is especially true for the measurements of the maximum achieved luminosities at various energies which were taken in a large part from the operations logbooks as achieved while running SPEAR for high energy physics.

REFERENCES

1. R. Belbeoch et al., Rapport Tech. 3-73 (1973).
2. Private communication from S. Tazzari. F. Amman et al., Proc. VIII Intern. Conf. on High Energy Accelerators, Geneva, 132 (1971).
3. I. B. Vasserman et al., Proc. All-Union Conf. on Charged Particle Accelerators, Dubna, USSR (1978).
4. J. M. Paterson and M. Donald, 1979 Particle Accelerator Conference, San Francisco, March 12-14, 1979. An Investigation of the Flip-Flop Beam-Beam Effect in SPEAR.
5. R. Belbeoch et al., Internal Report NI/38-77, ORSAY (1977).
6. M. Cornacchia, PEP Note 275, Stanford, (November 1978).
7. M. Sands, Physics with Intersecting Storage Rings, Ed. by B. Touschek, Academic Press, New York, 1971.
8. M. Bassetti, Proc. V Intern. Conf. on High Energy Accelerators, Frascati, 708 (1965).

9. B. Richter, Proc. Intern. Symp. Electron and Positron Storage Rings, Saclay, I-1-1 (1966).
10. SPEAR-Group, IEEE Trans. Nucl. Sci., NS-20, No. 3, 838 (1973).
11. H. G. Hereward, CERN ISR-DI 72-26 (1972).
12. J. R. LeDuff, ORSAY Rap. Tech. 6-72 (1972) and PEP Note 65 (1973).

REPORT ON SOME BEAM-BEAM

FUNCTIONAL DEPENDENCIES IN SPEAR

M. Cornacchia

Stanford Linear Accelerator Center*

I. INTRODUCTION

A considerable amount of experimental results on beam-beam effects in SPEAR is available. We have analyzed the results which give the functional dependences of some important machine parameters. The data have been taken from machine physics experiments carried out during the period December 1977 to October 1978, and from records of the operation runs.

II. DEFINITIONS

The experimental data cover the energy range 1.5 - 3.9 GeV. There are two circulating bunches (one electron, one positron) and consequently, two interaction regions. Unless otherwise specified, the symbols, definitions and values of the relevant parameters referred to in this paper are the following:

ν_{xo} = horizontal unperturbed betatron tune = 5.28

ν_{yo} = vertical unperturbed betatron tune = 5.18

β_x^* = horizontal beta function at the interaction points = 1.2 m

β_y^* = vertical beta function at the interaction points = 10 cm

η_x^* = horizontal dispersion at the interaction points = 0.00186 m

The luminosity is defined as[1]

$$\mathcal{L} = \frac{1}{4e^2 f} \frac{i^+ i^-}{A_{int}} , \qquad (1)$$

where

$A_{int} = \pi \sigma_x \sigma_y$

f = revolution frequency

e = electron charge

A_{int} = effective interaction area

i^+, i^- = beam currents.

---

*Current address: Brookhaven National Laboratory.

and where σ_x and σ_y are the standard deviations of the Gaussian distributions of the transverse density of the beam at the interaction points.

The 'space charge parameter', ξ, is the vertical linear tune shift due to the beam-beam forces if the vertical tune change between interaction points is not too close to an integer. It is a measure of the strength of the beam-beam interaction. For a Gaussian beam with head-on collision, it is given by[1]

$$\xi = \frac{r_e}{2\pi} \frac{N_B \beta_y^*}{\gamma \sigma_y (\sigma_x + \sigma_y)}, \qquad (2)$$

where

r_e = classical electron radius

N_B = number of particles per bunch

γ = (energy/rest energy).

The linear tune shift, $\Delta\nu_y$, is the tune shift due to the linear component of the space charge forces; it is related to the space charge parameter introduced earlier by the relationship:

$$\cos 2\pi(\nu_{yo} + \Delta\nu_y) = \cos 2\pi\nu_{yo} - 2\pi\xi \sin 2\pi\nu_{yo}. \qquad (3)$$

III. LUMINOSITY AS A FUNCTION OF BEAM CURRENT

Equation (1) predicts the dependence, if $i^+ = i^- = i$,

$$\mathcal{L} \simeq i^2.$$

Experiments performed at different times[2] indicate that, above a certain intensity, the luminosity at ~ 2 GeV does not scale like i^2 but, rather, like $i^{1.3}$. This is shown in Fig. (1); the luminosity starts departing from the quadratic law when the beam current reaches about 2 mA, at 1.95 GeV; this corresponds to a space charge parameter

$$\xi = 0.016.$$

Whereas the luminosity depends on many machine parameters in a way which is not completely understood, the functional dependence of the luminosity on the beam current is rather reproducible. It is, for instance, not very sensitive to the amplitude and shape of the residual closed orbit distortions and to the coupling between horizontal and vertical planes; it is also insensitive to the 'bunch lengthening cavity', which drastically modifies the beam density function in the longitudinal phase space. The same experiment[3] of luminosity versus current with a vertical beta function at the interaction points of 20 cm (instead of 10 cm) gave a dependence $\mathcal{L} \simeq i^{1.45}$ at 2 GeV.

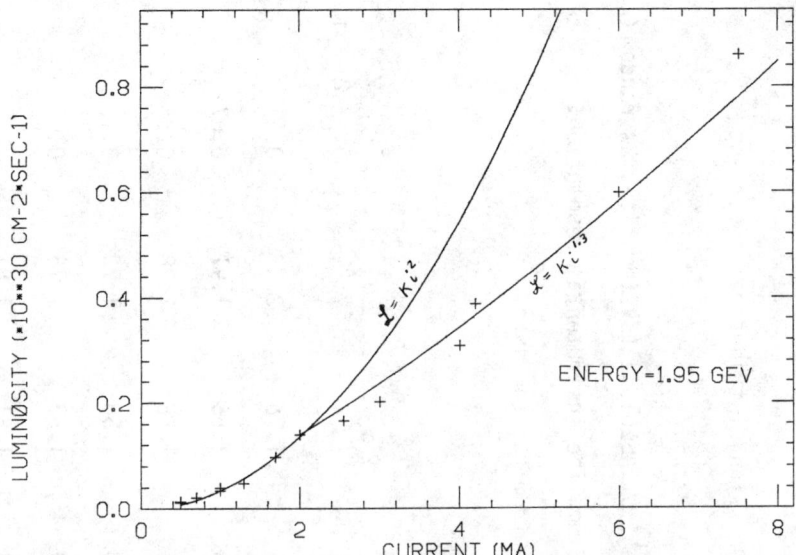

Fig. 1. Luminosity versus beam current.

Since the other factor intervening in the luminosity is the beam cross section, we postulate, in order to fit the experimental data, that the transverse beam size increases with current. In order to test the above hypothesis, the horizontal and vertical beam sizes, as given by the synchrotron light monitors, have been measured.[4] The results are given in Table 1 and 2 below. In these tables we have recorded the vertical and horizontal beam sizes in mm as a function of the colliding beams currents. The beam size in Table 1 is full width measured at 60.6% of the peak of the density distribution; if the current density is truly Gaussian, this beam dimension corresponds to two standard deviations.

Similarly, Table 2 gives the full beam sizes at 13.5% of the peak; for a Gaussian distribution, they should be 2 times the values listed in Table 1 (4 standard deviations). Any departure from this factor 2 is a qualitative indication of how much the distribution differs from a Gaussian.

The ratios between the beam sizes in Tables 1 and 2 are listed in Table 3.

The positron beam blows up much more than the electron beam, in the vertical plane. This is a phenomenon which is known to occur close to the beam-beam limit: depending on some machine parameters (phasing of the rf cavities, horizontal dispersion), in a not understood way, one or the other of the two colliding beams blows up.[6]

A least square fit of the logarithm of the vertical beam size against the logarithm of the current gave the following slopes (values of K for a dependence $\sigma \simeq i^K$):

TABLE 1

Beam size; 2 standard deviations for a Gaussian distribution of current density.

Estimated measurement error on beam size : 5% absolute (systematic), 2% random (rms)

Error in reading plotted output : ±0.03 mm. The instrumental resolution; 0.12 mm has been subtracted in rms fashion from the row data.

i^+ (mA)	i^- (mA)	$2\sigma_y^-$ (mm)	$2\sigma_y^+$ (mm)	$2\sigma_x^-$ (mm)	$2\sigma_x^+$ (mm)	$\mathscr{L}$ (10^{30} cm^{-2}×sec^{-1})
7.85	7.85	0.64	1.43	1.76	2.08	0.87
5.70	5.64	0.63	0.93	1.65	1.91	0.67
3.58	3.51	0.64	0.76	1.58	1.78	0.34
2.15	2.13	0.47	0.53	1.58	1.73	0.17
1.02	1.00	0.37	0.40	1.58	1.73	0.05
Theoretical (for 10% coupling)		0.28	0.28	1.81	1.81	

TABLE 2

Beam size; 4 standard deviations for a Gaussian distribution of current density.

i^+ (mA)	i^- (mA)	$4\sigma_y^-$ (mm)	$4\sigma_y^+$ (mm)	$4\sigma_x^-$ (mm)	$4\sigma_x^+$ (mm)	$\mathscr{L}$ (10^{30} cm^{-2} × sec^{-1})
7.85	7.85	1.57	3.10	3.53	4.05	0.87
5.70	5.64	1.48	2.05	3.19	3.69	0.67
3.58	3.51	1.38	1.47	3.06	3.51	0.34
2.15	2.13	.91	1.01	3.16	3.36	0.17
1.02	1.00	0.72	0.79	3.16	3.42	0.05
Theoretical (10% coupling)		0.56	0.56	3.62	3.62	

TABLE 3

Ratios between the values in Table 2 and the corresponding ones in Table 1 (the ratios should have the values 2 for Gaussian distributions).

i^+ (mA)	i^- (mA)	$\dfrac{4\sigma_y^-}{2\sigma_y^-}$	$\dfrac{4\sigma_y^+}{2\sigma_y^+}$	$\dfrac{4\sigma_x^-}{2\sigma_x^-}$	$\dfrac{4\sigma_x^+}{2\sigma_x^+}$
7.85	7.85	2.45 ±0.09	2.17 ±0.04	2.00 ±0.03	1.95 ±0.03
5.70	5.64	2.35 ±0.09	2.20 ±0.06	1.93 ±0.03	1.93 ±0.03
3.58	3.51	2.16 ±0.09	1.93 ±0.08	1.94 ±0.03	1.97 ±0.03
2.15	2.13	1.94 ±0.13	1.90 ±0.11	2.00 ±0.03	1.94 ±0.03
1.02	1.00	1.94 ±0.16	1.97 ±0.15	2.00 ±0.03	1.98 ±0.03

vertical; 2 standard deviations, electrons, K = 0.28 ± 0.08
vertical; 2 standard deviations, positrons, K = 0.59 ± 0.10
vertical; 4 standard deviations, electrons, K = 0.40 ± 0.08
vertical; 4 standard deviations, positrons, K = 0.65 ± 0.12

The least square fit of the logarithm of the luminosity versus the logarithm of the beam current gave the dependence $\mathcal{L} \simeq i^{1.41\pm0.09}$.

When two beams with Gaussian density distribution of different vertical standard deviations collide, the effective interaction dimension σ_y in the expression of the luminosity (Eq. 1) must be replaced by

$$\sqrt{(\sigma_y^+)^2 + (\sigma_y^-)^2}$$

where σ_y^+ and σ_y^- are the rms vertical beam sizes of the two beams. Thus, in Table 1, the luminosity at the higher current levels is largely determined by the beam size of the blown-up beam (positrons), which scales with the beam current like $\sigma_y^+ \simeq i^{0.59\pm0.10}$.

We find, therefore, good agreement between the dependence of luminosity and beam size with current: the empirical dependence for the luminosity was found to be $\mathcal{L} \simeq i^{1.41\pm0.09}$ which implies a vertical beam size dependence as $\sigma_y \simeq i^{0.59}$, which agrees with the measured $\sigma_y \simeq i^{0.59\pm0.10}$. We also observe in Table 3 that, at higher beam currents, the ratios of the beam sizes measured at 60.5% and 13.5% of the peak differ from the value 2; this implies that the density distribution departs from a Gaussian, and that the tails spread out more than the core. The horizontal beam sizes experience very little blow-up; note that the value of the horizontal dispersion function (η_x^*) at the light monitor positions is 0.98 m.

Another interesting result comes from the direct measurements of the tune shift with colliding beams;[7] the fit of the experimental data shows a dependence $\Delta\nu_y \simeq i^{1/3}$ at 2.4 GeV. Since (Eq. 2), $\xi \simeq i/\sigma_x\sigma_y$ (if $\sigma_y \ll \sigma_x$), there seems to be qualitative agreement with the proposed empirical law $\sigma_y \simeq i^{0.60}$.

We do not have recent data from machine physics experiments on the dependence of luminosity on current at different energies. We do have, however, operation data, where the luminosity and the current are recorded periodically. The range of recorded values covers a rather restricted interval, the upper limit being just below the beam-beam limit, for stable operation. We have selected most of the runs at 1.5, 2.5 and 3.7 GeV in the period July 1976 to June 1978. Each set of measurements consists of about 30 points (luminosity-current) at a given energy. A linear least square fit of the logarithm of the luminosity (in units 10^{30} cm^{-2} sec^{-1}) versus the logarithm of the current (in mA) was computed; the slope of the straight line gives the values of the exponent α, if the dependence $\mathcal{L} \simeq i^\alpha$ is admitted.

The results of the analysis are summarized in Fig. 2. In it, the exponent α versus the energy is plotted. Each point in the graph corresponds to one set of measurements. Figures 3, 4, and 5 show some examples of the original data (logarithm of the luminosity versus the logarithm of the beam current) at different energies.

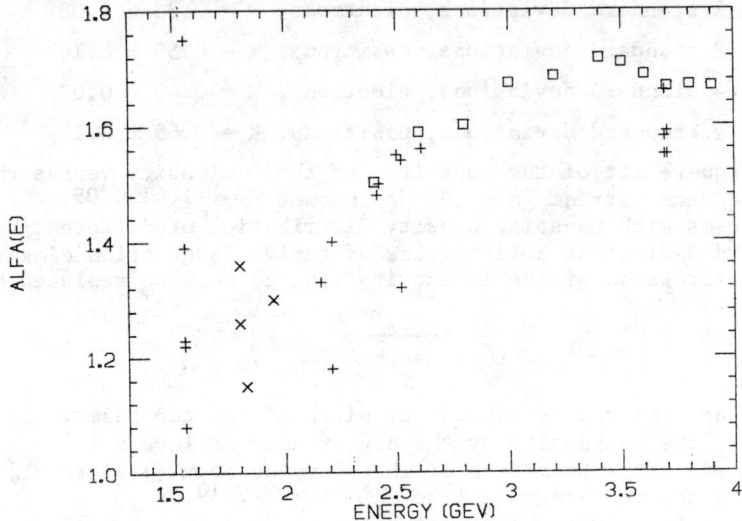

+ = OPERATION DATA
x = MACHINE PHYSICS EXPERIMENTS DATA
□ = CALCULATED FROM THE LUMINOSITY VERSUS ENERGY DATA

Fig. 2. Dependence of the coefficient α ($\mathcal{L} \simeq i^{\alpha}$) on energy.

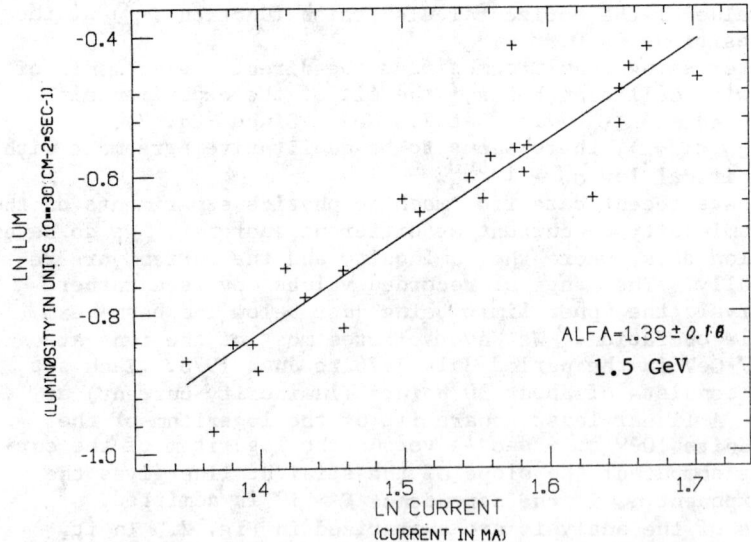

Fig. 3. Logarithmic dependence of the luminosity on the beam current.

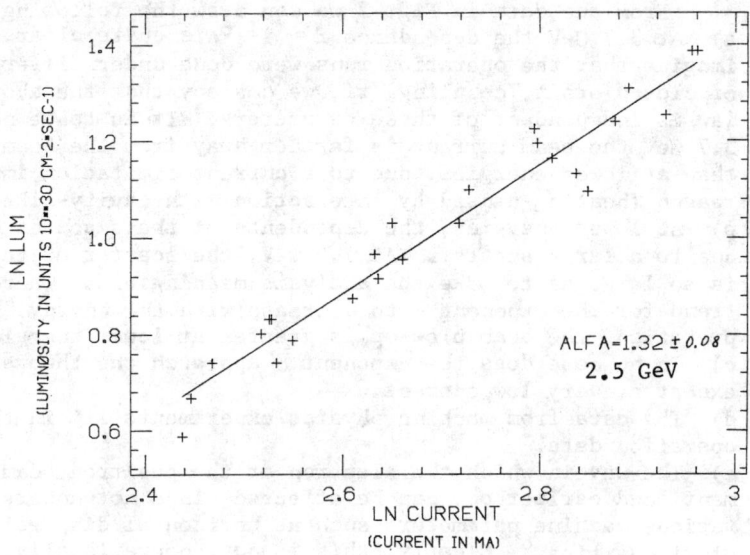

Fig. 4. Logarithmic dependnece of the luminosity on the beam current

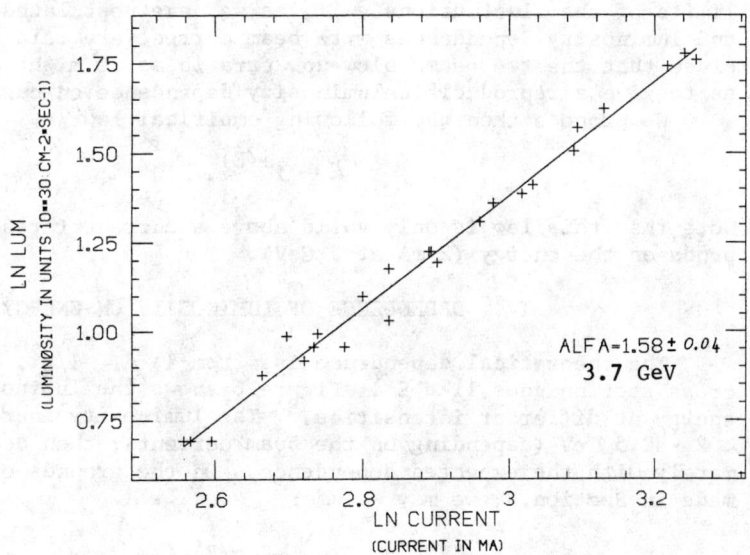

Fig. 5. Logarithmic dependence of the luminosity on the beam curren

From the data in Fig. 2 we can draw the following conclusions:
a) At 3.7 GeV the dependence $\mathcal{L} \simeq i^{1.6}$ is quite clear. Since we can imagine that the operation runs were done under different conditions of closed orbit, coupling, rf, we can say that the above empirical law is independent of these parameters. It is to be noted that at 3.7 GeV the beam current is farther away from the beam-beam limit than at lower energies, due to a current limitation imposed by another reason (heating caused by interaction with cavity-like objects).
b) At lower energies, the dependence of the exponent α is less clear, due to a large scatter. At 1.5 GeV, the scatter of the original data is so large as to make the analysis meaningless. There seems to be a trend for the exponent α to decrease with the energy. In our interpretation, the beam blow-up is greater at lower energies.
c) In no case does the exponent α approach the theoretical value 2, except at very low current.
d) The data from machine physics experiments fit in the trend of the operation data.
e) The way in which the electron or the positron beams blow-up, as mentioned earlier on, can be affected, in a not understood way, by various machine parameters such as horizontal dispersion and phasing of rf cavities.[6] Clearly, this is not, operationally, a very easily reproducible phenomenon, as the above parameters are often changed, either purposely or due to orbit errors. On the opposite side, the data from the operation runs analyzed in Fig. 2 show that the functional dependence of luminosity on beam current is reproducible within the limits of the fluctuations. If, as we have postulated, the beam size and luminosity dependences with beam current are related, we must conclude that the two beams blow-up, erratic as it might appear, is such as to give a reproducible luminosity dependence on current.

We propose then the following empirical law:

$$\mathcal{L} \simeq i^{\alpha(E)}. \qquad (4)$$

Note that this law is only valid above a current threshold which depends on the energy (2 mA at 2 GeV).

IV. DEPENDENCE OF LUMINOSITY ON ENERGY

The theoretical dependence is, from 1) $\mathcal{L} \simeq 1/E^2$, since the beam cross section goes like E^2. Figure 6 shows the luminosity versus energy at different intensities.[8] The luminosity increases up to $\sim 2 - 2.5$ GeV (depending on the beam current), then decays approximately with the expected dependence. On the grounds of observations made in Section 3, we may assume:

$$\mathcal{L} = \frac{k\, i^{\alpha(E)}}{E^2} \qquad (5)$$

where k is a constant.*

---
*The luminosity in Eq. (5) must not be confused with the maximum achievable luminosity, which scales like E^4

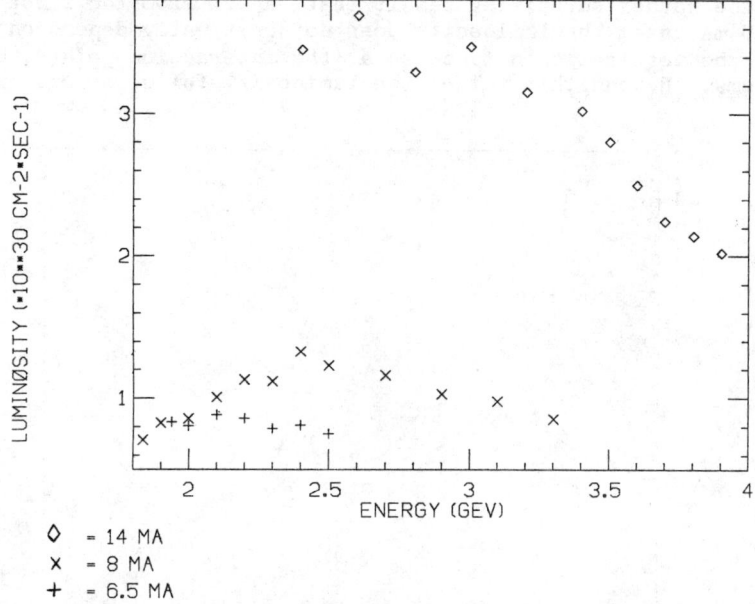

Fig. 6. Luminosity versus energy for three different values of the beam current.

Equation 5, solved for $\alpha(E)$, gives:

$$\alpha(E) = \frac{\ln \mathcal{L} + 2 \ln E - \ln k}{\ln i} \qquad (6)$$

The factor k depends on machine parameters, in particular, coupling. We have taken the data from Eq. 6 for $\alpha(E)$. For k we have taken an average value from the operation data, namely k = 0.37 if $\mathcal{L}$ is in 10^{30} cm^{-2} sec^{-1}, i in mA, E in GeV. The results for $\alpha(E)$ are shown as squares in Fig. 2, and they confirm the trend observed from the measurements of luminosity as a function of the beam current.

V. LUMINOSITY AND SPACE CHARGE PARAMETERS AS A FUNCTION OF β_y^*

From beam size and linear optics considerations, one should find, at low current

$$\mathcal{L} \simeq \frac{1}{\sqrt{\beta_y^*}} . \qquad (7)$$

At higher current, increasing β_y^* increases the beam-beam tune shift and the luminosity should fall more rapidly. The results of two experiments[9] are summarized in Figs. 7, 8, and 9. The two sets of points refer to two different conditions: in one case the beam current was 7 mA and the vertical tune ν_{yo} = 5.123; the second case had a beam current of 10 mA and the vertical tune ν_{yo} = 5.174. Figure

7 shows the rather surprising result that, apart from the first point of the 10 mA case, the luminosity does not critically depend on the value of the vertical beta function at the interaction points, up to $\beta_y^* = 20$ cms. Beyond this value, the luminosity falls, as one expects.

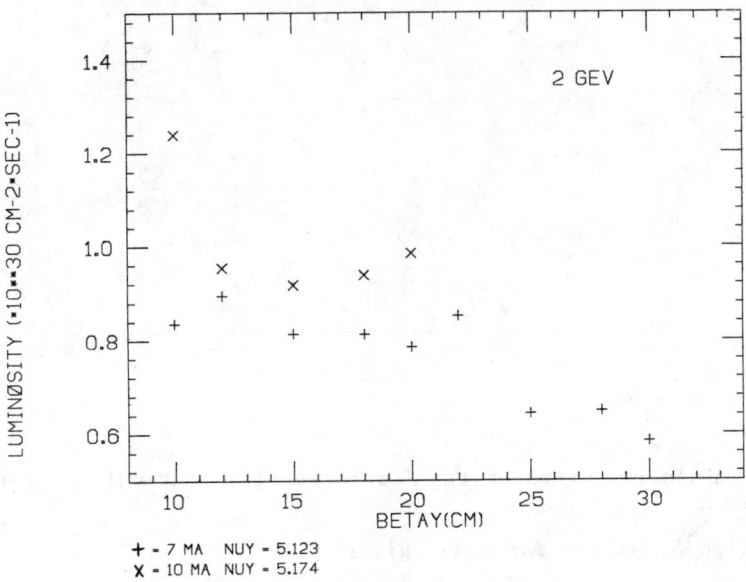

Fig. 7. Dependence of the luminosity on the vertical β-function at the interaction point.

In Fig. 8 we plot the value of the space charge parameter calculated from the luminosity for the two experiments described above; in Fig. 9 the real vertical tune shift is plotted. These results indicate that luminosity, for a fixed current, does not depend on the value of the beam-beam tune shift, if this is less than 0.045. Clearly, these surprising results require further investigation.

VI. DEPENDENCE OF LUMINOSITY ON THE UNPERTURBED VERTICAL TUNE

We have analyzed the results of three experiments in which the luminosity was measured as a function of the unperturbed vertical tune.[10]

Figure 10 shows that the luminosity decreases as the tune increases. At first sight this looks like a reasonable result, as one could argue that the luminosity decreases because the real tune shift increases with the vertical tune. In fact, the luminosity decreases with increasing tune faster than it would just to keep the real tune shift constant. This is shown in Fig. 11, where the total vertical tune shift (i.e., tune shift per interaction region times the number of interaction regions) is plotted as a function of the unperturbed

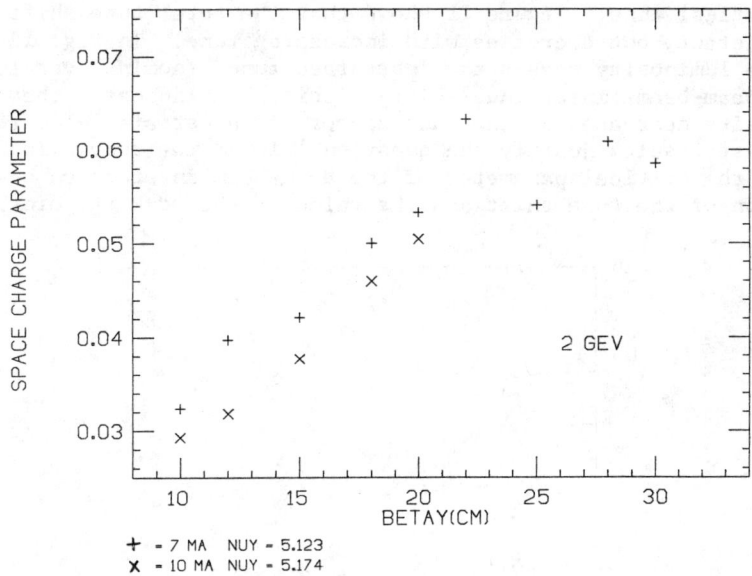

Fig. 8. Dependence of the space charge parameter on the vertical β function at the interaction point.

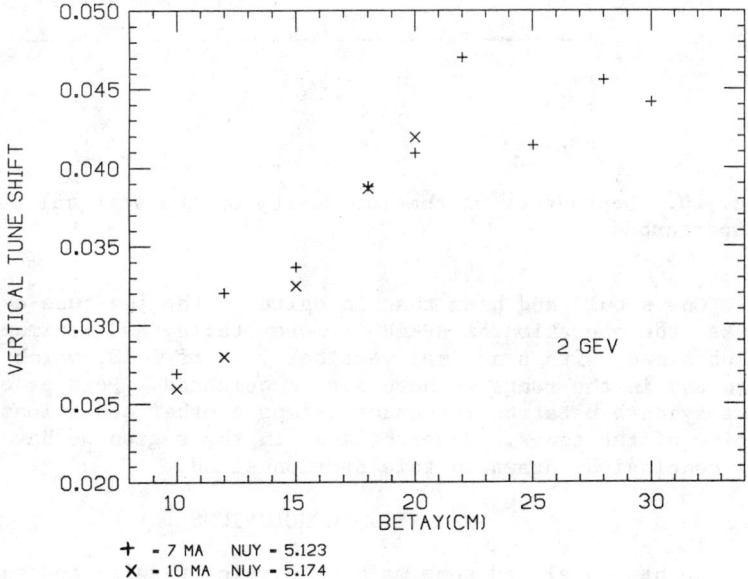

Fig. 9. Dependence of the vertical beam-beam tune shift on the vertical β function at the interaction point.

vertical tune. Figure 11 shows that the total tune shift is not constant, but decreases with increasing tune. In Fig. 12 we plot the luminosity versus the 'perturbed tune' (nominal vertical tune + beam-beam linear tune shift): this plot indicates that the luminosity decreases as the tune approaches a certain value (6 ν_y = 31?). These results justify the question: is it the tune shift alone which is the critical parameter of the beam-beam interaction or a combination of the tune shift and the value of the working point?

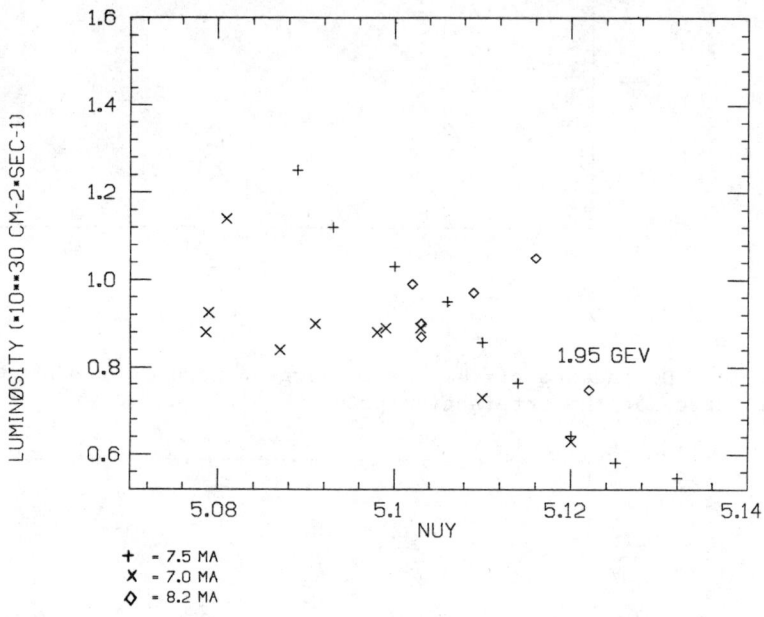

Fig. 10. Dependence of the luminosity on the vertical tune (unperturbed).

One should add here that in spite of the low-tune-experiment results, the operation of SPEAR is more stable, and maximum luminosity is obtained, with a nominal vertical tune of 5.18, which is higher than any in the range we have just considered. Perhaps other effects, like synchro-betatron resonances, impose other conditions on the choice of the tunes. Nevertheless, in the region we have considered, the conclusions drawn in this section stand.

VII. CONCLUSIONS

We have analyzed some machine physics results and operation data in SPEAR. There is reproducible evidence of a vertical beam blow-up, whose extent is a function of the beam current. The rate of the blow-up with current shows little dependence on machine parameters like closed orbit, coupling, rf conditions.

113

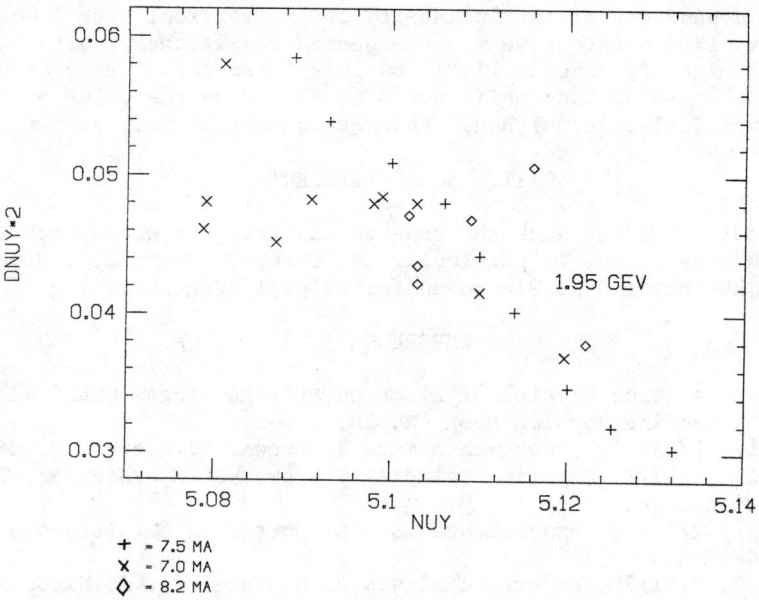

Fig. 11. Dependence of the total vertical beam-beam tune shift on the vertical tune (unperturbed).

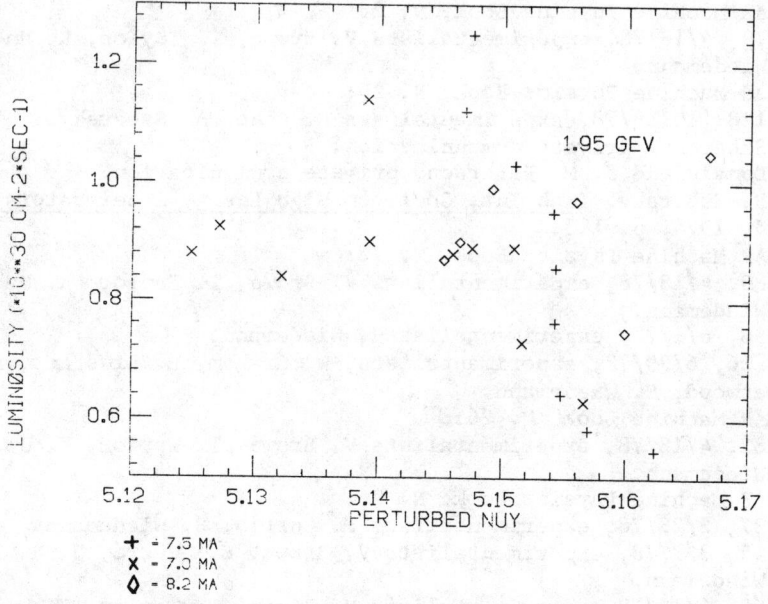

Fig. 12. Dependence of the luminosity on the vertical perturbed tune (nominal tune + beam-beam linear tune shift).

The dependence of the luminosity on the vertical beta function at the interaction points gave the unexpected result that, up to $\beta_y^* = 20$ cm, the luminosity depends little on this parameter. The maximum tolerable beam-beam tune shift seems to depend on the value of the nominal vertical tune, although this dependence is not, as yet, clear.

VIII. ACKNOWLEDGMENTS

I would like to thank the experimentalists, on whose work this report is based, and, in particular, A. Chao, P. Morton, J. M. Paterson, A. Sabersky and H. Wiedemann for helpful discussions.

REFERENCES

1. M. Sands, "The Physics of Electron Storage Rings", SLAC-121.
2. SPEAR Machine Physics Book, N. 26:
    p. 17, 1/25/78, experimentalists V. Brown, E. Guerra, H. Wiedemann.
    p. 29, 2/8/79, experimentalists V. Brown, J. M. Paterson, T. Taylor, H. Wiedemann.
    p. 31, 2/15/78, experimentalists S. Curtis, J. M. Paterson, H. Wiedemann.
    p. 32, 2/21/78, experimentalists J. M. Paterson, H. Wiedemann.
    p. 49, 3/8/78, experimentalists V. Brown, J. M. Paterson, H. Wiedemann.
    p. 78, 4/18/78, experimentalists V. Brown, T. Taylor, K. Underwood, H. Wiedemann.
3. SPEAR Machine Physics Book, N. 26:
    p. 79, 4/18/78, experimentalists V. Brown, T. Taylor, K. Underwood, H. Wiedemann.
4. SPEAR Machine Physics Book, N. 26:
    p. 128, 10/18/78, experimentalists A. Chao, A. Sabersky.
5. A. Sabersky, private communication.
6. M. Donald and J. M. Paterson, private communication.
7. A. P. Sabersky, <u>IXth Int. Conf. on High Energy Accelerators</u>, Stanford, 1974, p. 113.
8. SPEAR Machine Physics Book, N. 26:
    p. 80, 4/18/78, experimentalists V. Brown, T. Taylor, K. Underwood, H. Wiedemann.
    p. 96, 6/1/78, experimentalist H. Wiedemann.
    p. 116, 6/29/78, experimentalists, W. Graham, E. Linstadt, K. Underwood, H. Wiedemann.
9. SPEAR Machine Book, N. 26:
    p. 82, 4/18/78, experimentalists V. Brown, T. Taylor, K. Underwood, H. Wiedemann.
10. SPEAR Machine Physics Book, N. 26:
    p. 37, 2/28/78, experimentalists S. Curtis, H. Wiedemann.
    p. 45, 3/7/78, experimentalists V. Brown, E. Guerra, J. M. Paterson, H. Wiedemann.
    p. 82, 4/18/78, experimentalists V. Brown, T. Taylor, K. Undersood, H. Wiedemann.

RECENT RESULTS FROM DORIS AND PETRA

A. Piwinski

Deutsches Elektronen-Synchrotron DESY

I. INTRODUCTION

The limitation of the luminosity in the electron positron storage rings DORIS and PETRA has different reasons. The limitation in DORIS during its double ring operation was given by satellite resonances or synchro-betatron resonances.[1] These resonances are produced by the beam-beam interaction with a crossing angle and lead to a reduction of the life time of the stored particles. Since a splitting of the betatron and synchrotron frequencies between different bunches was needed to suppress multibunch single beam instabilities the satellite resonances could practically not be avoided. Analytical investigations of these resonances as well as computer simulations are in good agreement with measurements made at DORIS.[2]

The limitation in DORIS during its single ring operation up to 5 GeV was not given by the beam-beam interaction but by the available rf power and by a pressure rise with increasing bunch currents.[3]

The luminosity of the storage ring PETRA is limited up to present energies of 8.5 GeV, by the beam-beam interaction which leads here to a vertical blow up.[4,5] The blow up reduces the luminosity and the life time of the beams. Close to the limit the two beams are usually blown up differently. If the beam currents differ by more than a few percent the weaker beam always assumes a larger beam height than the stronger beam.

The observations in PETRA are still preliminary. The beam-beam effect will be further studied. Especially the maximum tune shift which will be obtained at higher energies and for different operating points will be investigated.

II. DORIS

II.I Analytical Investigation

The luminosity of the storage ring DORIS is limited by satellite resonances which are produced by the beam-beam interaction. The resonance frequencies are given by

$$Q_\beta = (p + r Q_s)/q \qquad (1)$$

where Q_β is the betatron wave number, Q_s is the synchrotron wave number and p, r and q are integers. It is shown that these satellite resonances are excited by the space charge forces of the two colliding beams, if the two beams cross at an angle or if the dispersion at the interaction point is not zero.[2]

Figure 1 shows two bunches which cross at an angle of 2Φ. A particle which has a distance s from the center of its own bunch does not pass through the center of the other bunch and hence gets a verticle kick.

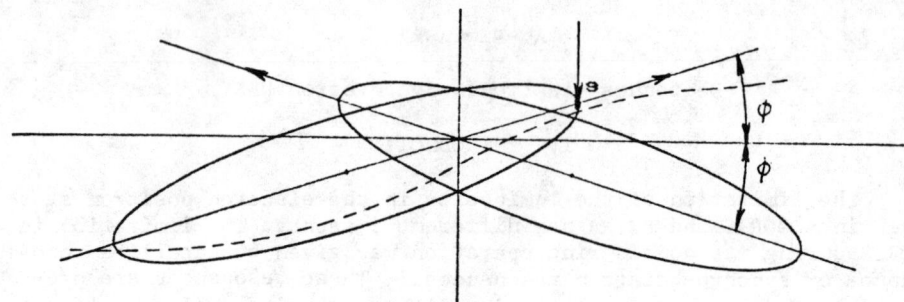

Fig. 1.

Because of this kick the closed orbit is distorted which is shown by the dotted line. For DORIS the maximum orbit distortion, which is caused by one interaction point, is 15% of a verticle standard deviation of the particle distribution. In the case of two interaction points the maximum orbit distortion is only 3% of the standard deviation, since the vertical Q-value is above an even integer and the two contributions to the orbit distortion are nearly compensated. So the displacement of the orbit is very small and has no influence on the luminosity. On a synchro-betatron resonance, however, the oscillation amplitude can increase so far that the particle is lost.

The verticle kick $\delta z'$ depends on the sum of z and s times $\tilde{\Phi}$:

$$\Delta z' = f(z + s\tilde{\Phi}) \tag{2a}$$

The linear part of the function f gives the tune shift or ΔQ. Equation (2a) shows the influence of the synchrotron oscillation on the betatron oscillation ($\delta z'$ depends on s). The synchrotron oscillation is also influenced by the betatron oscillation. This can be seen if one takes into account that the energy of a particle that crosses a bunch at an angle[6] is changed:

$$\frac{\Delta E}{E} = \tilde{\Phi} f(z + s\tilde{\Phi}). \tag{2b}$$

Equations (2a) and (2b) give the complete coupling between the betatron and synchrotron oscillation which is produced by the crossing angle.

If the betatron and synchrotron amplitudes are small one gets a linear coupling and one may apply the matrix formalism for the investigation of the coupled motion. A resonance occurs for $Q_\beta = p \pm Q_s$ and the phases of the eigenvalues are given by[2]

$$\mu = \mu_\beta + 2\pi \xi \pm 2\pi n \xi \, |\tilde{\Phi}| \sqrt{\frac{-\alpha_M C}{n\beta_o \sin(\mu_\beta/n)}} \tag{3}$$

with ξ = space charge parameter, n = even number of equidistant interaction points, α_M = momentum compaction factor, C = circumference, $\mu_\beta = 2\pi Q_\beta$.

Here all quadratic terms in ξ are neglected. Equation (3) shows that the amplitudes increase on the resonance exponentially if $\sin(\mu_\beta/n)$ is positive, that means if Q_β/n is above an integer.

The nonlinear satellite resonances, which occur for large amplitudes, can be investigated with help of approximation methods. Only one result should be given here. One can derive an invariant of motion which is determined by

$$\hat{z}^2 - \frac{q\alpha_M \beta C}{r 2\pi Q_s}\left(\frac{\Delta E}{E}\right)^2 = \text{const.} \tag{4}$$

with β = vertical amplitude function.

Since r can be positive or negative [Eq. (1)], the betatron and synchrotron amplitude can increase or decrease at the same time or they can exchange their oscillation energy periodically. So an instability occurs for a difference resonance, that means positive r, whereas in the case of a coupling resonance of horizontal and vertical betatron oscillations the difference is stable and the sum resonance is unstable. The reason is the assumption that the particle energy is above the transition energy.[7] Below the transition energy one has to replace the momentum compaction factor α_M by $(\alpha_M \gamma^2 - 1)/(\gamma^2 - 1)$, where γ is the relative particle energy. So below the transition energy ($\alpha_M \gamma^2 < 1$) the satellite resonances show a similar behavior as the coupling resonances of horizontal and vertical betatron oscillations.

II.2 Computer Simulations

A simulation of the betatron and synchrotron oscillation has been done on a digital computer. The nonlinear coupling of the oscillations due to the crossing angle has been taken into account. Since in DORIS the effective beam cross section is approximately circular, i.e. $\sigma_{zef} \approx \sigma_x = \sigma$, an exact expression for the change of the betatron angle due to the space charge forces could be used:

$$\Delta z' = \frac{8\pi \xi \sigma^2}{\beta_o (z+x\dot{\phi})}\left[e^{-\frac{(z+s\dot{\phi})^2}{2\sigma^2}} - 1\right] \tag{5}$$

with σ = standard deviation for the particle distribution.

The simulation showed that the change of the synchrotron coordinates due to the beam-beam interaction has no large influence on the results. To reduce the computer time this change as well as the nonlinearity of the synchrotron potential was neglected. With these simplifications the oscillations were simulated for 2000 revolutions. The initial phase between the betatron and synchrotron oscillation was varied in steps of $2\pi/100$, to look for the maximum amplitude which can occur during the 2000 revolutions. The maximum betatron amplitude was determined for more than 1000 Q_β-values between 6.02 and 6.48. The results for Q_β between 6.52 and 6.98 are mirror symmetric, since the synchrotron oscillation is constant.

The following parameters from DORIS were used $\xi = 0.01$, $\beta_0 = 1$ m, $\sigma = 0.23$ mm, $Q_S = 0.034$, $k = 480$, $\alpha_M = 0.018$, $\bar{R} = 45.8$ m, $\Phi = 12$ mrad, $\hat{z}_{ini} = 0.8$ mm = $4\,\sigma_z$, $(\Delta E/E)_{ini} = 1.8 \cdot 10^{-3} = 3\,\sigma_E$. Figure 2 shows the ratio of the maximum betatron amplitude over the minimum or initial amplitude. The simulation was done for 1 and 2 interaction points. In the case of 2 interaction points, the two crossing angles have opposite signs. The total width of most of the resonances is smaller than 0.001. This could not be shown in Fig. 2. Furthermore, only those resonances are shown which reached, either for 1 or for 2 interaction points, an increase of more than 50% of their initial amplitude.

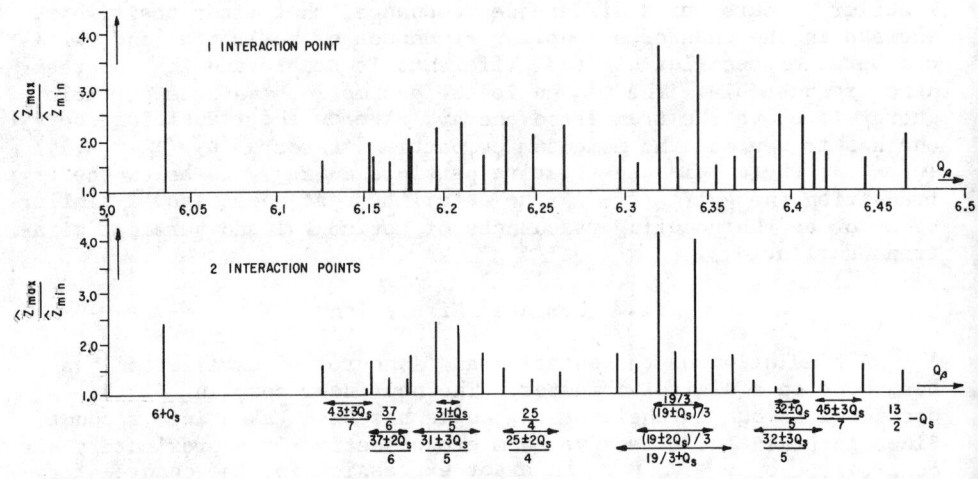

Fig. 2.

II.3 Measurements at DORIS

The measurements in the storage ring DORIS were done at 1.8 Gev. The behavior of weak positron bunches colliding with strong electron bunches was investigated. Without decoupling transmitter and without rf-quadrupole a maximum bunch current of about 1.3 mA could be stored in 30 electron bunches. The resulting ξ is 0.009 ($\sigma_x = 0.23$ mm, $\sigma_z = 0.2$ mm, $\sigma_s = 1$ cm).

In most of the measurements the vertical betatron frequency of the positrons was varied nearly continuously by a computer, whereas the synchrotron frequency was kept constant. The speed of the variation was 1 kHz per 10 sec. Simultaneously the current of the 30 positron bunches was plotted. The sharp beam losses yield exactly the frequencies of the resonances.

An example of such a measurement is shown in Fig. 3 for one interaction. Because of the very small width of the resonances and because of the speed of the variation only those resonances could be observed which lead to a life time less than 15 min. Table I shows a comparison of the measured and calculated frequencies ($Q_s = 0.024$).

TABLE I:

N	$Q_{\beta meas.}$	$Q_{\beta cal.}$	resonance
1	6.0932	-	-
2	6.0961	6.0960	$6 + 4Q_s$
3	6.1212	6.1190	$(49 - 2Q_s)/8$
4	6.1243	6.1250	$49/8$
5	6.1274	6.1280	$(49 + Q_s)/8$
6	6.1397	6.1394	$(43 - Q_s)/7$
7	6.1628	6.1627	$(37 - Q_s)/6$
8	6.1666	6.1667	$37/6$
9	6.1801	6.1808	$(31 - 4Q_s)/5$
10	6.1858	6.1856	$(31 - 3Q_s)/5$
11	6.1921	6.1904	$(31 - 2Q_s)/5$
12	6.1952	6.1952	$(31 - Q_s)/5$
13	6.1994	6.2000	$31/5$
14	6.2043	6.2048	$(31 + Q_s)/5$

In many measurements with one interaction point 25 vertical resonances were observed. The life time on a resonance was between a few seconds and 15 minutes. The strength of the resonances was not well reproducible because the variation of Q_β was not quite continuous. The power supplies of the quadrupoles are digitally controlled, so that Q_β is varied in very small steps which have the same order of magnitude as the width of the resonances.

The width of the satellite resonances was measured in several cases. It is about 0.0005. For satellites of coupling resonances it can be larger by a factor of 2 or 3. A horizontal satellite resonance, which was only produced by the beam-beam interaction, was not found.

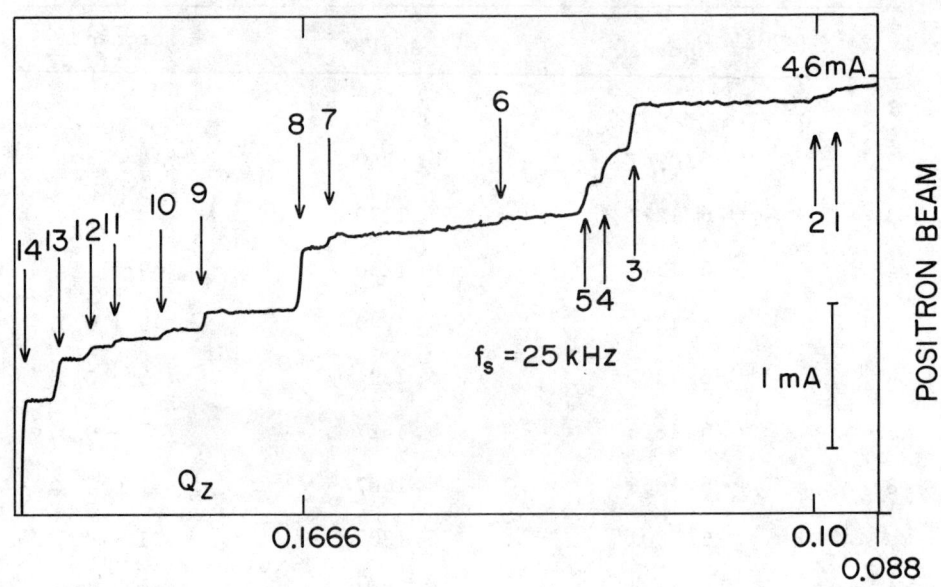

Fig. 3.

III. PETRA

III.I Behavior Of The Beams Near The Limit

During injection the two beams are separated vertically with an electrostatic field. After switching off the electrostatic field one beam or both are blown up vertically. For currents above the limit the life time is very short, it can be less than 1 sec. For smaller currents near the limit the beams are blown up but the life time is longer, it can be of the order of an hour. Figures 4 and 5 show the two beams on a television monitor. In Fig. 4 the beam currents are 0.3 mA and both beams are blown up. Figure 5 shows the beams at half the current, and they are flat.

Sometimes only one beam is blown up, and the other one is flat. If one then excites the two beams on their vertical betatron frequency it can happen that the second one blows up and the

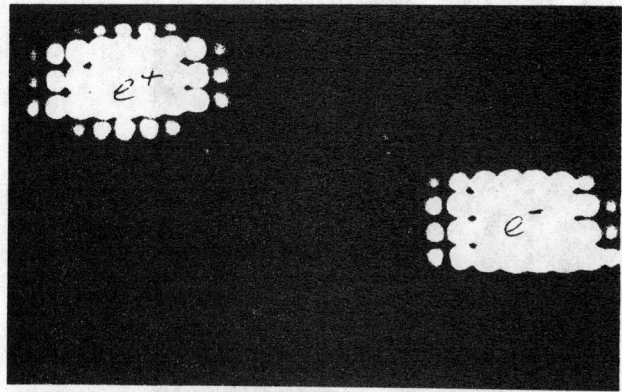

Fig. 4: Beam dimensions at $I^+ = I^- = 0.3$ mA.

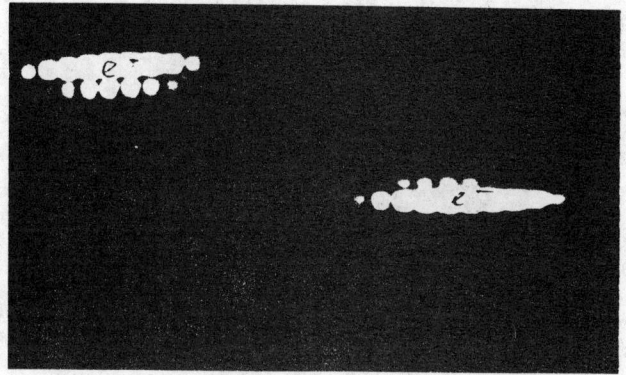

Fig. 5: Beam dimensions at $I^+ = I^- = 0.15$ mA.

first one becomes flat, i.e. the two beams may behave like a flip-flop system.

Figures 6 and 7 show the beams on a horizontal sixth-order resonance. This is a very special case which was obtained at 6.5 GeV after carefully tuning the working point on the resonance frequency. For the usual working point such a single resonance has never been observed. The time for the filling of the six islands is several minutes, and after a small change of the betatron frequency they vanish with the same time constant. Since the bunch currents or, more exactly, the currents in the central parts of the bunches are different, the tune shifts due to the space charge forces are different for the two bunches. Therefore the resonance does not appear for the two bunches at the same time, and the working point is different in Figs. 6 and 7.

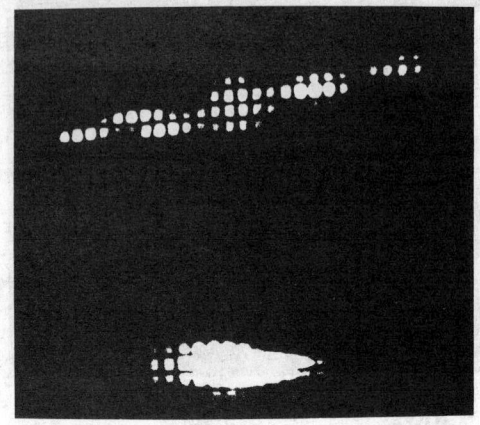

I^+ = 0.52 mA

I^- = 0.74 mA

Fig. 6.

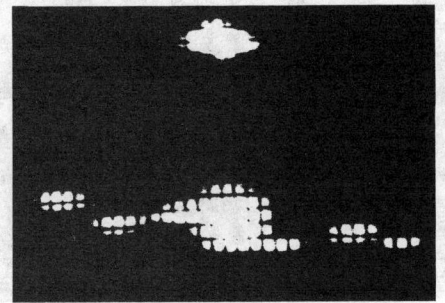

I^+ = 0.6 mA

I^- = 0.6 mA

Fig. 7.

III.2 Measurement Of The Coherent Tune Shift

Figure 8 shows the pickup signal of two colliding bunches (one in each beam) which are excited at their vertical betatron frequency. The currents of the bunches are below the limit. The excitation frequency increases from left to right, and there are two resonance frequencies where both beams are oscillating vertically. These frequencies are the two eigenfrequencies of the resonant system of the two bunches which are coupled by the space charge forces.

Figure 9 shows the response when the two bunches are separated by the electrostatic field. The small frequency splitting is due to "long range forces" and a small quadrupole component in the separating field. When only one electron or positron bunch is stored the measurement gives only the left peak in Fig. 8.

The distance of the two peaks in Fig. 8 gives the coherent tune

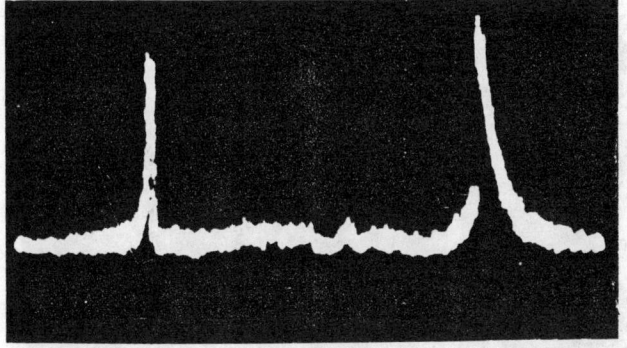

Fig. 8. Vertical eigenfrequencies of two colliding bunches.

Fig. 9. Vertical eigenfrequencies of two separated bunches.

shift per revolution, and that is twice the coherent tune shift per interaction, since we have two interaction points in this case. In linear approximation, the coherent tune shift is twice the incoherent tune shift.[8-10]

The measurement of the coherent tune shift has two practical difficulties. If the excitation is too strong the second resonance frequency is shifted, as shown in Fig. 10. With increasing excitation voltage the coherent amplitude increases also, the oscillation becomes smaller. Only the right corner of the triangle in Fig. 10 is independent of the strength of the excitation. It gives the largest frequency split, which occurs for the weakest excitation where the linear approximation is valid.

Another difficulty arises if the excitation and the observation of the two eigenmodes are not gated. In that case the excitation and the observation depend on their respective position in the ring. This can easily be understood if one assumes for instance that the antenna measures the electric field at the interaction point. Then

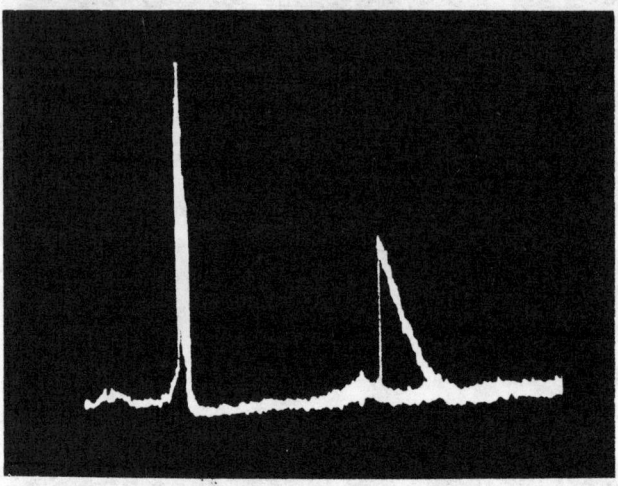

Fig. 10: Eigenfrequencies at strong excitation.

only the second eigenmode can be observed where the two bunches have opposite displacements at the interaction point. The first eigenmode where the two bunches have the same displacement cannot be observed since the signals of the two bunches compensate. Such positions occur periodically in the ring according to the betatron phase advance. Then there are other positions where only the first mode can be observed, and between these positions there are regions where both modes can be detected.

The same consideration holds for the excitation of the two modes which therefore can be done only in certain regions of the ring. For the measurements in PETRA only the excitation was gated whereas the position for the observation was optimized such that both modes could be observed.

The problem becomes more complex if there are two bunches in each beam. In that case one obtains four eigenfrequencies. If the bunch currents are equal and if the optics are symmetric, two eigenfrequencies coincide and the difference between the largest and the smallest frequency is eight times the incoherent tune shift per interaction.

By measuring both the horizontal and the vertical coherent tune shift one can, in linear approximation, independently calculate the horizontal and vertical beam dimensions at the interaction point. A comparison of the measured luminosity with the luminosity calculated from these beam dimensions shows good agreement.

The coherent tune shift has been measured as a function of several parameters. Figures 11 and 12 show the frequency splitting as a function of the bunch current for different energies. It is remarkable that the vertical splitting goes through a maximum. For larger currents at least one of two beams is blown up. The maximum vertical incoherent tune shift is about 0.016 in this case whereas the horizontal one increases up to 0.027.

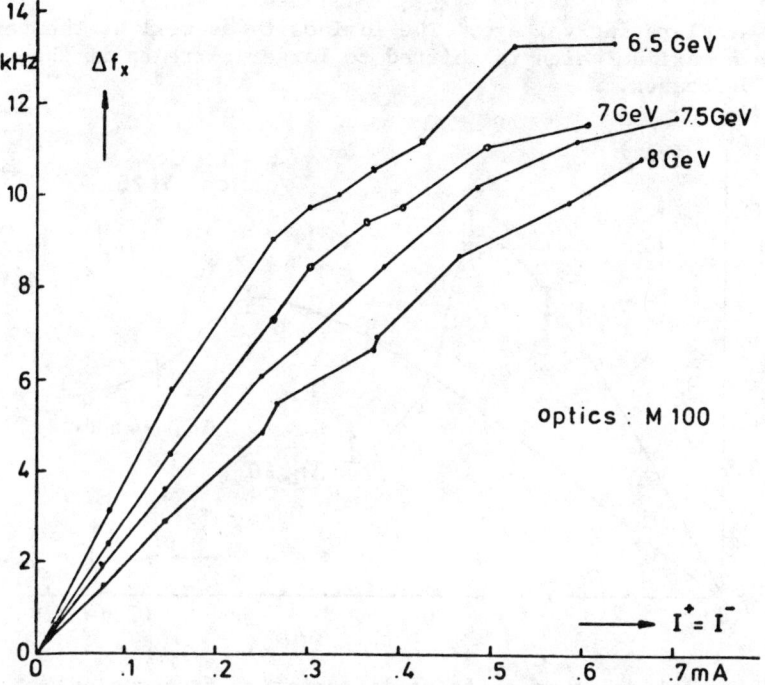

Fig. 11: Horizontal frequency splitting at different energies.

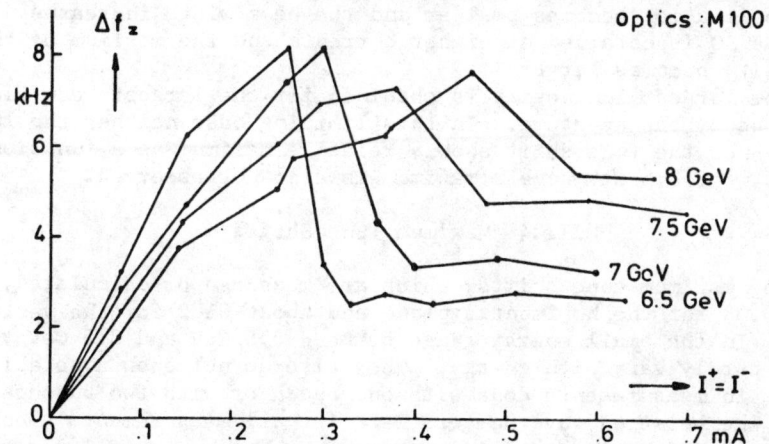

Fig. 12: Vertical frequency splitting at different energies.

III.3 Measurement Of The Luminosity

The vertical tune shift can be calculated from the measured luminosity since the beam height is small as compared to the beam width. Figure 13 shows an example of such a calculation. The luminosity was measured as a function of the current for two different frequencies of the accelerating voltage. The luminosity as well as the vertical ΔQ have a maximum which is shifted to larger currents if the rf frequency decreases.

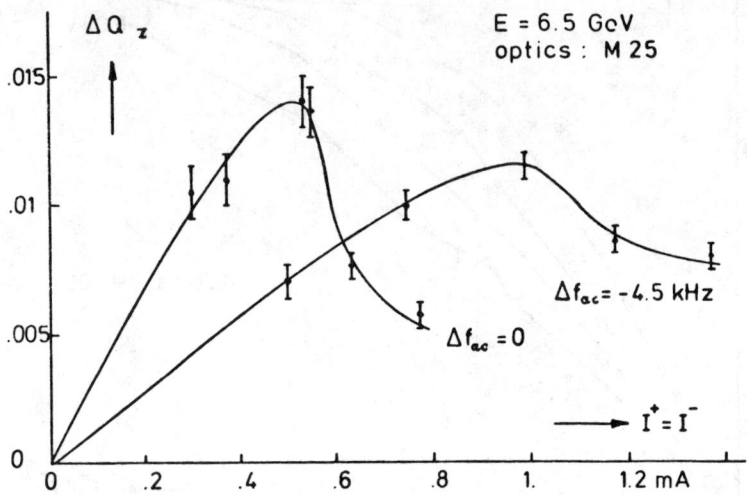

Fig. 13: Vertical tune shift at different rf frequencies.

For smaller rf frequencies the damping of the horizontal betatron oscillation becomes smaller and the beam width increases. Then the same ΔQ is obtained at higher currents and the maximum of the luminosity becomes larger.

The largest luminosity is obtained for the largest beam width permitted by the aperture. In this limiting case neither the luminosity nor the tune shift show a relative maximum as a function of current. At 8.5 GeV such a maximum is also not observed.

III.4 Maximum Tune Shifts

The maximum tune shifts, which are measured or calculated, are about 0.03 for the horizontal plane and about 0.02 for the vertical plane. In the small energy range between 6.5 GeV and 8.5 GeV these values hardly vary with energy. They also do not show a notable difference in measurements done with one bunch or with two bunches per beam. The betatron wave numbers were for all measurements about $Q_x = 25.18$ and $Q_z = 23.28$.

A comparison of the maximum tune shifts at different energies is difficult since the rf frequency was optimized for maximum luminosities. Thus the beam width and the horizontal damping have an energy dependence which is different from the usual scaling law.

REFERENCES

1. Wissenschaftlicher Jahresbericht DESY (1977).

2. A. Piwinski, <u>1977 Particle Accelerator Conference</u>, IEEE Trans. Nuclear Science, <u>NS-24</u>, No. 3, 1408 (1977).
3. The DORIS Storage Ring Group, DESY 79/8 (1979).
4. G. A. Voss, <u>1979 Particle Accelerator Conference</u>, San Francisco.
5. A. Piwinski, <u>1979 Particle Accelerator Conference</u>, San Francisco.
6. J. E. Augustin, Note Interne 35-69 (1969).
7. A. Piwinski, Internal Report, DESY PET-78/8 (1978).
8. The SPEAR-Group, <u>Proc. 9th Intern. Conf. on High Energy Accelerators</u>, Stanford, p. 37, (1974).
9. A. P. Sabersky, <u>Proc. 9th Intern. Conf. on High Energy Accelerators</u>, Stanford, p. 113, (1974).
10. A. Piwinski, Interner Bericht, DESY H2-75/3 (1975).

BEAM-BEAM EFFECTS AT THE 1.5 GeV, e^+e^- STORAGE RING ADONE

S. Tazzari

I.N.F.N.-Laboratori Nazionali di Frascati

C.P. 13-00044 Frascati (Italy)

I. INTRODUCTION

A summary of the experimental work done at Adone over the past years on beam-beam interaction effects is given.

In section II the qualitative behavior of two beams in interaction is described. In section III a discussion of the performance obtained over many hundreds of physics runs with two strong beams, given. In the last section the results of measurements on the behavior of a weak beam (approximating a single particle) colliding with a strong one are presented.

The main machine parameters are summarized in Table I.

TABLE I

$Q_x = Q_x = 3.05$ (unless otherwise specified)
$\beta_x^* = 8.87$ m
$\beta_z^* = 3.37$ m
$\eta^* = 2.085$ m
$\sigma_x^* = 1.29\, E_{(GeV)}$ mm (no coupling)
$\sigma_z^* = 0.444\, E_{(GeV)}$ mm (full coupling)
$\sigma_p = 3.82\, 10^{-4}\, E_{(GeV)}$
$R = 16.71$ m average radius

II. QUALITATIVE BEHAVIOR OF TWO INTERACTING BEAMS

In Adone, the beams are injected at around 300 MeV; the machine is then brought up to the operating energy and tune. During these preliminary operations the beams are kept separate at all interaction points by means of electrostatic plates. The voltage on the plates is then switched off to bring the beams into interaction. No special provision for switching the voltage off fast is needed. This may be connected to the fact that, due to the special way by which the separation is generated in our machine, the dipole attractive force of one beam on the other is important[1] and tends to bring the beams together much faster than one would expect from the voltage decay law only.

For every operating energy and tune there is a maximum current, I_{max}, that can be stored in each beam (we call this the space charge limit). If the value I_{max} is exceeded in either beam, the weakest beam size grows to very large values when the beams are brought into interaction (we refer to this as "flipping") and no useful luminosity is obtained. The growth is essentially in the vertical plane only, although, due to our machine parameters, the radial and vertical interaction strength parameters (ξ_x, ξ_z) are equal to within a few percent. The highest value of I_{max}, I_{max}^M, is obtained when the unperturbed machine tune is exactly on the main coupling resonance, $Q_x = Q_z$. The behavior is qualitatively shown in Fig. 1a).

Figure 1b) shows the qualitative behavior of luminosity, when the tune of the machine is changed along the path shown in Fig. 1a). The well-known "hysteresis" effect, also observed at SPEAR,[2] is evidenced.

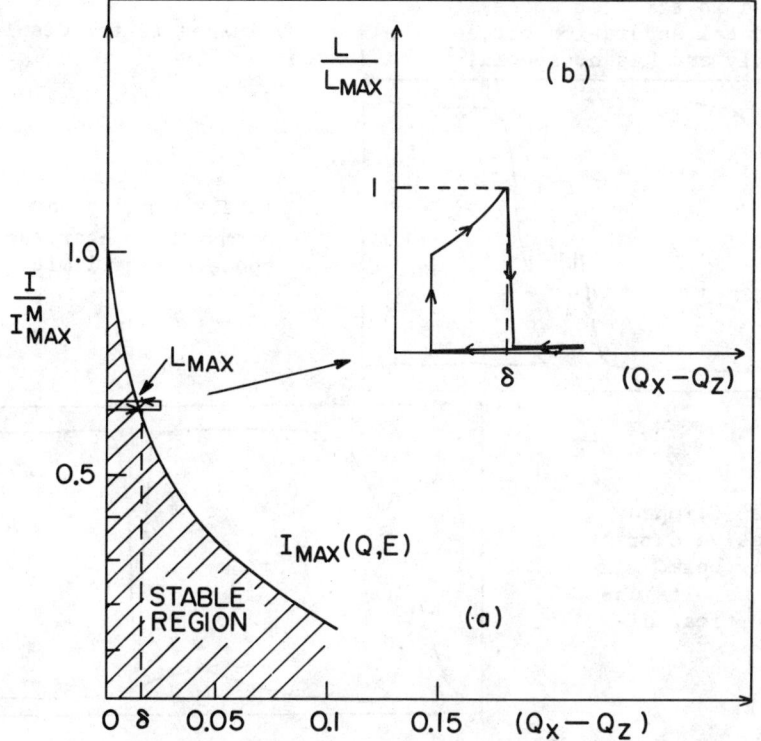

Fig. 1. Qualitative behavior of current and luminosity at the beam-beam limit. Operating tunes at the point $(Q_x - Q_z) = 0$ are $Q_x = Q_z \approx 3.05$.

It should be said here that there are two distinct energy regions in which the beams behave quite differently: at low energies (below ~ 0.7 GeV), if I_{max} is exceeded (e.g. by bringing the Q values too far apart) the weak beam "flips" and starts being lost with a lifetime that drops practically to zero as soon as the strongest current exceeds I_{max} by more than about 10%. On the contrary at high energies the "flipping" is not destructive, i.e. the lifetime of the flipped beam is not significantly different from that of the unflipped beam. Normal conditions can therefore be restored going around the hystereris curve shown in Fig. 1b) (provided, of course, $I_{max} \leq I_{max}^{M}$).

It may also be worth mentioning that near I_{max}^{M} the interacting beam system appears to be barely stable. Any perturbation, e.g. a small longitudinal instability, will cause the weakest beam to flip.

The behavior of the beam transverse dimensions (observed with a synchrotron light monitor) is qualitatively shown in Fig. 2.

The radial (R) and vertical (V) beam transverse dimensions are seen on a synchrotron light scanning-slit monitor. The ordinate is proportional to the beam local density, and equal for both pictures; the abscissa is time (or space).

Horizontal separation between the R and V images is for display purposes only and has no special significance.

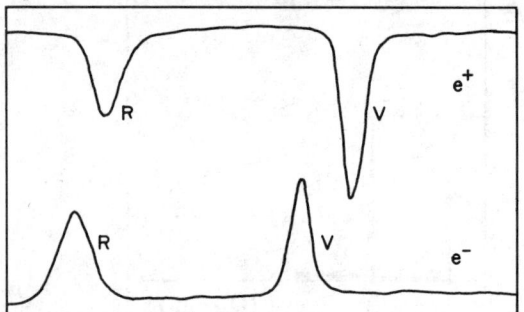

Interacting strong beams at or near the space charge limit.

Weak beam is "flipped" due to excessive current in the strong beam. The strong beam has its unperturbed vertical dimension.

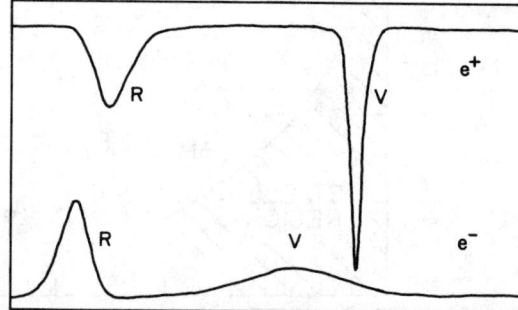

Fig. 2. Behavior of the beam transverse dimensions.

III. PERFORMANCE WITH TWO STRONG BEAMS IN HEAD-ON COLLISION

Figure 3 shows the dependence of the maximum luminosities obtained during physics runs, and of the corresponding currents (per beam, in three bunches), on energy. The dashed area around the current curve indicates the region where severe longitudinal instabilities occur, preventing higher currents from being reliably stored. Above approximately 1 GeV the maximum shown luminosity is obtained with reduced emittance ($Q_x \neq Q_z$).

The resulting linear tune shift per crossing, δQ, and interaction strength parameter ξ, can be computed from measured quantities[3] and are shown on the same figure.

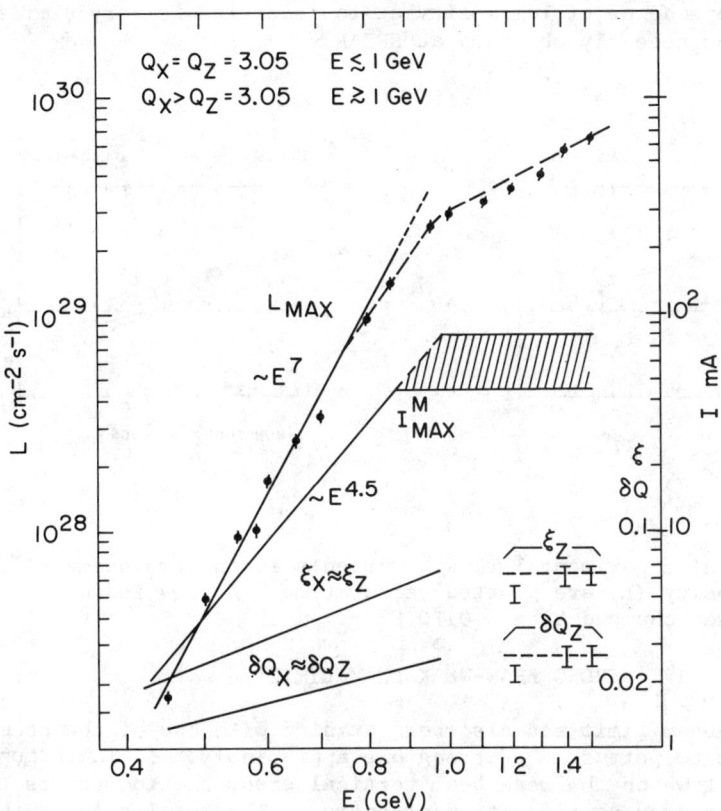

Fig. 3. Maximum luminosity, L_{max} and maximum storable current (in three bunches), I_{max}^M, versus energy. The corresponding interaction strength parameters, $\xi_{x,z}$, and the small amplitude Q shifts, $\delta Q_{x,z}$, are also shown.

All our results are consistent with the assumption that the beam-beam limit is determined by a maximum obtainable Q-shift per crossing, which is a function of energy and of the number of crossing around the ring. Given the errors involved, this does not necessarily contradict the SPEAR results of Ref. 5 and 6. In particular it has been shown that: -higher values of the interaction strength parameters, ξ_x, ξ_z, can be tolerated by tuning the machine as near as possible to the integer; the maximum value of δQ is however insensitive to tune, and depends on the number of bunches only[3,4]; -the vertical cross section blow-up factor is approximately the same for all energies at the beam-beam limit, so that the beam cross section maintains its natural E^2 dependence[3,4]; -the beam-beam limit can be controlled by varying the beam transverse emittance (through coupling). As an example two experimental runs are shown in Fig. 4 where luminosity is kept proportional to current (rather than $\propto I^2$) by varying $(Q_x - Q_z)$ as the current decreases.

The E^7 dependence of luminosity at low energies is very similar to the behavior recently observed at SPEAR.[5]

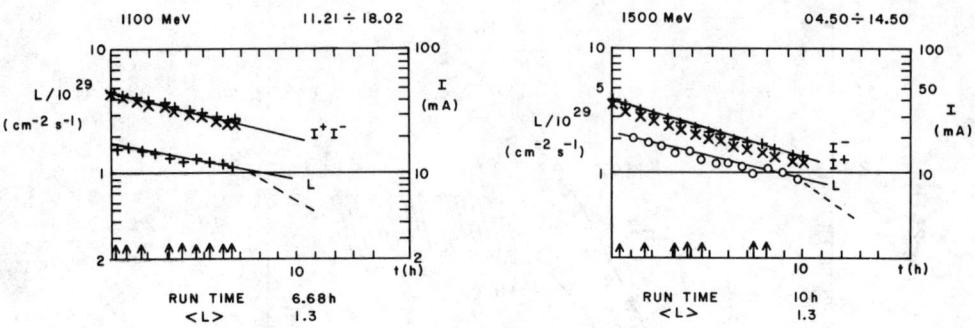

Fig. 4. Typical experimental runs. Currents in the two beams (I^+, I^-) and luminosity (L) are plotted versus time. Arrows indicate that $Q_x - Q_z$ was changed by $\sim 0.01/0.02$.

IV. STRONG BEAM-WEAK BEAM LIMIT

The beam-beam limit has also been studied with one of the beams so weak as not to perturb the strong one appreciably.[6,7] The strong beam current at which the weak beam vertical cross section starts to increase is measured as a function of energy. The results for both three bunches and a single bunch in the strong beam are shown in Fig. 5.

To check that the strong beam is actually unperturbed, its cross section is obtained by measuring luminosity and currents. It can be seen from Fig. 7 that, within the experimental errors, the strong beam cross section is equal to the unperturbed computed one.

The beam-beam limit has, to within the experimental errors, the same behavior as in the case of two strong beams; the resulting δQ values are plotted versus energy in Fig. 6. It also appears that the limiting δQ depends roughly on the square root of the number of crossing per turn.

With three bunches in the strong beam the maximum measured total Q-shift ($\delta Q \times$ number of crossing per turn) is $\Delta Q \cong 0.19$, at 1 GeV.

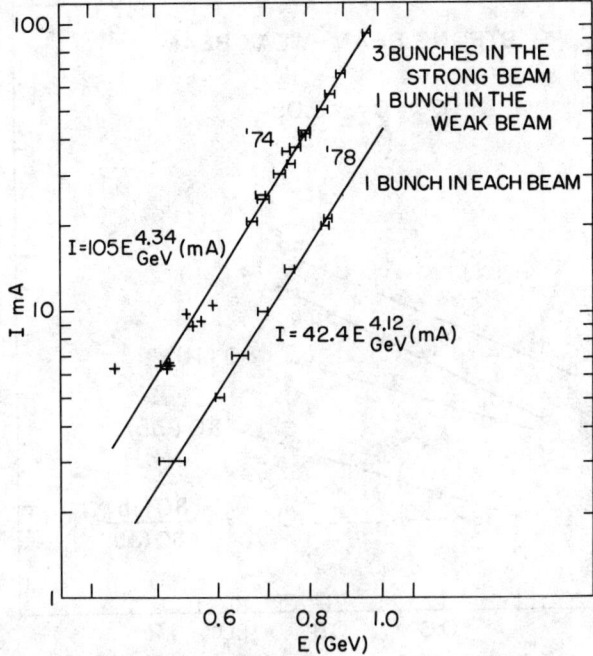

Fig. 5. Strong beam-weak beam. Maximum current versus energy, at the beam-beam limit. Three bunches (6 crossings) and 1 bunch (2 crossings) data.

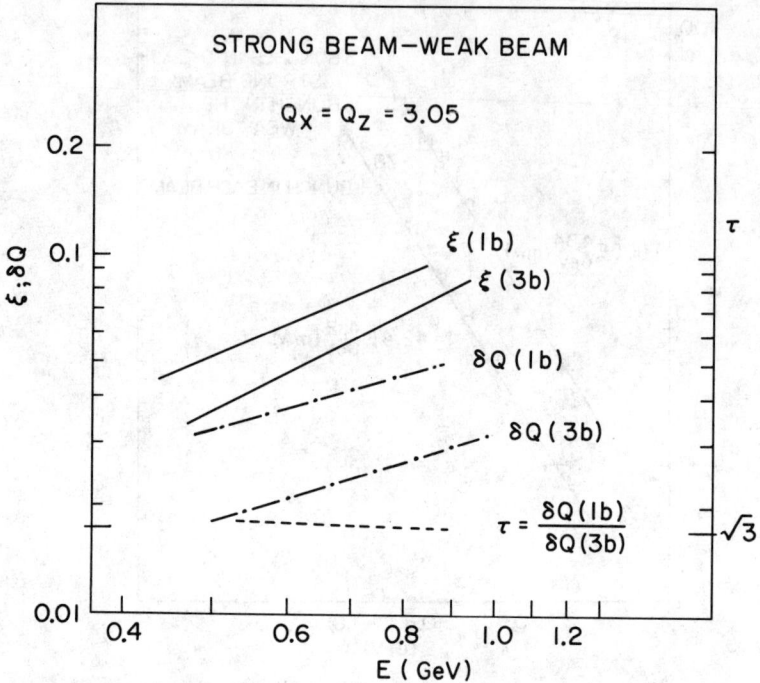

Fig. 6. Strong beam-weak beam. Interaction strength parameter ($\xi = \xi_x \cong \xi_z$) and linear tune shift (δQ) as derived from the measurements of Fig. 5.

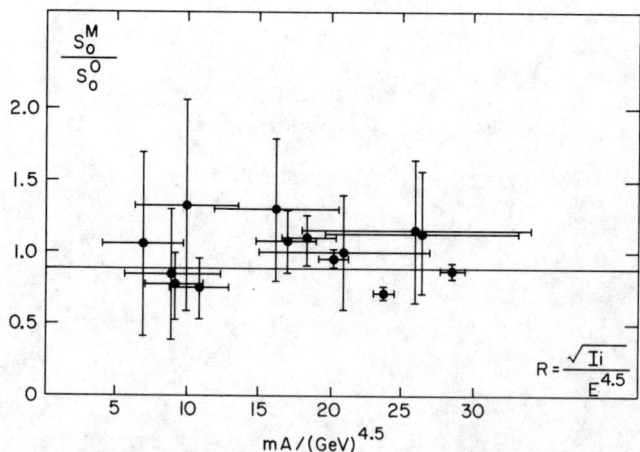

Fig. 7. Strong beam-weak beam. Ratio of the measured beam effective cross section (S_o^M) to the radiation cross section (S_o^O) as a function of energy and current. Parameter R, a function of the strong current I, the weak current, i, and energy, E, was chosen for comparison with the results presented in Ref. 3.

REFERENCES

1. S. Tazzari, "Closed orbit with separation electrodes and dipole beam-beam effect." DM Int. Memo T-99 unpublished.
2. M. H. R. Donald and J. M. Paterson, "Investigation of the flip-flop beam-beam effect in SPEAR. SLAC-PUB-2273 (79).
3. F. Amman et al., "Remarks on the two beam behavior of the 1.5 GeV electron positron storage ring Adone"-<u>Proc. VIII Intern. Conf. on High Energy Accelerators</u>-Geneva 1971, p. 132.
4. M. Bassetti et al., "Adone: Present status and experiments." <u>Proc. IX Intern. Conf. on High Energy Accelerators</u>-Stanford 1974, p. 104.
5. H. Wiedemann, "SPEAR Results," in these Proceedings.
6. "SPEAR Storage Ring Group-Operating results from SPEAR," <u>Proc. of the 1973 Particle Accelerator Conference</u>-S. Francisco 1973, p. 752.

BEAM-BEAM EFFECT

A REVIEW OF THE OBSERVATIONS MADE AT ORSAY

H. Zyngier

Laboratoire de l'Accélérateur Linéaire
91405 ORSAY (France)

This review is based on the results reported at different international conferences (Cambridge 1967, Erevan 1969, Geneva 1971, San Francisco 1974) or Erice Schools (1973, 1976), and a few other data obtained recently on DCI.

We have in Orsay an experience of over 10 years on beam-beam interactions with several different machines. The normal configuration of ACO is a lattice of 4 symmetric cells, but it was eventually modified to a 2 cell and a 1 cell structure. DCI is a 2 symmetric cell lattice in which we can perform e^+e^- collisions in either ring, e^+e^- collisions as well as three or four beam collisions. A magnetic detector has been implemented both on ACO and DCI.

The beam-beam limit is currently defined as the point where the lifetime of a beam (usually the weaker one) becomes short, say less than an hour. For stored currents approaching this limit, the interaction cross-section sometimes increases, but we never observed a saturation of the luminosity. The situation of DCI is illustrated by Fig. 1. On ACO, we had no beam enlargement at high energies, and a small blow-up at low energy, but less marked than with DCI.

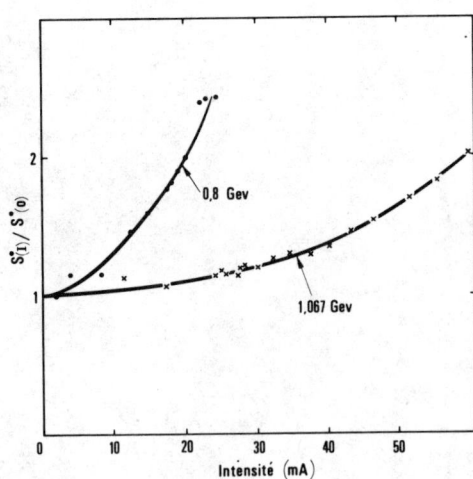

Fig. 1. Interaction Cross Section of e^+e^- Colliding Beams in DCI.

There is a clear relationship between beam-beam effects and rational values of the tune. Figure 2 shows the beams as seen on a TV screen for different tunes. The stronger beam is affected only by the coupling resonance $\nu_x - \nu_z = 2$ and the vertical nonlinear resonance $\nu_z = 2/3$. The weaker beam is sensitive to many nonlinear resonances.

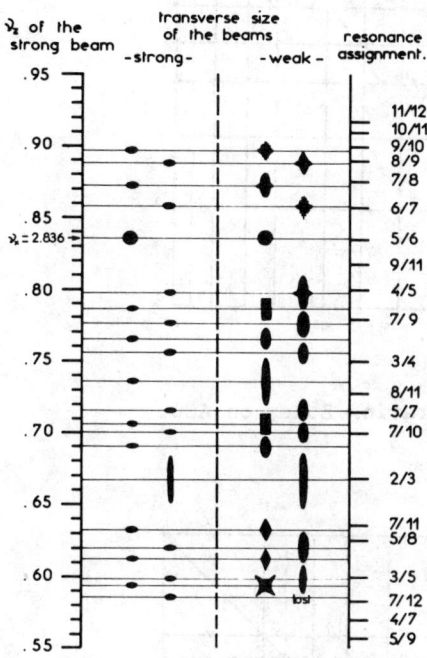

Fig. 2. Crossing of Nonlinear Resonances for Strong-Weak Counter Rotating Beams.

For higher intensities, the two beams are stable only for definite values of the tunes as is shown by Fig. 3. Islands of stable interaction can be found in the vicinity of the coupling resonance, and are separated by rational values of the tunes. The points situated far away from the resonance were not looked for on ACO. But it should be noted that the coupling strength is easily tunable on DCI, and not on ACO.

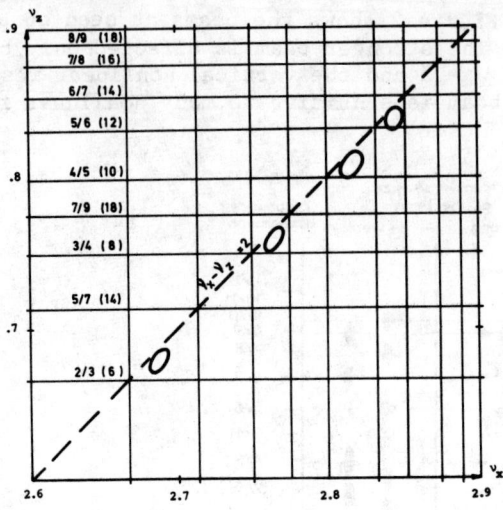

Fig. 3-a. Operating Zones on ACO.

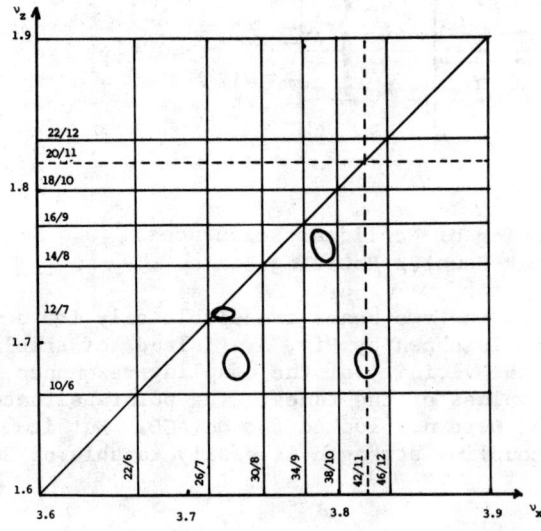

Fig. 3-b. Operating Zones on DCI.

When intensities are raised towards the limit, the islands decrease in size. This can be seen on the $\nu_x \nu_z$ diagram of Fig. 4, and the ν_z, I diagram of Fig. 5. The latter shows that the stable zones are limited by two rational tunes. For the central point ($\nu_x = 2.817$), data have been taken both with two and four crossing points per turn. The upper limit should then be related to either 10/12 or 20/24.

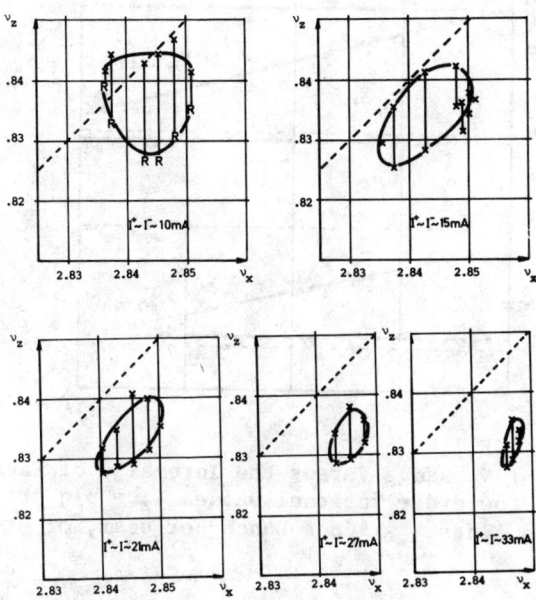

Fig. 4. Range of ν_z Versus the Radial Wave Number ν_x for Different Intensities. The coupling resonance $\nu_x - \nu_z = 2$ is indicated by a dotted line. Enlargement of one beam at an end point of the ν_z range is indicated by R.

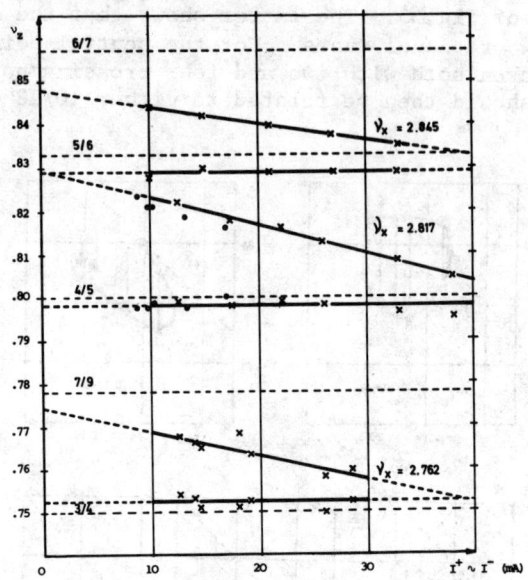

Fig. 5. Limits of ν_z Range Versus the Intensity of Each Beam and for 3 Values of ν_x. Low order rational values $\nu_z = p/q$ are indicated by horizontal dotted lines. x: one bunch per beam, o: two bunches per beam.

The three examples shown on Fig. 5 correspond to slightly different tunings. Table I gives a summary of the results obtained with radically different structures, and at different times. The parameter ranges from 0.01 to 0.045. There is a definite improvement of the attainable ξ with time, as well with ACO as with DCI. Implementation of a magnetic detector gave a loss in luminosity for ACO and a gain for DCI. In some cases, coherent oscillations have been detected with frequencies corresponding either to ξ or 2ξ per crossing.

TABLE I

	E (MeV)	(cm^{-2}s^{-1})	L/γ^4	ξ	
ACO 4-1967	385	1. × 10^2	3 × 10^{16}	0.03	
				0.02	(4 crossings)
ACO 4-1969	535	3.6 × 10^{28}	3 × 10^{16}	0.03	
ACO 4-1971	510	6. × 10^{28}	6 × 10^{16}	0.04	
ACO 2-1971	380	1.5 × 10^{28}	5 × 10^{16}	0.01	(2-cell structure)
ACO 4-1974	510	10. × 10^{28}	10 × 10^{16}	0.045	
ACO 1-1974	360	5. × 10^{28}	20 × 10^{16}	0.01	(1-cell structure)
DCI-1978	1000	2.5 × 10^{29}	5 × 10^{16}	0.02	
DCI-1979	1000	5. × 10^{29}	10 × 10^{16}	0.03	

As the energy is varied, the luminosity goes like E^4, and the intensity like E^3 (ACO) or E^2 (DCI). These dependences can be related to a constant ξ and a cross section proportional to E^2 (like the natural cross section) for ACO. For DCI, the cross section is constant, an ξ propotional to E.

Some efforts have been devoted to the study of the behavior of the beam-beam effect when the orbits of the two beams are slightly separated. In ACO, the separation was made by electric fields, and the beam-beam limit dropped drastically for separations of the order of $\sigma/10$ or less. In DCI, the two rings can be separated by magnetic corrections, and no sharp effect could be detected neither in e^+e^+ collisions nor with three or four beams.

A new domain has been open in the beam-beam studies with the space charge compensation with DCI. The results are summarized by Fig. 6. On the right side, we can see the increase of the interaction cross section in the normal e^+e^- collisions, with two crossings per turn. On the left side, a similar curve is drawn for e^+e^+ collisions, with one crossing per turn. Then, for three beam collisions, that is two strong companion beams against a weak beam, there is essentially no beam enlargement up to higher currents. With four beams of nearly equal intensities, there is no enlargement either. All these measurements have been made at the same tune, slightly higher than the normal e^+e^- tune, in order to allow for the negative tune shift of e^+e^+ collisions. To be complete, I must add the following remarks:
- e^+e^+ collisions with two crossing points per turn are limited about 15 mA.
- Three beams are stable when the small beam has the sign of the weaker of the companion beams.
- Four beam operation is also limited at 15 mA.

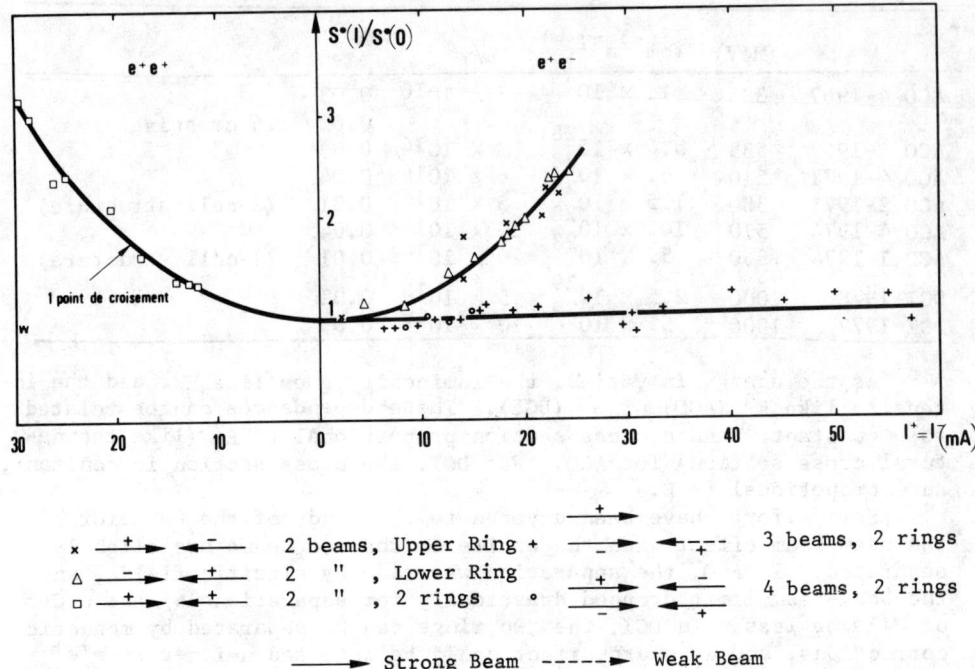

Fig. 6. Beam-Beam Effect on DCI (E = 800 MeV)

We are now investigating new operating tunes to overcome the four beam limit.

TRANSFER MAP APPROACH TO THE BEAM-BEAM INTERACTION

Alex J. Dragt

Los Alamos Scientific Laboratory, Los Alamos, New Mexico 87545

and

University of Maryland, College Park, Maryland 20742

ABSTRACT

A study is made of a model for the beam-beam interaction in ISABELLE using numerical methods and the recently developed method of Transfer Maps. It is found that analytical transfer map calculations account qualitatively for all the features of the model observed numerically, and show promise of giving quantitive agreement as well. They may also provide a kind of "magnifying glass" for examining numerical results in fine detail to ascertain the presence of small scale stochastic motion that might lead to eventual particle loss. Preliminary evidence is presented to the effect that within the model employed, the beam-beam interaction at its contemplated strengths should not lead to particle loss in ISABELLE.

I. INTRODUCTION

The purpose of this paper is to explore the model of Herrera, Month, and Peierls[1] for the ISABELLE beam-beam interaction with the aid of the recently developed method of Transfer Maps[2] and its associated Lie algebraic techniques.[3] The model employed for the beam-beam interaction is "weak-strong".[1,4] One beam, the strong beam, is taken to be fixed, and the other beam, the weak beam, is treated as a collection of particles that are affected by their passage through the strong beam but not by each other. The strong beam is assumed to be an unbunched ribbon in the horizontal plane whose vertical charge distribution is well described by a Gaussian shape. The weak beam also lies in the same horizontal plane and crosses the strong beam at a fixed angle. Only vertical deflections of the weak beam by the strong beam are taken into account.

In the strong beam-weak beam limit, the net motion of a particle in the weak beam can be viewed as the continual repetition of two sequential motions: passage through the storage ring followed by passage through the strong beam. (See Fig. 1.) The equations of motion for each of these two passages (through the ring and through the strong beam) are derivable from Hamiltonians, and therefore each passage is described by a symplectic (Poisson bracket preserving) transfer map.[2,3]

By design, the passage through a storage ring is well described by a linear map. Upon restricting attention only to vertical motion and making a suitable choice of coordinates, the "ring" transfer map can be written as

$$q' = q \cos(2\pi w) + p \sin(2\pi w)$$
$$p' = -q \sin(2\pi w) + p \cos(2\pi w). \tag{1}$$

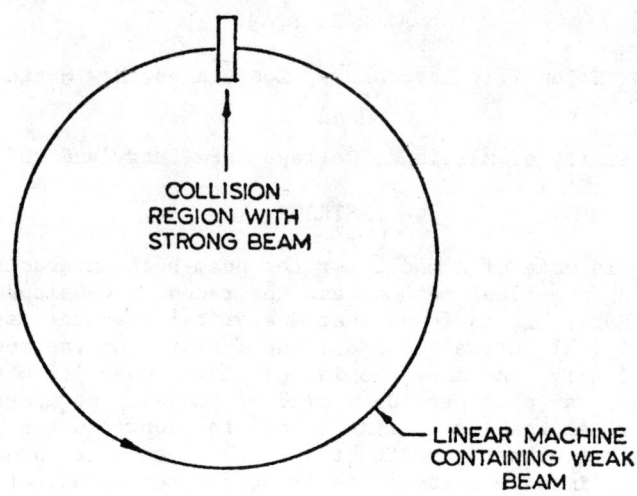

Fig. 1. Schematic representation of particle motion in a storage ring and a colliding beam region.

Here q is proportional to the vertical coordinate of a particle in the weak beam, p is a suitably chosen canonically conjugate momentum, and w (modulo an integer) is the tune of the storage ring.[1] The unprimed variables q,p specify the particle orbit just as it enters the ring, and the primed variables q′,p′ describe the orbit upon exit.

The effect of passage through the strong beam is more complicated. To find the "beam" transfer map exactly, it is necessary to integrate the nonlinear equations of motion for a particle passing through the strong beam. However, a good approximation to this map is given by assuming that the particle suffers a vertical momentum change depending only upon its initial vertical position, and that the vertical position itself remains unaffected:

$$q'' = q'$$
$$p'' = p' + u(q'). \tag{2}$$

This impulse approximation becomes exact in the limit that the interaction region becomes a point and/or the transit time through the region approaches zero. In any case, the beam mapping (2) is symplectic, and therefore its use will produce no qualitative error.

The function u is proportional to the electrostatic force exerted by the strong beam. In the coordinates and Gaussian model employed, u is given by the relation[1]

$$u(q) = 4\pi D/\sqrt{3} \int_0^{q\sqrt{3}} dt\, e^{-t^2}. \tag{3}$$

Here D is the beam-beam strength parameter that typically[4] has values ranging from 10^{-3} to 10^{-2}. It is normalized in such a way that the beam-beam interaction depresses the tune for infinitesimal betatron oscillations by an amount D when D is small.

There is one last caveat to be made. According to the current design, ISABELLE will actually have 6 collision regions separated by 6 identical lattice sections. Thus, in reality, the maps 1 and 2 must be iterated 6 times to simulate the effect of one complete turn. Correspondingly, Eq. (1) is the transfer map for one lattice section and w (modulo an integer) is actually 1/6 of the total machine tune.

In this paper the effect of repeated iteration of the maps (1) and (2) are studied using Transfer Map methods and results are compared with numerical calculations.

II. TRANSFER MAP RESULTS

To treat q and p on an equal footing, it is notationally convenient to introduce variables z_1 and z_2 by the relations

$$z_1 = q$$
$$z_2 = p. \tag{4}$$

Employing this notation, let $f(z)$ be any function of the phase-space variables z. With each such function f these is an __associated Lie operator__ F. This operator acts on functions and is defined by the rule

$$Fg = [f,g]. \tag{5}$$

Here g is any function of the phase-space variables, and the square bracket [,] denotes the Poisson bracket operation familiar from classical mechanics.

Next, consider the object exp(F), called a __Lie transformation__, defined by the exponential series

$$\exp(F) = I + F + F^2/2! + F^3/3! + \ldots. \tag{6}$$

More explicitly, the action of exp(F) on an arbitrary function g is given by the expression

$$\exp(F)g = g + [f,g] + [f,(f,g)]/2! + \ldots. \tag{7}$$

Now consider the operator $\exp(F_2)$ where F_2 is the Lie operator associated with the quadratic polynomial

$$f_2 = -\pi w(z_1^2 + z_2^2). \tag{8}$$

It is easily verified that

$$F_2 z_1 = [f_2, z_1] = 2\pi w z_2$$
$$F_2 z_2 = [f_2, z_2] = -2\pi w z_1. \tag{9}$$

Consequently, use of (7) and (9) gives the relation

$$\exp(F_2) z_1 = z_1 + z_2 (2\pi w) - z_1 (2\pi w)^2/2!$$
$$- z_2 (2\pi w)^3/3! + \ldots \tag{10a}$$
$$= z_1 \cos(2\pi w) + z_2 \sin(2\pi w).$$

Similarly, it can be checked that

$$\exp(F_2) z_2 = -z_1 \sin(2\pi w) + z_2 \cos(2\pi w). \tag{10b}$$

Therefore the ring transfer map (1) can be written in the compact form

$$z' = \exp(F_2) z. \tag{11}$$

A similar Lie transformation representation can be found for the beam transfer map (2). Let $f_b(z)$ be the function defined by the relation

$$f_b(z) = \int_0^{z_1} u(q) \, dq. \tag{12}$$

The Poisson bracket relations analogous to (9) are

$$F_b z_1 = [f_b, z_1] = 0$$
$$F_b z_2 = [f_b, z_2] = \partial f_b / \partial z_1 = u(z_1) \tag{13}$$
$$F_b^2 z_2 = [f_b, (f_b, z_2)] = [f_b, u(z_1)] = 0, \text{ etc.}$$

Consequently the infinite sum (7) is trivial to evaluate in this case because it terminates. One finds the result

$$\exp(F_b) z_1 = z_1$$
$$\exp(F_b) z_2 = z_2 + u(z_1). \tag{14}$$

Therefore the beam transfer map (2) can be written in the form

$$z'' = \exp(F_b) z'. \tag{15}$$

Combing the two results (11) and (15), one finds that the net transfer map M for passage through the ring followed by passage through the strong beam is given by the product

$$M = \exp(F_2) \exp(F_b). \tag{16}$$

The observant reader may be worried about the order in which the two factors appear in (16). It can be verified that the above order indeed is correct because Lie transformations have the property

$$\exp(F_2) \, g(z) = g \, \exp(F_2) \, z = g(z') \tag{17}$$

for any function $g(z)$.[3]

The problem at hand is to evaluate M^n for large n in order to compute the effect of many passages through the ring and the strong beam. The computation of M^n would be easy if a Lie operator H could be found such that M could be reexpressed in the form exp(H), for then M^n would be simply given by exp(nH). The determination of such an H is a standard problem in the theory of Lie algebras that is solved by using the Campbell-Baker-Hausdorff formula.[3] This formula gives H in terms of F_2 and F_b, and their multiple commutators. In addition, there is an analogous formula that gives the function h associated with H in terms of f_2 and f_b, and their multiple Poisson brackets. It also can be shown that the computation of exp(nH) is equivalent to the integration of a "trajectory" in "z space" for n units of "time" using -h as an "effective" Hamiltonian. Consequently, the function h(z) is formally invariant under the map. This means that the function h(z) generalizes the Courant-Snyder invariant to the case of nonlinear motion.

For the problem under consideration, h is given by the formal operator formula,

$$h = f_2 + F_2 \left[1 - \exp(-F_2)\right]^{-1} f_b + \ldots. \tag{18}$$

The terms not shown in the series involve Poisson brackets with more than one f_b, and therefore are quadratic and higher order in the beam-beam strength parameter. Consequently, as it stands, Eq. (18) is correct through first order in the beam-beam strength.

The computation of the effect of the operator F_2 and the functions of F_2, such as occur in (18), is facilitated by the introduction of "polar" coordinates in phase space and the use of Fourier series. This can be achieved in a canonical way by using action angle variables a, Φ defined by the relations

$$q = z_1 = (2a)^{\frac{1}{2}} \sin \Phi$$
$$p = z_2 = (2a)^{\frac{1}{2}} \cos \Phi. \tag{19}$$

It is evident from (5) and (8) that F_2 annihilates any function of a. By contrast, use of (1), (11), and (17) shows that

$$\exp(F_2)\, a^{n/2} \exp(in\Phi) = \exp(i2n\pi w)\, a^{n/2} \exp(in\Phi). \tag{20}$$

Consequently, the functions $\exp(in\Phi)$ are eigenfunctions of F_2 with eigenvalues $i2n\pi w$:

$$F_2 \exp(in\Phi) = i2n\pi w \exp(in\Phi). \tag{21}$$

This result can also be obtained by direct evaluation of the Poisson bracket $[f_2, \exp(in\Phi)]$.

The determination of h as given in (18) is now straightforward. Inserting (19) into (12) and making a Fourier expansion, one finds

$$f_b = \sum_{-\infty}^{\infty} c_n(a) \exp(i2n\Phi) \tag{22}$$

where

$$c_n = 4\pi D a \int_0^1 \int_0^1 du\, dv\, v \exp(-3au^2v^2) \\ \times [I_n(3au^2v^2) - I_n'(3au^2v^2)]. \tag{23}$$

Here the symbols I_n denote modified Bessel functions, and use has been made of the standard relations[5]

$$\exp(x \cos y) = \sum_{-\infty}^{\infty} I_n(x) \exp(iny) \tag{24a}$$

$$I_{n+1} + I_{n-1} = 2I_n'. \tag{24b}$$

Now insert (22) into (18) and use the eigenfunction property (21). The result is that h is given in complex form by the expression,

$$h = -2\pi wa + \sum_{-\infty}^{\infty} c_n(a)\, [2n\pi w/\sin(2n\pi w)] \exp[2in(\Phi + \pi w)], \tag{25a}$$

and in real form by the expression,

$$h = -2\pi wa + c_0(a) + 2 \sum_1^{\infty} c_n(a)[2n\pi w/\sin(2n\pi w)]\cos[2n(\Phi + \pi w)]. \tag{25b}$$

The expressions (25) provide a generalization of the Courant-Snyder invariant through first order in the beam-beam interaction strength. Upon inspecting them, several points are immediately evident:

(a) Resonances occur and the formulas diverge whenever the tune w is of the form

$$w = k/(2N) \tag{26}$$

where k and N are integers. Thus there are resonances at half-integer tunes, quarter-integer tunes, sixth-integer tunes, etc. This was to be expected because u(q) as given by (3) contains no even powers and all odd powers of q.

(b) The strengths of the various order resonances are proportional to $nc_n(a)$. Using the large n expansion,[5]

$$I_n(x) \sim \exp[-n \log(2n/ex)], \qquad (27)$$

one finds from (23) that the strengths of the various resonances fall off faster than exponentially as n is increased. Therefore the sizes of various resonance features in phase space should decrease in size according to their proximity to the origin in phase space.

It also follows that (25) converges rapidly at all tune values for which w is badly approximated by rationals. Indeed, the points in tune space where (25) fails to converge are of measure zero.[6]

(c) With the aid of time reversal invariance it can be shown from (16) that the locations of various features in phase space as fixed points and separatrices must be symmetric about the line $\Phi = \pi/2 - \pi w$ for all beam-beam interaction strengths. (Note that according to (19), the line $\Phi = 0$ corresponds to the p axis.) Because f_b as given by (12) is even in z_1, there is also symmetry in phase with respect to inversion through the origin. Examination of (25) shows that both of these symmetries are present in h.

To calculate the behavior of M exactly at and near resonance, it is necessary to work with powers of M. For example, consider m'th order resonances. Then m = 2N and tunes near an m'th order resonance value can be written in the form

$$w = k/m + \delta \qquad (28)$$

where δ measures departure from exact resonance. Moreover, it can be shown that there is a Lie operator H_r such that M^m can be written in the exponential form $\exp(mH_r)$ at and near resonance without divergence difficulties. Finally, there is again an effective Hamiltonian h_r corresponding to H_r that is given in this case by the formula

$$h_r = (\delta/w)f_2 + (\delta/w)F_2 \{1 - \exp[-m(\delta/w)F_2]\}$$
$$\times \{1 + \exp[-F_2] + \exp[-2F_2] + \ldots + \exp[-(m-1)F_2]\}f_b + \ldots \qquad (29)$$

Upon inserting (22) into (29), one finds

$$h_r = -2\pi\delta a + c_o(a) + 2\sum_1^\infty c_n(a)[2\pi\delta/\sin(2n\pi w)]\cos[2n(\Phi + \pi w)]. \qquad (30)$$

It is evident that the expression for h_r is well behaved nearby and exactly at the resonance value $\delta = 0$.

As a specific example, consider the case of fourth order resonances. Near a one-quarter tune k = 1, N = 2, and

$$w = 1/4 + \delta. \qquad (31)$$

Thus one finds for small δ that

$$2n\pi\delta/\sin 2n\pi w = 2n\pi\delta/\sin(n\pi/2 + 2n\pi\delta)$$
$$= O(\delta) \text{ for n odd} \qquad (32)$$
$$= (-1)^{n/2} + O(\delta^2) \text{ for n even.}$$

Consequently, neglecting terms of order δD and δ^2, one has in this case for h_r the expression

$$h_r = -2\pi\delta a + c_o(a) + 2 \sum_{n \text{ even}} (-1)^{n/2} c_n(a) \cos[2n(\Phi + \pi w)]. \qquad (33)$$

Because $-h_r$ acts as an effective Hamiltonian, the fixed points of M^4 are the equilibrium points of h_r. These points are therefore the solutions to the equations

$$0 = \partial h_r/\partial a = -2\pi\delta + c'_o(a) - 2c'_2(a) \cos 4(\Phi + \pi w) + \ldots \qquad (34a)$$

$$0 = \partial h_r/\partial \Phi = 8c_2(a) \sin 4(\Phi + \pi w)$$
$$\qquad\qquad - 16c_4(a) \sin 8(\Phi + \pi w) + \ldots . \qquad (34b)$$

The solutions to (34b) are readily found to be

$$\Phi + \pi w = 0, \pi/4, 2\pi/4, \ldots, 7\pi/4. \qquad (35a)$$

When these solutions are inserted into the "radial" Eq. (34a), it takes the simple form

$$0 = -2\pi\delta + c'_o(a) \pm 2c'_2(a) + 2c'_4(a) + \ldots . \qquad (35b)$$

Thus, as expected, there are 8 fourth-order fixed points when the tune is near a quarter.

The nature of these fixed points can be obtained by expanding h_r about them. At the fixed points one finds the results

$$\partial^2 h_r/\partial a^2 = c''_o(a) \pm 2c''_2(a) + 2c''_4(a) \ldots$$

$$\partial^2 h_r/\partial a \partial \Phi = 0 \qquad (36)$$

$$\partial^2 h_r/\partial \Phi^2 = \pm 32c_2(a) - 128 c_4(a) + \ldots .$$

It follows that if the Eq. (36) are dominated by their first terms, then the fixed points are alternately elliptic and hyperbolic (stable and unstable), as also expected,[7] because the quadratic form corresponding to (36) is either definite or mixed.

Note, moreover, that Eq. (36) and all the higher order terms in the Taylor series expansion about a fixed point are linear in the beam-beam interaction strength. Consequently, for small beam-beam interaction strength, the size and shape of resonant islands and their associated separatrix structure are <u>independent</u> of the beam-beam interaction strength, and are dependent <u>only</u> on their location in phase space. Only the width of the resonance in tune space, i.e., the rate at which various features move as δ is changed, depends on the interaction strength. This latter dependence can be inferred from (34a) and (35b).

III. NUMERICAL RESULTS

A proper study of the usefulness of h and h_r involves the numerical integration of the trajectories that they generate, or at least a determination of their level lines, and a comparison of these results with points generated by iterating M and M^m numerically. Such a comparison has been made in a similar but simpler problem involving the insertion of a short sextupole element into a ring.[2] In that case the quantitative agreement proved to be excellent, and similar agreement is expected for this problem as well.

However, because of the complexity of evaluating the coefficients $c_n(a)$, the equivalent study has not yet been carried out for the present problem. Instead, a preliminary exploration of the nature of M has been made by studying the points obtained by iterating M numerically. In this section it will be shown that M indeed does have all the qualitative properties that were predicted in the previous section.

Figures 2 through 5 show phase-space plots generated by successive iterates of M for various initial conditions and tune values. The phase-space coordinates range over (-2, 2), and the scale is chosen so that the beam lies within (-1, 1).[1] The tunes are near the resonant values 1/2, 1/4, 1/6, and 1/8 respectively, and the beam-beam interaction strength is 10^{-2}. Observe that the size of resonance features, e.g., island dimensions, indeed do decrease with increasing order of the resonance. Symmetry about the line $\Phi = \pi/2 - \pi w$ and inversion symmetry are evident. The number and nature of fixed points is as anticipated.

Figure 6, which appears to be almost identical to Fig. 3, was obtained by running at a tune w = 0.253 and a beam-beam interaction strength of 5×10^{-3}. It shows, as predicted, that the size of resonant features is independent of the size of the interaction strength provided the tune is suitably adjusted so as to make the features a appear in the same region of phase space. Note that according to (35b), when the size of the c_n is halved, δ should also be halved to keep the radial location of fixed points the same. Examination of the tune values for Figs. 3 and 6 shows that this is indeed the case. When the tune is thus adjusted, there is a slight change in the angular location of the fixed points in accord with (35a).

Figure 7 shows a tenth-order resonance obtained by running near a tune of 1/10. It was not shown as part of the sequence of Figs. 2 through 5 because the island structure becomes too small to see when (by adjusting the tune) it is located closer to the origin. This example verifies that the sizes of resonant features decrease with proximity to the origin, and in fact the higher the order of the

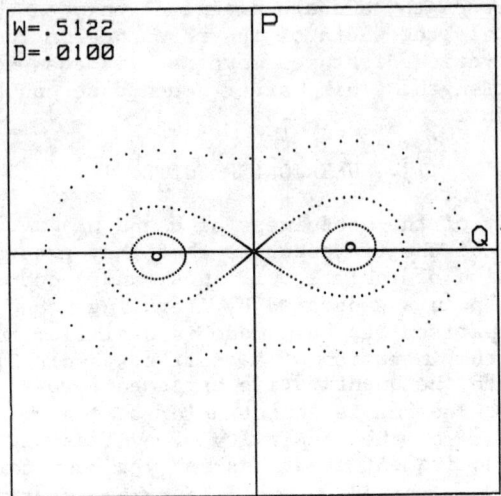

Fig. 2. Phase-space plot generated by successive iterations of the transfer map M for various initial conditions. The tune is near one half. The coordinates extend from - 2 to 2, and are normalized in such a way that the beam will be within the unit circle under actual operating conditions. The beam-beam interaction strength is 10^{-2}.

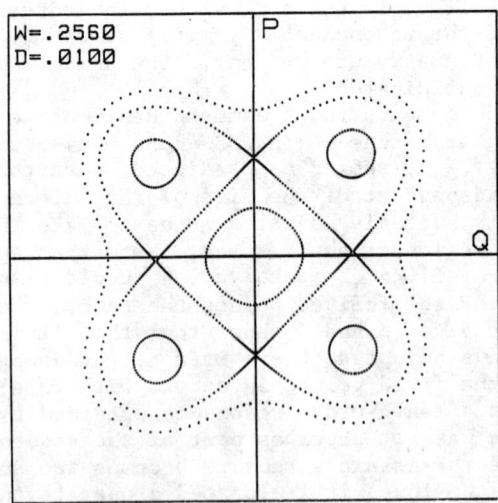

Fig. 3. Phase-space plot when the tune is near one fourth.

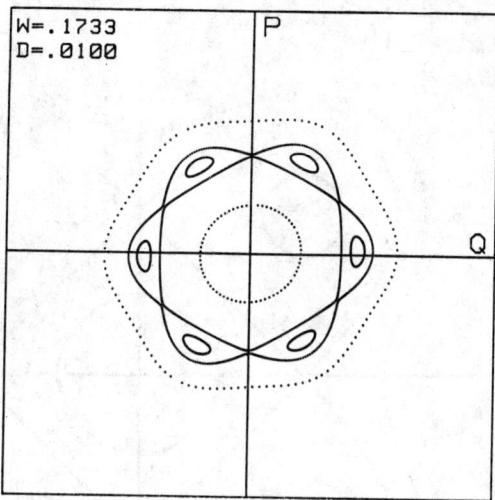

Fig. 4. Phase-space plot when the tune is near one sixth.

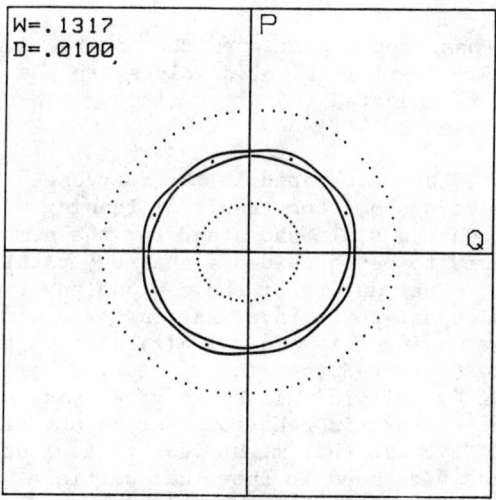

Fig. 5. Phase-space plot when the tune is near one eighth.

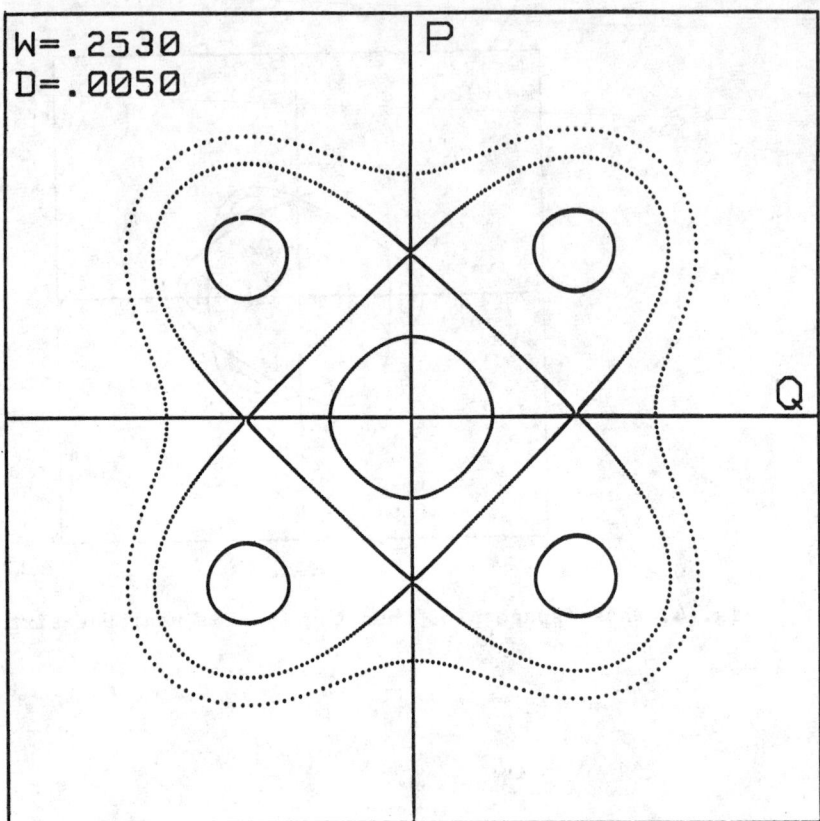

Fig. 6. Phase-space plot when the interaction strength is half that of Fig. 3. The initial p, q values are the same as in Fig. 3, and the tune is adjusted to make various phase-space features match those in Fig. 3.

resonance, the more rapid is the decrease.

Figure 8 shows the result of running with a nonresonant tune of 77/100. On the scale shown and for the number of iterations made, there seems to be no evidence that any points will leave the beam envelope. The nature of the map and any tendency for points to move off what appear to be invariant curves could be examined in finer detail by studying the value of h(z) at each point. Because h generalizes the Courant-Snyder invariant, its variations could be used as a kind of "magnifying glass" to give evidence for small scale homoclinic or stochastic behavior that is not otherwise discernible to the naked eye and that might lead to eventual particle loss. This method has been used to show that particle motion in the Van Allen radiation belts is not integrable.[8]

Figure 9 illustrates that stochastic behavior indeed can occur for the beam-beam problem if the interaction strength is large enough and the tune (taking into account its depression by the beam-beam

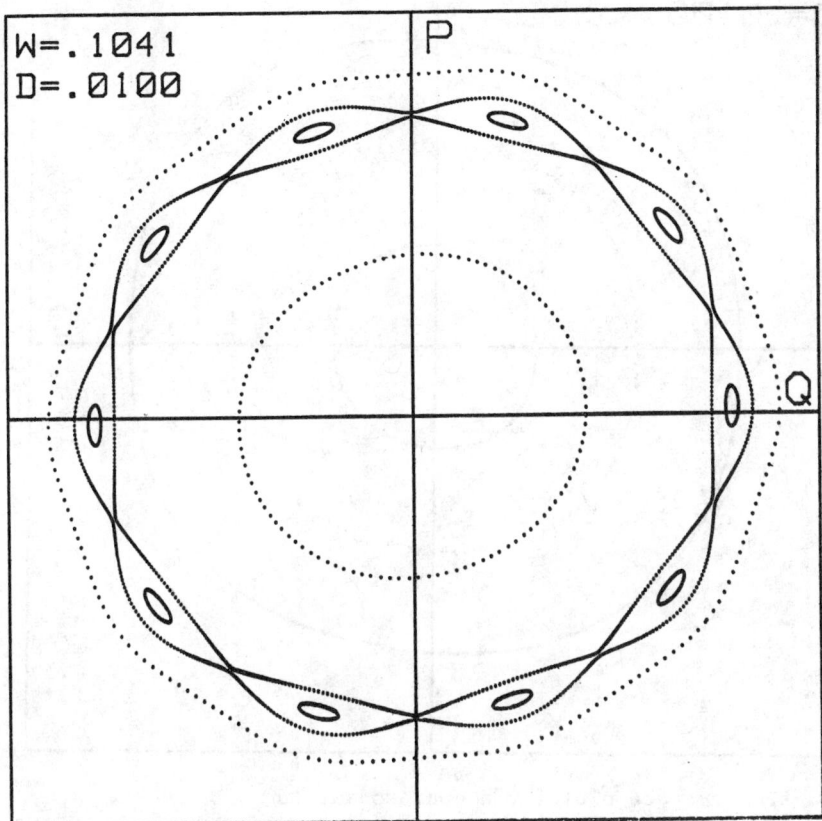

Fig. 7. Phase-space plot near a tune of one tenth.

interaction) is sufficiently close to a resonant value. The stochastic behavior in this case leads to particle losses within a few hundred turns.

IV. CONCLUDING REMARKS AND COMMENTS

Operation of ISABELLE with each 1/6 lattice section having a tune near a multiple of 1/2, 1/4, 1/6, or 1/8 corresponds to operating the total ring near an integer, half integer, or quarter integer tune. Because operation of the total ring near any of these tunes is probably already precluded by structure resonances in the ring, the first beam-beam interaction resonance of significance is at least of tenth order. Figure 7 illustrates that the tenth-order resonance structure is small even when it is far from the phase-space origin, and consequently it is even smaller when it is within the beam. This observation, and the regular behavior found in the nonresonant case of Fig. 8, give preliminary evidence that within the model employed, the beam-beam interaction at its contemplated strengths should not lead to particle loss. However, in accord with our earlier comment, it would be worthwhile to examine the behavior of $h(z)$ and $h_r(z)$ for evidence of small scale stochastic behavior.

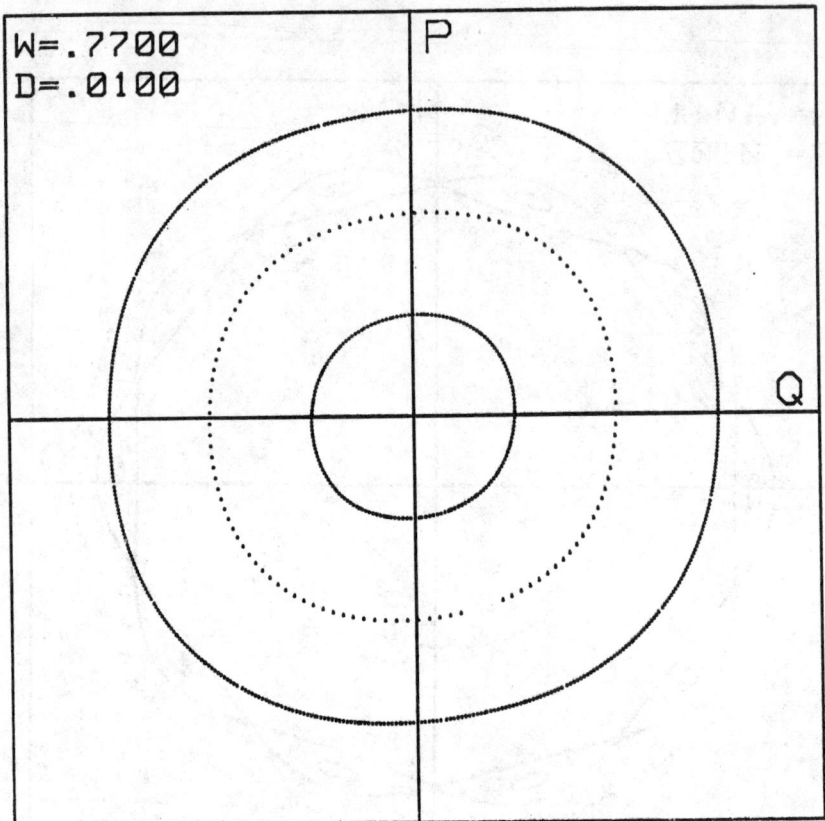

Fig. 8. Phase-space plot for a nonresonant tune.

The conclusion that resonances below tenth order are not significant depends on the assumption that all 6 interaction regions and all 6 lattice periods are identical. The validity of this assumption should be examined, and the effect of lower order resonances should be reexamined when the 6 interaction regions are all slightly different.

Finally, consideration should be given to the possible effect of adding nonlinearities to the transfer map for the ring. It is anticipated that the addition of suitable nonlinearities, perhaps by the use of octupoles, would lead to a reduction in the size of beam-beam interaction resonance structures. In particular, it would then no longer be the case that the size of resonant structures would depend only on their location in phase space and not on the interaction strength. It might turn out, of course, that the ring nonlinearities required to achieve a significant effect would be difficult to obtain or would be undesirable for other reasons.

ACKNOWLEDGMENT

The author wishes to thank Dr. Richard K. Cooper for many helpful conversations.

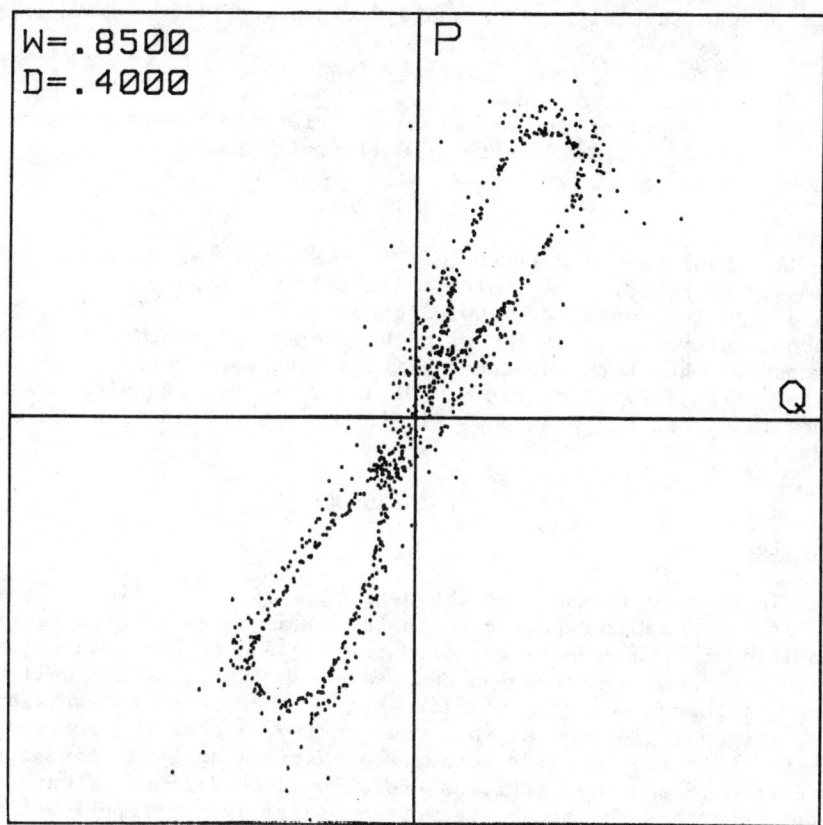

Fig. 9. Phase-space plot showing stochastic behavior for large beam-beam interaction strength. The reader is invited to draw the symmetry line $\Phi = \pi/2 - \pi w$.

REFERENCES

1. J. C. Herrera, M. Month, and R. F. Peierls, Brookhaven National Laboratory Report BNL 25703 (1979).
2. A. Dragt, "A Method of Transfer Maps for Linear and Nonlinear Beam Elements," To appear in IEEE Transactions on Nuclear Science, NS $\underline{26}$ (1979).
3. A. Dragt and J. Finn, J. Math. Phys. $\underline{17}$, pp. 2215-2227 (1976).
4. J. C. Herrera, Brookhaven National Laboratory Report BNL 25703 (1979).
5. M. Abramowitz and I. Stegun, Eds., <u>Handbook of Mathematical Functions</u>, National Bureau of Standards Applied Mathematics Series 55 (1966).
6. S. Sternberg, <u>Celestial Mechanics</u>, part II, (W. A. Benjamin, New York, 1969), p. 18.
7. V. Arnold and A. Avez, <u>Ergodic Problems of Classical Mechanics</u>, (W. A. Benjamin, New York, 1968).
8. A. Dragt and J. Finn, J. of Geophys. Res. $\underline{81}$, pp. 2327-2339 (1976).

The Instability Threshold for Bunched Beams in ISABELLE

Jeffrey Tennyson

Dept. of Electrical Engineering and Computer Science
University of California, Berkeley

ABSTRACT

An absolute upper limit on the tune shift is derived for bunched beams in ISABELLE. The limit is defined by the onset of stochastic phase flow in a model with two degrees of freedom. The appearance of stochasticity is identified with the overlap of synchro-betatron resonances when both the tune shift and the synchrotron period exceed certain critical values ($\Delta \nu = 0.01$ and $P_s = 60$). Results are based on theoretical and computational studies of a simple difference equation model.

I. INTRODUCTION

Overview

In storage rings, when the densities of the colliding beams exceed a certain value, the particles in the beams begin to diffuse radially, resulting in the expansion and/or loss of one or both beams.[1] The cause of this phenomena, known as the beam-beam limit, is not well understood, although several theories have been proposed to explain it. One such theory suggests that when the density exceeds the critical value, the particle trajectories become stochastic: an invariant of the motion is destroyed and the particles are allowed to diffuse outward. This paper describes the stochasticity mechanism and its application to the problem of the beam-beam limit. In particular, the resonance overlap criterion of Chirikov[2] is used to analyze the stability of bunched beams in ISABELLE.

Outline

The paper is divided into four sections. The first section describes betatron oscillations in a storage ring, how the diffusion of the betatron amplitude determines the lifetime of the beam, and how stochastic phase flow can contribute to this diffusion. The second section introduces a simple two dimensional model for the betatron oscillation and discusses briefly the dangers of both linear and non-linear resonance. It also describes how and where stochastic motion appears in the phase space and the consequences of resonance overlap. The third section presents the difference equations and the corresponding Hamiltonian. It shows that beam bunching can lead to the appearance of resonance sidebands and that the sidebands are more likely to overlap when the synchrotron frequency is small rather than large. Section four shows computer generated surface-of-section plots which illustrate the ideas of the previous sections.

The Betatron Oscillation

The cross-sectional profile of a storage ring beam is roughly elliptical. As the particles travel around the ring, they oscillate transversally about the beam center (see Figure 1). For each particle, the oscillation amplitude is approximately constant over many oscillation periods. The size of the elliptic cross section is thus determined by the average oscillation amplitude of the particles in the beam. For an e^+-e^- machine, radiation damping reduces the average oscillation amplitude to an equilibrium value at which the damping is balanced by a weak diffusion. For p-$\bar{\text{p}}$ or p-p machines, the radiation damping is insignificant because of the higher particle mass. Consequently, only diffusion of the oscillation amplitude results in beam growth. The only way to preserve the beam shape in a proton machine, is to periodically "clean" the beam with a mechanical "scraper". The scraper removes particles whose amplitudes have diffused beyond the desired beam width. This results in a steady loss of particles and the eventual decay of the beam.

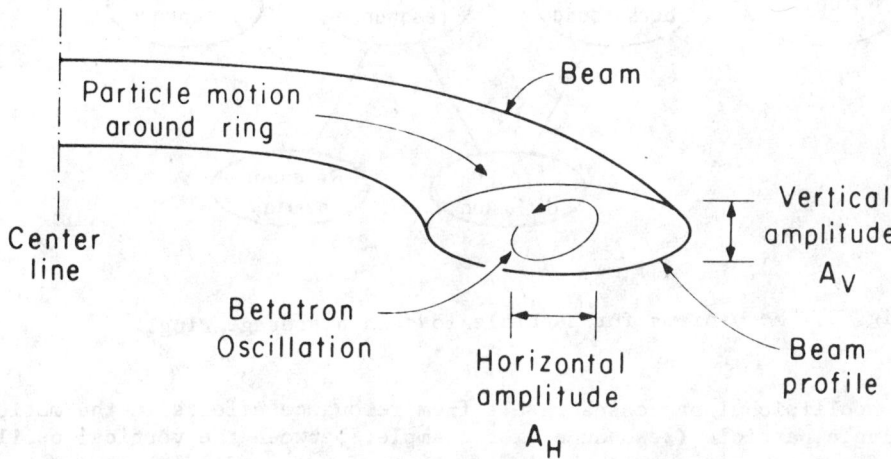

Fig. 1. Transverse oscillations about the beam center are called betatron oscillations. Here, the transverse motion is projected onto a cross section of the beam for the case where the horizontal and vertical frequencies are equal.

Beam Diffusion

The lifetime of a proton beam is determined by the rate at which the amplitudes of the transverse oscillations diffuse. There are several different mechanisms for this diffusion (see Figure 2). These can be divided into two main catagories: collisional and noncollisional.
   Collisional processes include intra-beam coulomb interactions and scattering off of low energy background neutrals.

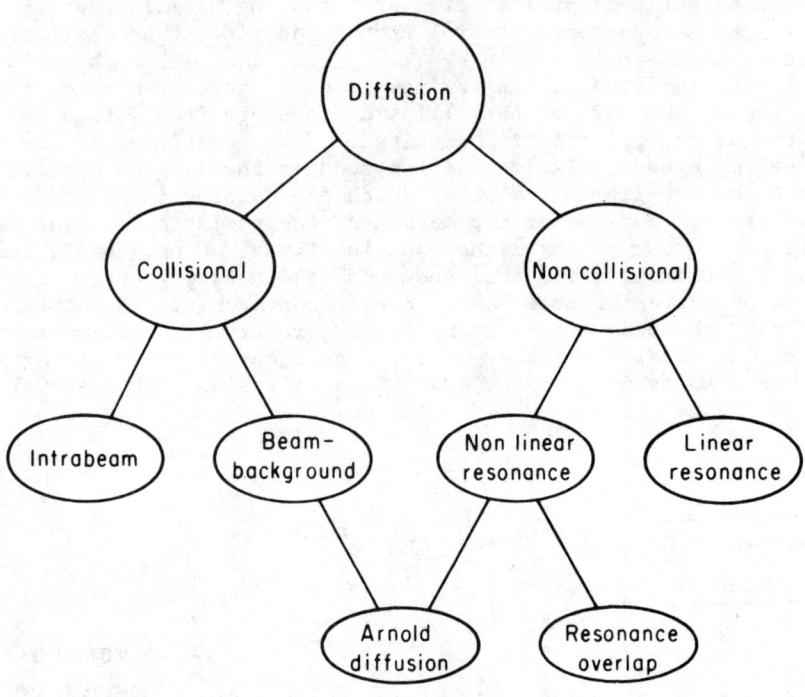

Fig. 2. Mechanisms for particle loss in a storage ring.

Noncollisional processes result from resonance effects in the motion of a single particle (resonance, for example, between the vertical oscillation frequency and the revolution frequency). Accordingly, the importance of noncollisional processes increases with the coupling strength between the two or more degrees of freedom. Since these processes constitute the major concern of this paper, they are now discussed in detail.

Single Particle Dynamics

The phase flow of a Hamiltonian system may be qualitatively described as either regular or stochastic. This is a local property of the flow in the sense that some regions of the phase space may be predominantly regular while others are predominantly stochastic.[3] The motion is regular if it is confined (for all time) to an N dimensional submanifold of the 2N dimensional phase space (i.e., if there are N invariants of the motion). It is stochastic if it is confined to a submanifold of dimensionality greater than N. Stochastic trajectories are also characterized by a macroscopic irreversibility. Neighboring stochastic trajectories diverge exponentially with time while regular trajectories diverge linearly. This means that the smallest change in the initial conditions of a stochastic trajectory will drastically effect its future behavior. In this respect, the distinction between regular and stochastic motion is very similar to the difference between laminar and turbulent flow in a fluid.

For conservative systems with only one degree of freedom, and for linear systems, the motion is always regular. For nonlinear conservative systems with two or more coupled degrees of freedom, the coupling can cause stochastic flow in certain resonant regions of the phase space.

Of systems that exhibit stochastic flow, those with two degrees of freedom (N = 2) have a special property: the stochastic region of the phase space is divided into an infinite number of unconnected open sets by the surfaces of regular motion. This follows directly from the fact that the stochastic region is a three dimensional space, while the regular submanifolds are closed two dimensional surfaces. Because the system cannot jump between unconnected stochastic regions, there is little difference in practice between a trajectory that is confined to a very thin stochastic region and one that is truly regular. It is only when the unconnected stochastic regions begin to merge and become large that the stochasticity becomes important. In many cases, this merging occurs quite suddenly when the coupling exceeds a certain critical value. From a practical point of view, this means that there is a "margin of grace" in which small imperfections in the symmetries of a system will not destroy the invariants of the motion derived from those symmetries. This result seems very fortuitous when one considers the imperfect nature of all physical devices.

The properties of stochastic flow are somewhat different if the system has more than two degrees of freedom (N > 2). In this case, the surfaces of regular motion cannot divide the stochastic region (they differ in dimensionality by (N - 1) > 1. Consequently, a stochastic trajectory is not necessarily confined to any particular region of the phase space, even when the coupling parameters are very small. Furthermore, V.I. Arnold has proved[4] that for a specific, but typical system with three degrees of freedom, at least some trajectories will link any two finite regions of the energy surface, regardless of how far apart they are or how small the coupling is. This exploration has accordingly been given the name "Arnold Diffusion". Because the stochastic layers are typically thin and cover only a small portion of the phase space, only a very small class of particles fall inside them and are effected by Arnold diffusion. Consequently, this phenomena is usually not important unless it is complemented by a collisional diffusion that is weaker than the Arnold diffusion, but strong enough to transport particles from nearby regular regions into the stochastic layers at an appreciable rate. (Systems with N > 2 will not be discussed further in this paper. For more details see ref. 5, 6, or 7).

II. A SIMPLE STORAGE RING MODEL

The Model

The ideas of stochastic and regular motion may be illustrated with a simple example (Fig. 3). A single particle in a storage ring describes a conservative system with two degrees of freedom.*

$$\text{revolution } (V_\theta, \theta)$$
$$\text{vertical oscillation } (V_Y, Y)$$

The phase space is four dimensional. Because the Hamiltonian is an invariant, the motion is confined to a three dimensional "energy surface" in the phase space. By studying the qualitative nature of the trajectories on the energy surface, it is possible to determine a threshold for beam expansion due to coupling between the revolution and the vertical oscillation.

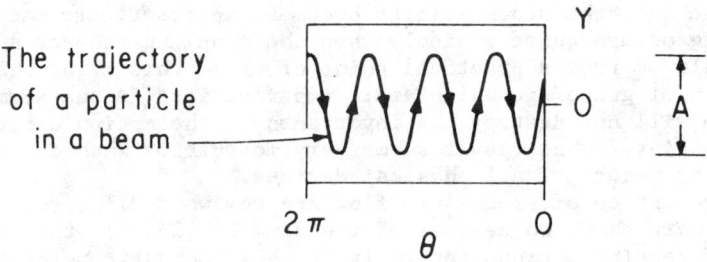

Fig. 3. The model for ISABELLE has two degrees of freedom, represented here by the coordinates Y and θ. The particle experiences libration (betatron oscillation) in Y and rotation in θ.

---

*For some storage rings (those with crossing beams) the energy in the horizontal oscillation is well conserved. Although the horizontal motion is ignored here, it would have to be included if one were studying Arnold diffusion rather than simple resonance overlap.

Linear Resonance

In the first approximation, the oscillations of a single particle in a non-intersecting storage ring are linear[8] (the frequencies of the transverse oscillation and the revolution are independent of the transverse and longitudinal energies, respectively). Consequently, even a very small coupling between the two degrees of freedom can be disastrous if the frequencies of revolution Ω, and the transverse oscillation ω_ρ are resonant, i.e. if $\Omega/\omega_0 = m/n$ where n and m are integers. If this condition is met, the revolution can pump energy steadily into the transverse oscillation until the amplitude becomes so large that the particle is removed by the scraper. For this reason, linear resonances must be carefully avoided.

Nonlinear Resonance: The Beam-Beam Force

The vertical oscillation becomes nonlinear when the two beams are made to intersect one another. The trajectory of a particle is deflected slightly as it passes through the opposing beam.* This effect, known as the "beam-beam interaction", is dependent upon the displacement Y of the particle at <u>the moment of crossing</u>. The interaction has two important effects. First, it introduces a substantial coupling between the revolution and the oscillation, and second, it makes the vertical oscillator "nonlinear", i.e. the frequency becomes dependent upon the betatron amplitude A. The linear frequency normalized to the revolution frequency $\nu_0 = \omega_0/\Omega$ is called the vertical "tune". The change in the tune due to the beam-beam interaction is called the "tune shift" $\Delta\nu(A)$.

The nonlinearity introduced by the beam-beam interaction tends to stabilize the resonances. A particle that is in resonance at one amplitude moves out of resonance as soon as its amplitude changes slightly. This prevents the type of unbounded growth seen in the linear case. The introduction of the beam-beam nonlinearity would be very beneficial to beam stability if it did not also result in the appearance of orbital instability (stochastic motion).

Regular and Stochastic Motion in the Storage Ring Model

Figure 4a shows the three dimensional energy surface for the uncoupled linear system. In this limit, all trajectories are regular and are confined by a second energy invariant to two dimensional tori in the phase space (these tori are really energy surfaces in their own right). The oscillation amplitude, given by the minor radius of the torus, is constant for all initial conditions.

---

*Relativistic effects result in an interaction between a particle and the opposing beam, but not between the particle and its own beam.

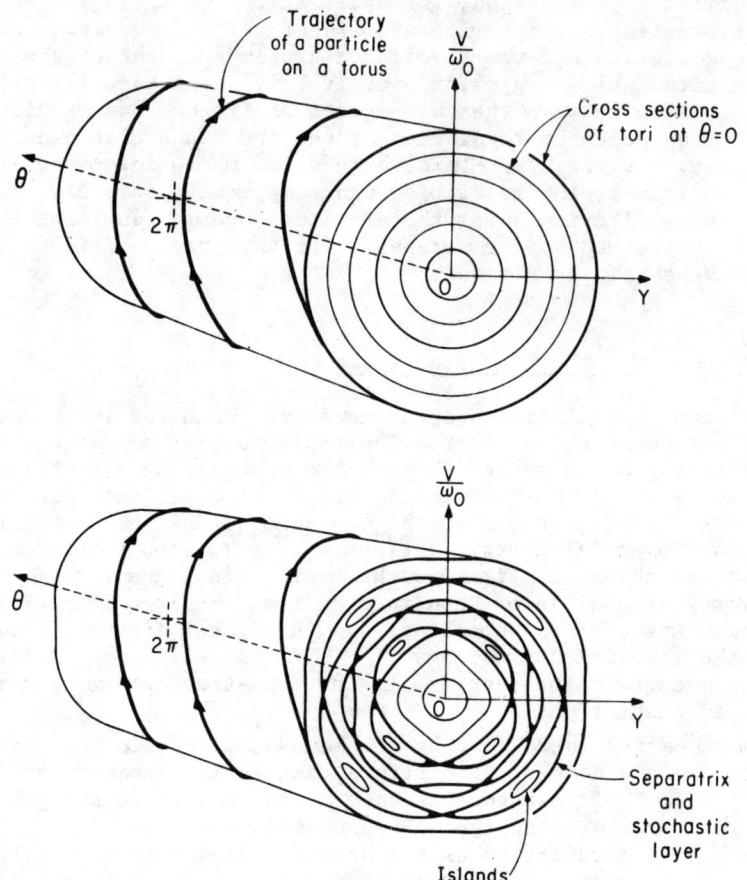

Fig. 4a. Uncoupled linear motion. As the particle rotates in the θ direction, it also rotates in the γ,V plane. The particle is confined to a closed cylinder (or torus) in the energy surface.

Fig. 4b. Coupled non-linear motion. The coupling distorts the tori of Fig. 4a. Where there are resonances between the θ and γ motion, thin stochastic layers form about the separatrices.

Figure 4b shows the same energy surface with a small beam-beam nonlinearity and coupling. Surprisingly, most of the tori are not destroyed, but only somewhat distorted.* Because motion that begins on a certain torus cannot leave the torus, and because no trajectory may cross a torus, the small coupling does not lead to beam expansion, even over very long periods of time. Although most of the tori in Fig. 4b remain intact, a few (those very close to the resonances) are either destroyed or altered. Each resonance is characterized by a separatrix, a surface whose cross section looks something like a chain of footballs. It is the tori in the immediate vicinity of the separatrix that are destroyed and replaced by a three dimensional stochastic layer.

As the beam-beam force increases, both the resonance width and the thickness of the stochastic layer increase. At some critical strength, the stochastic layers of neighboring resonances touch each other and form a stochastic bridge, allowing the particle to diffuse from one resonance to the next (see Fig. 5). Beam expansion is then expected to occur when neighboring resonances overlap in the region of phase space occupied by the beam. Because resonance overlap can create an extremely rapid diffusion (the beam could be lost in 10^4 revolutions), it represents an absolute upper limit on the beam-beam force and hence on the density of the beam. Below this threshold, the beam lifetime might be determined by other resonance effects not considered here, by collisional effects, or possibly by a combination of the two (Arnold diffusion).

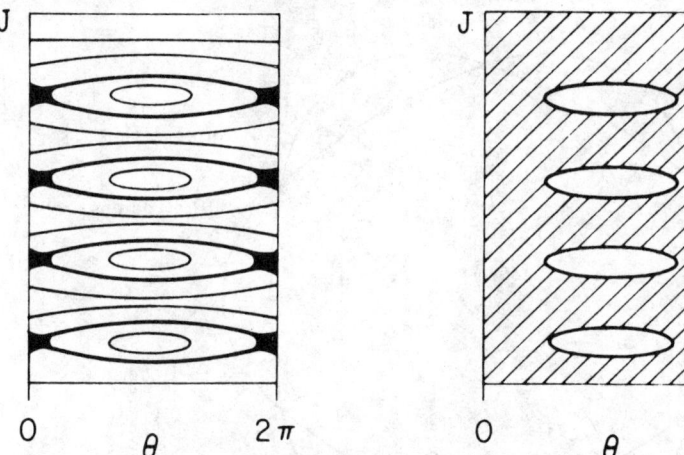

Fig. 5. In Fig. 5a, the stochastic regions surrounding the separatrices are isolated from one another by the tori of regular motion.. In Fig. 5b, the resonances have grown wide enough to overlap. Only some of the island tori remain intact (the hatched region is tochastic). Although these are only representative sketches, they are quite similar to iteration plots of the standard mapping with $K < 1$ and $K > 1$, respectively.

---

*This system satisfies the conditions of the KAM theorem (see for example Ref. 6). The stability of such a system is not destroyed by small perturbations. The invariant tori are only gradually dissolved as the perturbation is increased.

The remainder of this paper examines the phenomenon of resonance overlap for the simple model introduced above, with one additional feature: the beam-beam force is modulated at a very low, fixed frequency by synchrotron oscillations. In a real storage ring, this is an important effect when the beams are bunched by an rf field. Although not necessary in a proton ring, bunching increases the luminosity, and is therefore a desirable option.

III. THE EQUATIONS OF MOTION

The Difference Equations

In the ISABELLE ring, there are six identical sectors with a single intersection in each sector (see Fig. 6). The change in the betatron phase between intersections is $\omega_o = 2\pi\nu_o$ where $\nu_o = 3.770$. Assuming a linear harmonic oscillation in the region between intersections, the integrated change in the vertical position and velocity, Y and V, is

$$Y_{n+1} = Y_n \cos(\omega_o) + (V_n/\omega_o)\sin(\omega_o) \tag{1}$$

$$V_{n+1} = -Y_n \omega_o \sin(\omega_o) + V_n \cos(\omega_o), \tag{2}$$

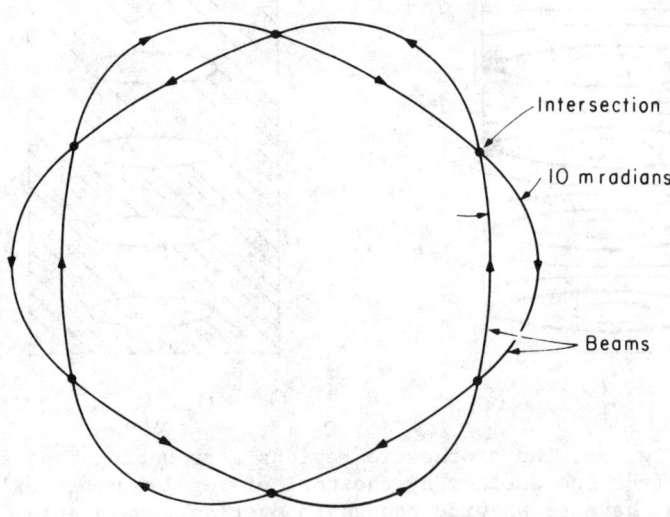

Fig. 6. A schematic view of the ISABELLE rings showing the six identical sectors and their intersection points. The crossing angle is actually very small, $\alpha = 10$ mradians.

where Y_n, V_n are the position and velocity just after the nth intersection, and $\underline{Y}_{n+1}, \underline{V}_{n+1}$ are the position and velocity just before the (n + 1)th intersection. As the particle moves through the intersection, it receives a vertical impulse ΔV (see Fig. 7) due to the collective electromagnetic field of the crossed beam. The impulse depends only upon Y and may be approximated with

$$\Delta V = \xi \, Y \, (r_o^2 + Y^2)^{-1/2} \, , \qquad (3)$$

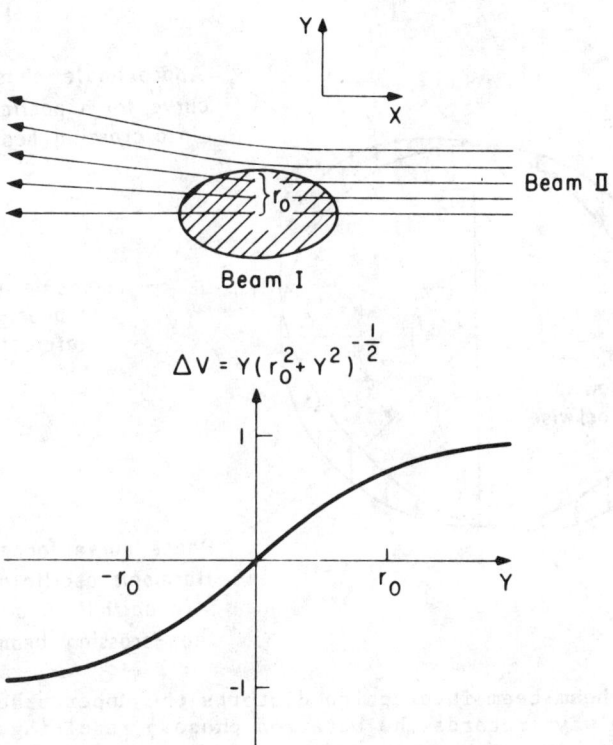

Fig. 7a. The beam-beam interaction. When a particles trajectory crosses the opposing bea, it receives a small kick, changing its vertical velocity V. The trajectories shown here have initial velocities V = 0.

Fig. 7b. The magnitude of the kick ΔV is dependent upon the vertical displacement Y. The actual dependence depends on the density profile of the beam. The hyperbolic potential $\Delta V = Y(r_o^2+Y^2)^{-1/2}$ is chosen here as a qualitatively representative model ($\Delta V(Y=0)=0$, $\Delta V(Y=\pm\infty)=\pm 1$).

where ξ is the beam-beam strength parameter and r_o is the beam radius. The integrated motion through the intersection is

$$Y_{n+1} = Y_{n+1} \tag{4}$$

$$V_{n+1} = V_{n+1} + \xi Y_{n+1} (r_o^2 + Y_{n+1}^2)^{-1/2} \tag{5}$$

where Y_{n+1}, V_{n+1} are the position and velocity just after the $(n+1)$th intersection. The jump in V at the intersection causes the betatron phase to make a small step backwards (see Fig. 8). The average effect is to reduce the effective tune ν of the vertical oscillation.

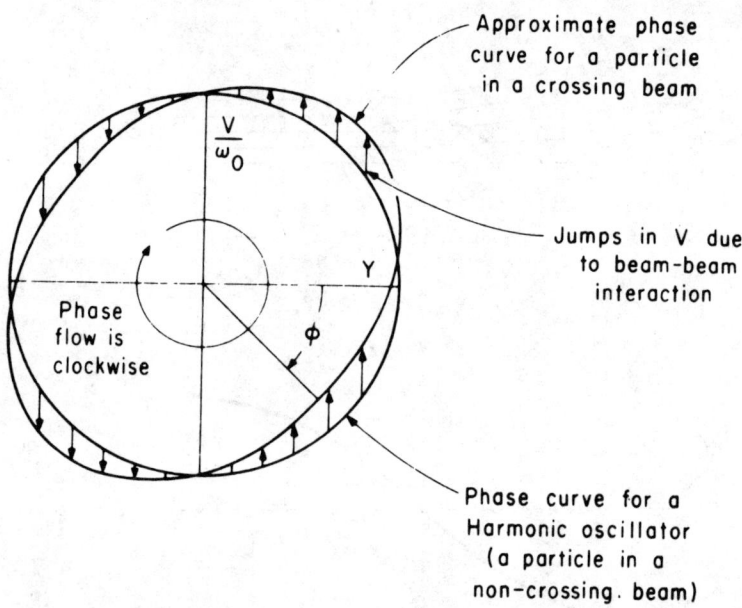

Fig. 8. The beam-beam interaction distorts the unperturbed tori. The jump ΔV always retards the betatron phase ϕ resulting in a negative tune shift $\Delta \nu$. The arrows represent the magnitude and direction of jumps ΔV as a function of the vertical displacement Y at the intersection.

<u>The Hamiltonian</u>

The difference equations (1), (2), (4) and (5) can be derived exactly from a "kicked" Hamiltonian

$$H(Y, V; n) = \frac{V^2}{2} + \omega_o^2 \frac{Y^2}{2} - \xi \sum_{m=-\infty}^{\infty} \delta(n+m) \sqrt{r_o^2 + Y^2} \;. \tag{6}$$

This is a non-autonomous Hamiltonian with one degree of freedom. The discrete index n has become a continuous parameter equivalent to time. The delta-function sum can be put into a Fourier series form

$$\sum_{m=-\infty}^{\infty} \delta(n+m) = \sum_{m=-\infty}^{\infty} \cos(2\pi mn). \tag{7}$$

The m = 0 term is independent of n and can be removed from the sum. Equation (6) then becomes

$$H(Y, V; n) = \frac{V^2}{2} + \omega_o^2 \frac{Y^2}{2} - \xi\sqrt{r_o^2 + Y^2} - \xi 2 \sum_{m=1}^{+\infty} \cos(2\pi mn)\sqrt{r_o^2 + Y^2}. \tag{8}$$

By assigning ξ the role of small parameter, Eq. (8) can be divided into an unperturbed integrable system, plus a small perturbation. The first three terms on the right describe a conservative nonlinear oscillator with one degree of freedom

$$H_o(Y, V) = \frac{V^2}{2} + \omega_o^2 \frac{Y^2}{2} - \xi\sqrt{r_o^2 + Y^2}. \tag{9}$$

The remaining terms describe a time dependent perturbation

$$H_1(Y; n) = \xi 2 \sum_{m=1}^{\infty} \cos(2\pi mn)\sqrt{r_o^2 + Y^2}. \tag{10}$$

Equation (8) can then be written

$$H(Y, V; n) = H_o(Y, V) + H_1(Y; n). \tag{11}$$

Tune Shift

The frequency of the unperturbed oscillator H_o is a function of the amplitude and can be derived using action angle variables (see Fig. 9). The frequency of small oscillation, defined by the condition

$$H_o \ll \omega_o^2 \frac{r_o^2}{2} \tag{12}$$

can be obtained directly from Eq. (9) by expanding the third term on the right hand side in a power series about Y = 0.

$$H_o(Y, V) = \frac{V^2}{2} + \omega_o^2 \frac{Y^2}{2} - \xi\left[r_o + \frac{1}{2r_o}Y^2 + \ldots\right]. \tag{13}$$

The frequency of small oscillation is then

$$\lim_{A \to 0} \omega^2(A) = \left(\omega_o^2 + \frac{\xi}{r_o}\right). \tag{14}$$

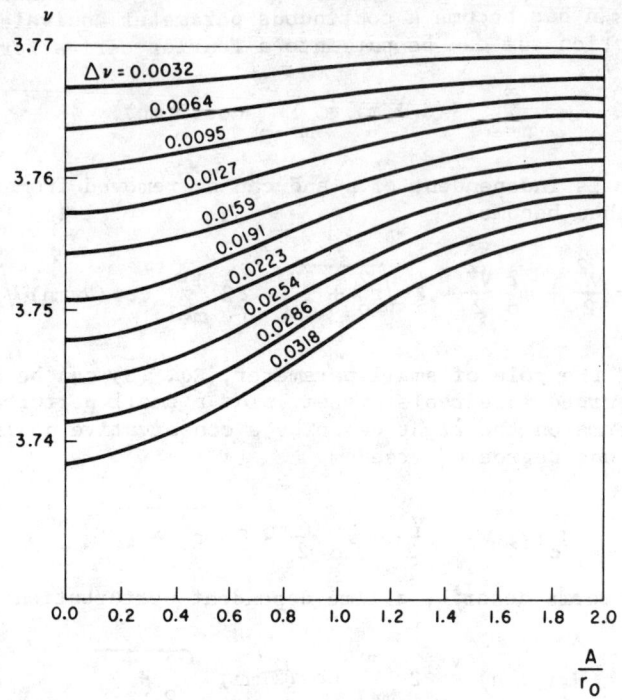

Fig. 9. Nonlinearity in ISABELLE. The betatron frequency varies with amplitude due to the effects of the beam-beam interaction. The beam-beam strength for each contour is proportional to the zero amplitude tune shift $\Delta\nu = \nu(A=0)-3.770$. The graphs are experimentally derived from computer mappings.

For a small beam-beam force $\xi \ll \omega_o^2 r_o$, the tune shift is given by

$$\Delta\nu(0) = \frac{1}{2\pi}[\omega(0) - \omega_o] = -\frac{1}{2\pi}\frac{\xi}{2r_o\omega_o} . \quad (15)$$

Coupling and Resonance

The perturbation terms Eq. (10) provide the coupling between the vertical oscillation and the revolution. The square root can be expressed in terms of the action-angle variables J, θ of the unperturbed system H_o.

$$H(J, \theta; n) = H_o(J) + \xi \sum_{m=1}^{\infty} 2\cos(2\pi mn) \sqrt{r_o^2 + Y^2(J, \theta)} . \quad (16)$$

If it is then expanded in a Fourier series, the result is

$$\sqrt{r_o^2 + Y^2}(J, \theta) = \sum_{k=0}^{\infty} A_k(J)\cos k\theta . \tag{17}$$

The perturbation Eq. (10) becomes

$$H_1(J, \theta; n) = -\xi \sum_{m=1}^{\infty} \sum_{k=-\infty}^{\infty} A_k(J)\cos(2\pi mn + k\theta) . \tag{18}$$

The change in the action of the vertical oscillation is given by

$$\dot{J} = -\xi \sum_{m=1}^{\infty} \sum_{k=-\infty}^{\infty} A_k(J)k\sin(2\pi mn + k\theta) . \tag{19}$$

The action is roughly a constant unless initial conditions put the system in a resonance, i.e. for some integers m, k

$$2\pi mn + k\theta = \text{constant} . \tag{20}$$

Using the first order approximation $\dot{\theta} \approx \omega n$, the resonance condition Eq. (20) is equivalent to

$$2\pi m + k\omega = 0 . \tag{21}$$

As mentioned in section I, resonances do not individually threaten the stability of the motion. It is only when resonances interact with one another (or overlap) that they present a problem. To evaluate the stability of a theoretical system, it is therefore necessary to determine the width of, and the spacing between, the strongest resonances in the operational region of the phase space. These may be calculated using only the resonance strength A_k and the nonlinearity $\Lambda = \partial\omega/\partial J$. If Λ is approximately a constant near some resonance m_1, k_1, then

$$\omega(J) \sim \omega(J_r) + \Lambda(J - J_r) , \tag{22}$$

where $\omega(J_r)$ satisfies

$$2\pi m_1 + k_1 \omega(J_r) = 0 . \tag{23}$$

For $J \approx J_r$, Eq. (19) becomes

$$\dot{J} \sim -\xi A_{k_1} k_1 \sin(2\pi m_1 n + k_1 \omega(J)n) \tag{24}$$

where θ has been replaced with $\omega(J)n$. Using Eq. (22) and (23), this becomes

$$\dot{J} \sim -\xi A_{k_1} k_1 \sin\varphi \tag{25}$$

where

$$\varphi = k_1 \Lambda (J - J_r) n \qquad (26)$$

$$\dot{\varphi} = k_1 \Lambda (J - J_r) . \qquad (27)$$

Defining $I = (J - J_r)k_1\Lambda$, Eq. (25) and (27) became

$$\dot{I} \sim - k_1^2 \Lambda \xi A_{k1} \sin\varphi \qquad (28)$$

$$\dot{\varphi} \sim I . \qquad (29)$$

These are the equations of motion for a pendulum. The width of the m, k resonance is given by the pendulum separatrix width

$$\Delta_\omega J = 4\sqrt{\frac{\xi A_{k1}}{\Lambda}} . \qquad (30)$$

The spacing between resonances m_1, k_1 and m_2, k_2 is from Eq. (22) and (24)

$$\Delta_s J = \frac{2\pi}{\Lambda k_1} . \qquad (30)$$

If the two resonances have the same width, they will overlap when

$$\frac{\Delta_\omega J}{\Delta_s J} = \frac{4}{2\pi} \sqrt{\xi A_{k1} k_1^2 \Lambda} > 1 . \qquad (32)$$

Equations (28) and (29) are the continuous counterparts to the so-called standard mapping[5]

$$I_{n+1} = I_n + K(J)\sin\varphi_n \qquad (33)$$

$$\varphi_{n+1} = \varphi_n + I_{n+1} . \qquad (34)$$

The function $K(J)$ is related to Eq. (32) by

$$K(J) = \left[\frac{2\pi}{4} \frac{\Delta_\omega J}{\Delta_s J}\right]^2 = \xi A_{k1}(J) k_1^2 \Lambda . \qquad (35)$$

More rigorous derivations[5,9,10] and experimentation have shown that the resonances of the standard mapping overlap not at $K = \Lambda^2/4$, but at $K \approx 1$. This may be attributed to the finite width of the stochastic layers and the presence of intervening higher order resonances. The overlap condition Eq. (32) then becomes

$$\xi A_{k1}(J) k_1^2 \Lambda > 1 \ . \tag{36}$$

The Resonance Spectrum

In order to explicitly calculate the stability limits from an overlap condition such as Eq. (36), it is necessary to know the Fourier coefficients $A_k(J)$. Using the approximation

$$Y(J, \theta) \sim \sqrt{\frac{2J_o}{\omega_o}} \cos \theta \ , \tag{37}$$

The coefficients of Eq. (17) are given by the integral

$$A_k(J) = \frac{1}{\pi} \int_0^{2\pi} d\theta \cos(k\theta) \sqrt{r_o^2 + \frac{2J}{\omega_o} \cos^2 \theta} \ . \tag{38}$$

Only the k = even terms are non-zero. Numerical integration yields the first six coefficients

$$\begin{aligned}
A_0 &= 1.22 & A_6 &= 7.55 \times 10^{-4} \\
A_2 &= 2.06 \times 10^{-1} & A_8 &= -8.09 \times 10^{-5} \\
A_4 &= -8.82 \times 10^{-3} & A_{10} &= 9.71 \times 10^{-6} \ .
\end{aligned} \tag{39}$$

where $2J/\omega_o = r_o^2 = 1$. These fall off very quickly: each A_k is an order of magnitude smaller than the preceding one.

Even with a knowledge of the resonance spectrum, the overlap condition Eq. (36) cannot be applied to the storage ring model described above for two reasons. The first is that the range of frequencies spanned by the J space is very small ($\nu(J = 0) = 3.74$ and $\nu(J = \infty) = 3.77$ for a tune shift of $\Delta\nu = 0.03$). The two strongest resonances in this range are m, k = 15,4 and m, k = 98,26. The second is negligible, so the first is essentially alone. Since it has no neighbors, and in particular no m = 4 neighbors, there is no porsiblity of overlap. The second problem is that even if another resonance were available for overlap, Eq. (36) would still not be applicable unless the nonlinearity Λ was roughly a constant over the region between the two resonances.

But the derivation of Eq. (36) will not be wasted. With some modifications, it can be applied to the problem of bunched beams, where the synchrotron oscillation produces a dense set of potentially overlapping sideband resonances.

Bunched Beams: Resonance Splitting

The model described above represents storage rings with unbunched, crossing beams. It is extremely stable, even at unrealistically high tune shifts ($\Delta \nu > 0.03$). The situation is somewhat different if the beams are bunched by a locally imposed radio frequency field. Opposing bunches are synchronized to reach the intersection point at exactly the same time. Since the particle density is greatest at the center of a bunch and tapers off near the ends, particles close to the center receive a stronger kick than those near the ends (see Fig. 10). The particles oscillate very slowly back and forth within each bunch, and thus experience a modulated beam-beam force.

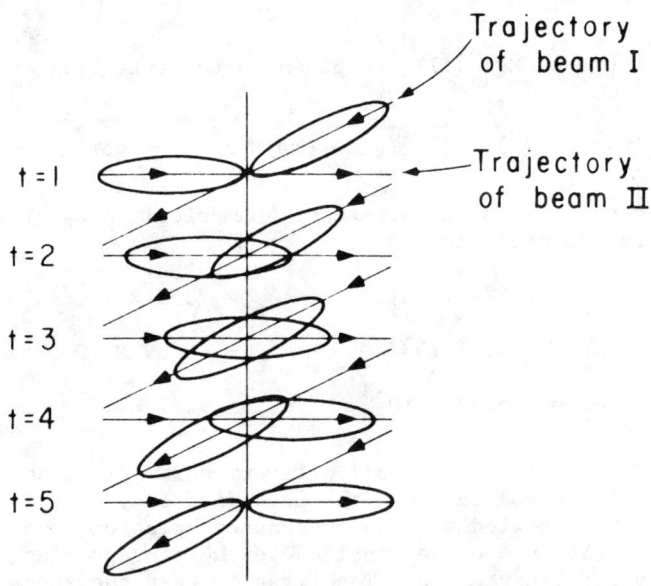

Fig. 10. The crossing of bunched beams. The two bunches reach the intersection point at exactly the same time. Particles near the ends of each bunch feel a weaker deflection than those in the center.

Because the synchrotron oscillation has an effect on, but is not affected by the betatron oscillation, it does not constitute an additional degree of freedom for the problem. It is introduced into the difference equations as a simple n-dependent modulation of the beam-beam force parameter ξ. For the bunched beam model, the velocity difference equation (5) is rewritten

$$V_{n+1} = \underline{V}_{n+1} + \xi_o (1 + \cos 2\omega_s (n+1)) \underline{Y}_{n+1} (r_o^2 + \underline{Y}_{n+1}^2)^{-\frac{1}{2}}, \tag{40}$$

where ξ_0 is the average beam-beam force parameter, and ω_s is the synchrotron frequency (a typical ω_s for ISABELLE is 2.5×10^{-6}). The Hamiltonian for the bunched beam system is obtained from Eq. (6) by replacing ξ with $\xi_0[1 + \sin(2\omega_0 n)]$:

$$H^B(Y,V;n) = \frac{V^2}{2} + \omega_0^2 \frac{Y^2}{2} - \xi_0(1 + \cos 2\omega_s n)\sqrt{r_0^2 + Y^2}$$

$$- \xi_0(1 + \cos 2\omega_s)2 \sum_{m=1}^{\infty} \cos(2\pi mn)\sqrt{r_0^2 + Y^2} \,. \quad (41)$$

The unperturbed Hamiltonian Eq. (9) becomes

$$H_0^B(Y,V;n) = \frac{V^2}{2} + \omega_0^2 \frac{Y^2}{2} - \xi_0(1 + \cos 2\omega_s n)\sqrt{r_0^2 + Y^2} \,. \quad (42)$$

This is a nonlinear oscillator in one dimension with a slowly oscillating parameter ξ. Since $\omega_s \ll \omega_0$, there is an adiabatic invariant of the motion described by the instantaneous action angle variables

$$\begin{aligned} J &= J(Y,V;\xi) \\ \theta &= \theta(Y,V;\xi) \end{aligned} \quad (43)$$

When Eq. (42) is rewritten in terms of J and θ, the θ dependence may be relegated to terms of order ω_s or smaller[11]

$$H_0^B = H_{00}^B(J;\xi) + O(\omega_s) \,. \quad (44)$$

The equations of motion are

$$\dot{J} = O(\omega_s) \quad (45)$$

$$\dot{\theta} = \omega(J;\xi) + O(\omega_s) \,. \quad (46)$$

In what follows, the instantaneous frequency $\omega(J,\lambda)$ is approximated by the average frequency plus the fundamental harmonic

$$\omega(J;\xi) = \overline{\omega}(J) + \overline{\Delta\omega}(J)\cos 2\omega_s n \quad (47)$$

where $\overline{\omega}(J)$ is the betatron frequency at J averaged over one synchrotron period, and $\overline{\Delta\omega}(J) = \omega_0 - \overline{\omega}(J)$. This is integrated over n for fixed J to obtain the angle

$$\theta(n) = \overline{\omega}(J)n - \frac{\overline{\Delta\omega}(J)}{2\omega_s}\sin 2\omega_s n \,. \quad (48)$$

The full Hamiltonian Eq. (41) can now be written in terms of J and θ

$$H^B(J,\theta;\xi) = H^B_{oo}(J;\xi) - \xi_o(1 + \cos 2\omega_s n)$$

$$\times 2 \sum_{m=1}^{\infty} \cos(2\pi mn) \sum_{k=0}^{\infty} A_k(J)\cos(k\theta) + O(\omega_s). \quad (49)$$

The rate of change of J is

$$\dot{J} = \xi_o(1 + \cos 2\omega_s n) 2 \sum_{m=1}^{\infty} \cos(2\pi mn) \times \sum_{k=0}^{\infty} A_k(J) k \sin(k\theta), \quad (50)$$

Using Eq. (48), the $\sin(k\theta)$ term becomes

$$\sin(k\theta) = \sin\left[k\bar{\omega}n + k\frac{\overline{\Delta\omega}}{2\omega_s}\sin(2\omega_s n)\right] = \sum_{\ell=0}^{\infty} J_\ell\left(\frac{k\overline{\Delta\omega}}{2\omega_s}\right)\sin(k\bar{\omega}n + 2\omega_s \ell n), \quad (51)$$

where J_ℓ are the Bessel functions. Using Eq. (51) and some trigonometric manipulation, Eq. (50) becomes

$$\dot{J} = \xi_o k \sum_{m=-\infty}^{\infty} \sum_{k=0}^{\infty} \sum_{\ell=0}^{\infty} A_k(J) B_\ell\left(\frac{k\overline{\Delta\omega}}{2\omega_s}\right) \times \sin(k\bar{\omega}n + \ell 2\omega_s n + m 2\pi n), \quad (52)$$

where $B_\ell = J_\ell + 1/2(J_{\ell-1} + J_{\ell+1})$. The new resonance conditions are

$$(k\bar{\nu} + \ell 2\nu_s m) = 0. \quad (53)$$

Each of the k,m resonances has been split into a multiplet of k,l,m sideband resonances.[6,12] The multiplet width is determined by the Bessel functions which fall off quickly when $\ell > (k\overline{\Delta\nu}/2\nu_s)$ (see Fig. 11). Physically, this width corresponds to the depth of the tune shift modulation. The m,k resonance moves back and forth through the phase space like a mixing spoon*, with frequency ω_s (see Fig. 12). As the mixing spoon passes the system's phase point, the system receives a small nudge. It either gains or loses energy, depending on the betatron phase when the spoon goes by. If the phase is approximately the same each time the spoon passes, the system is in a sideband resonance.

The sidebands, because they are more closely spaced, tend to overlap long before the k,m resonances overlap. Consequently, synchrotron modulation is potentially a major threat to the dynamical stability of the beam.

---

*This is a subtle version of the bucket analogy in which the particles are actually trapped in the resonance as it goes by (see the section on trapping).

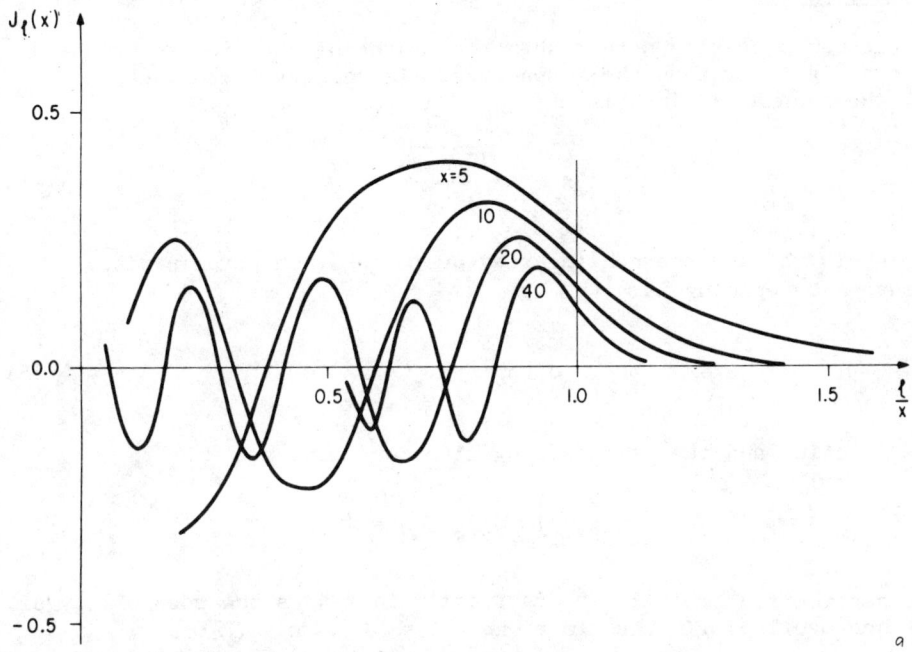

Fig. 11. Bessel functions. The Bessel functions $J_\ell(x)$ ($x = k\overline{\Delta\nu}/2\nu_s$) are plotted as a function of ℓ for fixed x. The edge of the sideband multiplet is approximately defined by $\ell/x=1$. When $\ell/x>1$, the $J_\ell(x)$ fall off rapidly to zero.

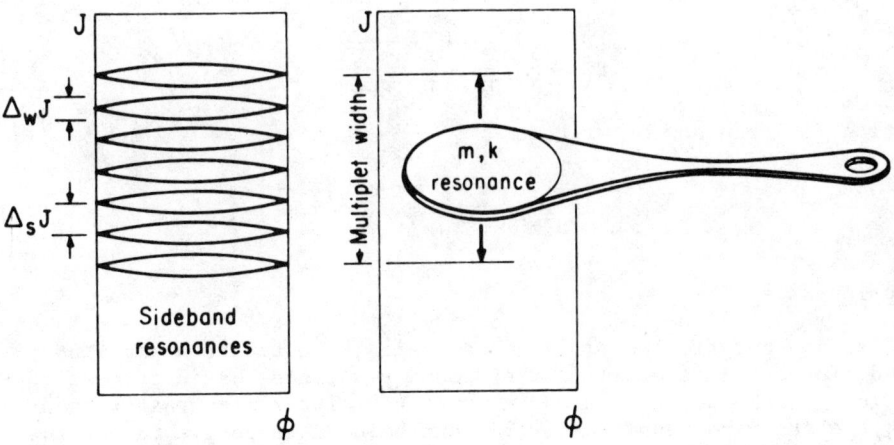

Fig. 12. Mixing spoon analogy. Slow oscillations of the parameter ξ cause the m,k resonances to move back and forth in the J space. This action results in the formation of a multiplet of sideband resonances.

Sideband Overlap

Resonance overlap occurs when the widths of the sideband resonances are greater than their spacings. In analogy to the unbunched case, the resonance width is

$$\Delta_\omega J = 4\sqrt{\frac{\xi_o A_k B_\ell}{\Lambda_2}} \,, \tag{54}$$

where Λ_2 is the average nonlinearity over one synchrotron period. The resonance spacing is

$$\Delta_s J = \frac{2\pi}{k\Lambda_2} 2\nu_s \,. \tag{55}$$

The K function and the resonance overlap condition are

$$K^B = \left(\frac{k}{2\nu_s}\right)^2 \xi_o A_k B_\ell \Lambda_2 > 1 \,. \tag{56}$$

Note that the factor $1/(2\nu_s)^2$ drastically increases the possibility of resonance overlap. In the limit where $1 \ll \ell \ll x = (k\overline{\Delta\nu}/2\nu_s)$

$$B_\ell \sim 2\, J_\ell(x) \sim 2\sqrt{\frac{2}{\pi x}} \cos\left(x - \frac{\ell\pi}{2} - \frac{\pi}{4}\right), \tag{57}$$

or even more approximately (see Fig. 11)

$$B_\ell \sim \frac{4}{\pi}\sqrt{\frac{2}{\pi x}} \sim \frac{4}{\pi}\sqrt{\frac{4\nu_s}{\pi k \Delta\nu}} \,. \tag{58}$$

Equation (56) can now be written

$$K^B = \left[k^2 \xi_o A_k \Lambda_2 \frac{2}{\pi}\sqrt{\frac{1}{\pi k \Delta\nu}}\right] \nu_s^{-3/2} > 1 \,. \tag{59}$$

Diffusion Rates

If the condition Eq. (59) is satisfied, the betatron amplitude will diffuse. Since overlap of sideband resonances has destroyed correlations in the betatron phase φ_n over time intervals greater than $P_s/2$, the standard mapping Eq. (33) can be used to derive the diffusion coefficient. The relation between J and I is now

$$I = (J - J_r) P_s k \Lambda_2 / 2 \,. \tag{60}$$

The diffusion coefficient from Eq. (35) for uncorrelated φ_n is[5]

$$D(I) = \frac{\langle \Delta I^2 \rangle}{2T} = \frac{(K^B)^2}{2P_s} , \qquad (61)$$

or

$$D(J) = \frac{2(K^B)^2}{k^2 P_s^3 \Lambda_2^2} . \qquad (62)$$

Substitution for K^B [Eq. (59)] gives

$$D(J) = \left(\frac{2}{\pi}\right)^3 \frac{k \xi_o^2 A_k}{\Delta \nu} . \qquad (63)$$

The P_s has cancelled out. By changing the variable again, Eq. (63) can be rewritten to describe the diffusion in amplitude directly. Using

$$\Delta J = \Delta A (2\omega_o A) , \qquad (64)$$

equation (63) becomes

$$D(A) = \frac{\langle \Delta A^2 \rangle}{2T} = \left(\frac{2}{\pi}\right)^3 \frac{k \xi_o^2 A_k^2}{\Delta \nu 4 \omega_o^2 A^2} . \qquad (65)$$

The Trapping Limit

The lower limit on the validity of Eq. (65) is approximately the stability border, at $P_s = P_b$. There is also an upper limit, called the trapping limit $P_s = P_t$, beyond which the diffusion is approximately proportional to $\nu_s = 1/P_s$. Trapping occurs when P_s is so large that some particles actually become trapped inside the separatrix of the mixing spoon resonance[13] (see Fig. 13). The trapped particles are forced to follow the resonance on its excusions through the action space. Under these conditions, the standard mapping and the linear integration of the equations of motion from which it was derived, are invalid. Since the particles cannot be transported faster than the speed at which the resonance itself moves, the diffusion rate is forced to drop off when $P_s > P_t$.

The stability border P_s and the trapping limit P_t are evaluated in the next section for some typical parameter values. The trapping limit can be roughly calculated by setting the maximum possible jump in action ΔJ^P due to a single kick from the beam-beam interaction equal to the maximum distance ΔJ^R traveled by the mixing spoon resonance in a time $n = 1$. From Eq. (25)

$$\Delta J^P = k \xi A_k . \qquad (66)$$

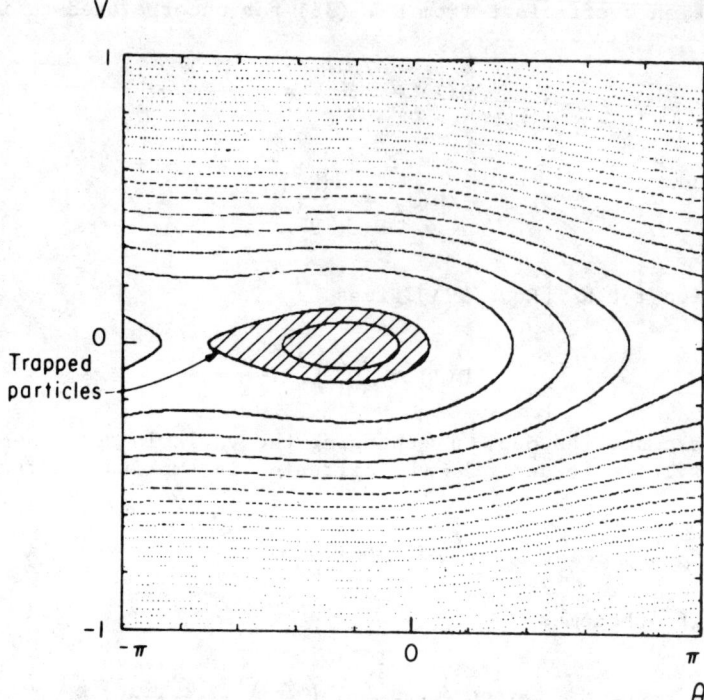

Fig. 13. Trapping. In a constantly accelerating wave (fixed k, $\dot{\omega}$=constant) particles can be trapped and accelerated with the wave, providing $\dot{\omega}$ is small enough. The contours shown here represents the phase flow in the phase space V, θ with respect to the accelerating frame of reference.

Assuming harmonic oscillation of the resonance in a J space with uniform nonlinearity

$$\Delta J^R = 8\pi^2 \overline{\Delta \nu} \, \nu_s / \Lambda_2 \, , \tag{67}$$

the trapping limit is then

$$P_t = \frac{8\pi^2 \overline{\Delta \nu}}{\Lambda_2 k \xi A} \, . \tag{68}$$

<u>Evaluation of P_b, $D(A)$, and P_t</u>

The stability border for the model described in section III is calculated here from Eq. (59), and later compared to more accurate values derived from computer generated surface of section plots. The values of m and k are 15 and 4 since this is the only significant resonance in the vicinity of the operating tune $\nu = 3.770$. To make a plot of the stability border in $\overline{\Delta\nu}$, $P_s = 1/\nu_s$ space it is necessary to express Eq. (59) as an explicit function of these variables. Using Fig. 9, it is possible to approximate the maximum nonlinearity Λ_o as a linear function of $\overline{\Delta\nu}$.

$$\Lambda_2^{max} \sim \frac{1}{\nu_o A} \frac{d\nu}{dA} \sim (0.1) \overline{\Delta\nu} , \qquad (69)$$

where $J = \omega_o A^2$, $r_o = 1$, $\nu_o = 3.770$ and $A = 1$. The beam-beam strength parameter is given by Eq. (15)

$$\xi_o = 2(2\pi)^2 r_o \nu_o \overline{\Delta\nu} = 298 \overline{\Delta\nu} . \qquad (70)$$

The Fourier coefficient A_4 was calculated previously

$$A_4(A = 1) = -8.82 \times 10^{-3} . \qquad (71)$$

Equation (59) can now be written

$$K^B = 0.755(\overline{\Delta\nu}_s P_s)^{3/2} > 1 \qquad (72)$$

The theoretical stability border is plotted in Fig. 14 along with the experimentally determined border. The horizontal line defines the multiplet edge. Below this line, the approximation Eq. (57) no longer holds, and the resonance widths fall off much faster than $\nu_s^{1/4}$ (see Fig. 11). Considering the large number of approximations made in this calculation, the agreement between experiment and theory is quite good.

Equations (70) and (71) can also be used to evaluate the diffusion coefficient Eq. (65) for $A = r_o = 1$,

$$D = 0.00318 \overline{\Delta\nu} . \qquad (73)$$

For an average tune shift of $\overline{\Delta\nu} = 0.015$ (see Fig. 15),

$$D = 4.8 \times 10^{-5} , \qquad (74)$$

where the units are r_o^2 per intersection period. Apparently resonance overlap can destroy the weaker of the two beams in about 10^4 intersections (or 20 msec). Although this result is derived for an imaginary ISABELLE with an unrealistically large tune shift, such an effect would not be unlikely in an $e^+ - e^-$ ring where the tune shifts are an order of magnitude larger. Furthermore, the diffusion rate given by Eq. (73) would compete quite successfully with the radiation damping in these machines (the damping time at SPEAR is about 4 msec).

The trapping limit Eq. (69) can be similarly evaluated. When we use Eq. (70) and (71), the limit is given by

$$0.013(P_t \overline{\Delta\nu}) = 1 . \qquad (75)$$

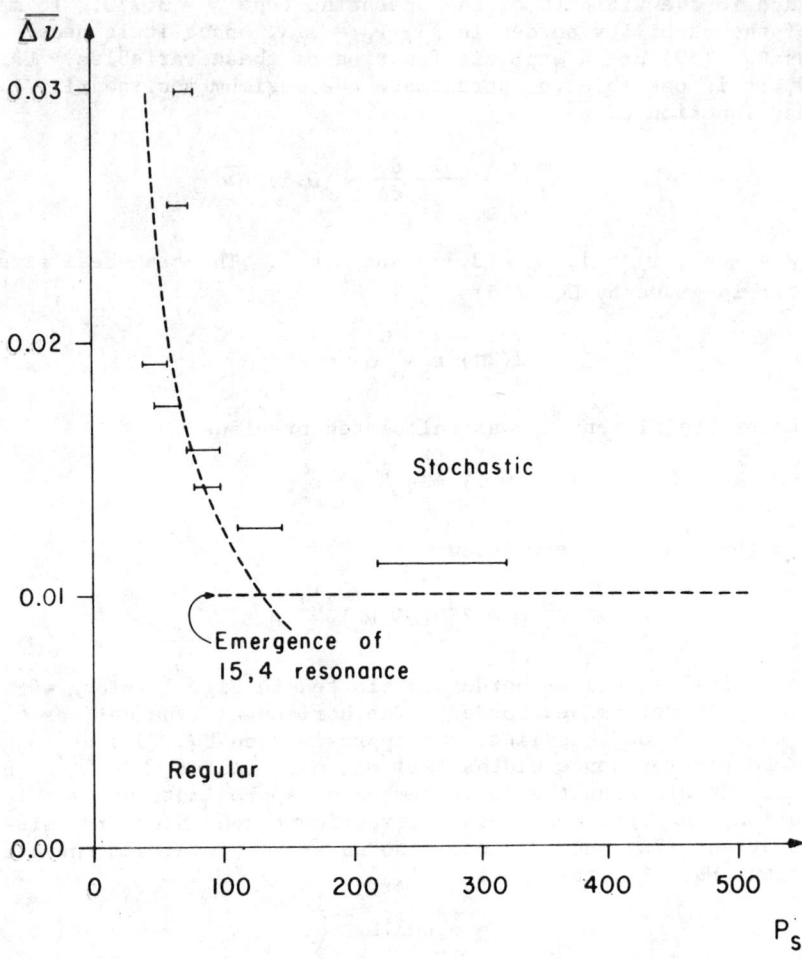

Fig. 14. The stability border. The dotted lines represent the theoretical border: the vertical line corresponds to Eq. (63), and the horizontal line to the multiplet edge. The error bars represent the experimentally determined stability border. This was obtained from a number of computer generated surface-of-section plots. For each measurement, $\overline{\Delta\nu}$ was held fixed and P_S was varied until the stochastic regions became large enough to visually resolve. A number of runs were done at very long modulation periods (P_S = 3000), but stochasticity did not appear for any case in which $\overline{\Delta\nu} < 0.01$.

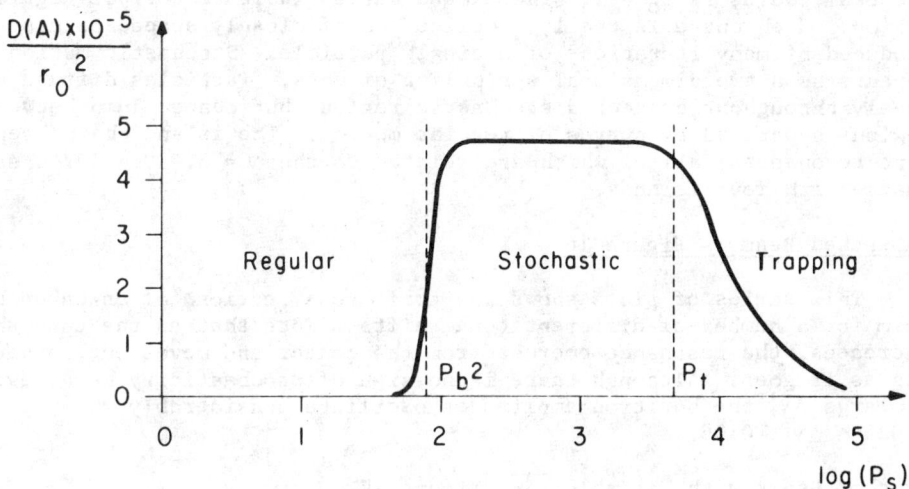

Fig. 15. The diffusion coefficient as a function of the synchrotron period. There are three diffusion regimes. When $P_s < P_b$ the motion is regular and there is no diffusion. When $P_b < P_s < P_t$, nonlinear resonances overlap to create a strong diffusion. When $P_s > P_t$, resonance trapping may occur. This is a rough theoretical curve derived for ISABELLE with a tune shift $\overline{\Delta \nu} = 0.015$.

When $\overline{\Delta \nu} = 0.015$ (Fig. 15), then

$$P_t = 5128 \ . \tag{76}$$

IV. COMPUTER SIMULATIONS

The following plots are surface of section mappings generated by the difference equations Eq. (1), (2), (4), and (40). For the unbunched beam, the system returns to the surface of section every iteration, i.e. the Hamiltonian is periodic in n with period $\Delta n = 1$. For the bunched beam, the period is $\Delta n = P_s/2$ and the phase position (Y,V) is plotted once every $P_s/2$ interations*.

---

*For this reason, P_s is always taken to be an even integer. There is no reason to believe that the motion would be qualitatively different if P_s were not an even integer.

The problem was run on a CRAY-1 computer using 13 digit precision. Each particle was plotted between 1000 and 5000 times. In every case, the beam radius is $r_o = 1$. The closed curves indicate tori of regular motion. Each curve is really a collection of closely spaced points produced by many iterations of a single particle. Stochastic motion appears as a two dimensional sprinkling of dots. Particles diffuse freely throughout connected stochastic regions but cannot jump between regions separated by curves of regular motion. The island chains represent resonances, all of which are related to the $\nu = 3.750 = 15/4$ resonance with four islands.

Unbunched Beams: Figure 16

This series of plots shows the tori cross sections of an unbunched beam for a number of different tune shifts. Note that as the tune shift increases, the resonance emerges from the center and moves out, expanding as it goes. Although there is no sign of stochasticity here, even at large $\Delta\nu$, the betatron amplitudes oscillate considerably when $0.025 < \Delta\nu < 0.06$.

Bunched Beams with Variable $\overline{\Delta\nu}$: Figure 17

These plots show tori cross sections of the bunched beam system at $P_s = 400$ and a variety of tune shifts. All of the islands observed in this series are sidebands of the 15,4 resonance. The locations of the sidebands are given by

$$\overline{\nu} = \frac{15 - 2\nu_s \ell}{4} . \qquad (77)$$

If ℓ is divisible by four, only one island appears in the plot. If it is divisible by two but not four, two islands appear. If ℓ is odd, four islands appear. Using this formula*, Eq. (64) and Fig. (9), the sidebands can be labeled with considerable confidence (see Fig. 18). Note that stochasticity appears first in the region of strongest nonlinearity and that its appearance at $\overline{\Delta\nu} = 0.01$ is quite sudden.

Bunched Beams with Varying P_s: Figure 19

If the average tune shift $\overline{\Delta\nu}$ is held constant and only the synchrotron period is varied, the resonance overlap when P_s exceeds a critical value dependent on $\overline{\Delta\nu}$. The plots shown here are for $\overline{\Delta\nu} = 0.0115$. P_s varies from 50 to 600 and overlap occurs between $P_s = 200$ and $P_s = 300$. Plots d) and e) show the multiplet edge clearly

---

*The formla only works for P_s divisible by eight.

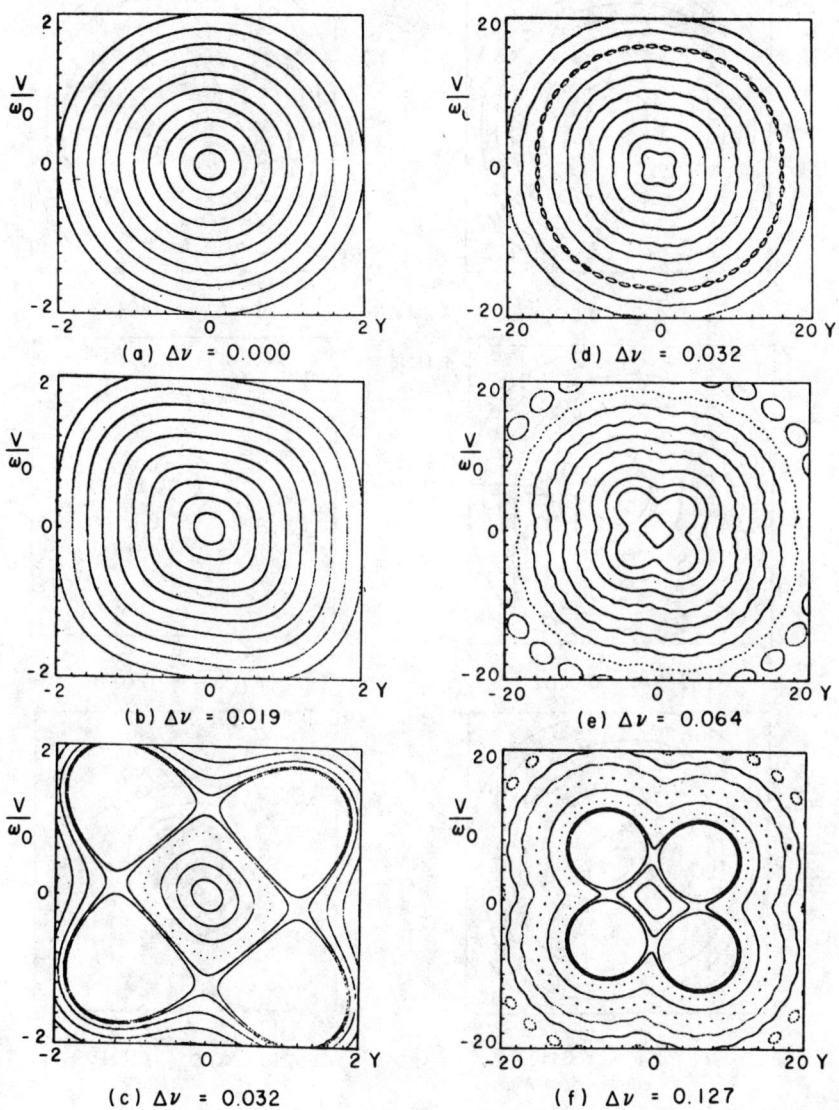

Fig. 16. Section plots for unbunched beams.

186

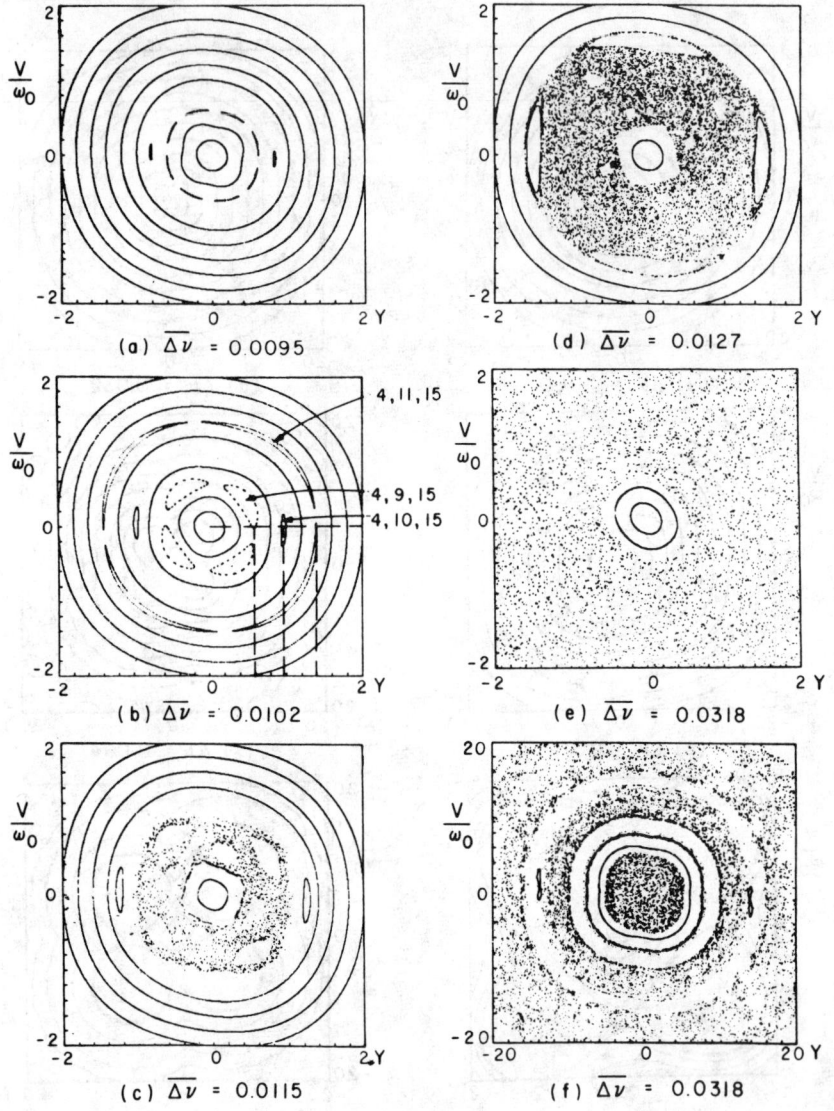

Fig. 17. Section plots for bunched beams.

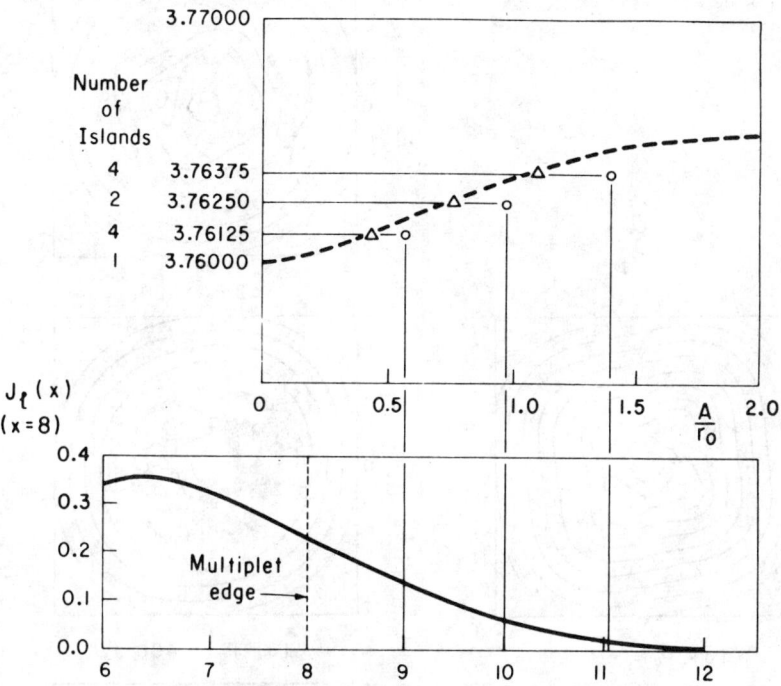

Fig. 18. Labeling the resonances in Fig. 17b. The three resonances are labeled as follows: the amplitude-dependent tune $\nu(A)$ is graphed using Fig. 9 for $\Delta\nu = 0.0102$ (dotted line). The resonances calculated from Eq. (77) are plotted on the ν axis using $\nu_s = 1/400$. The predicted locations of the three resonances (triangles) are then compared with the actual positions (circles) as measured from 15b. Ambiguities are resolved using the number-of-islands formula described in the text. These ambiguities are probably due to slight differences between the functions $\nu(A)$ and $\bar{\nu}(A)$. A plot of the $x = 8$ Bessel functions shows that these are "tail" resonances, beyond the multiplet edge.

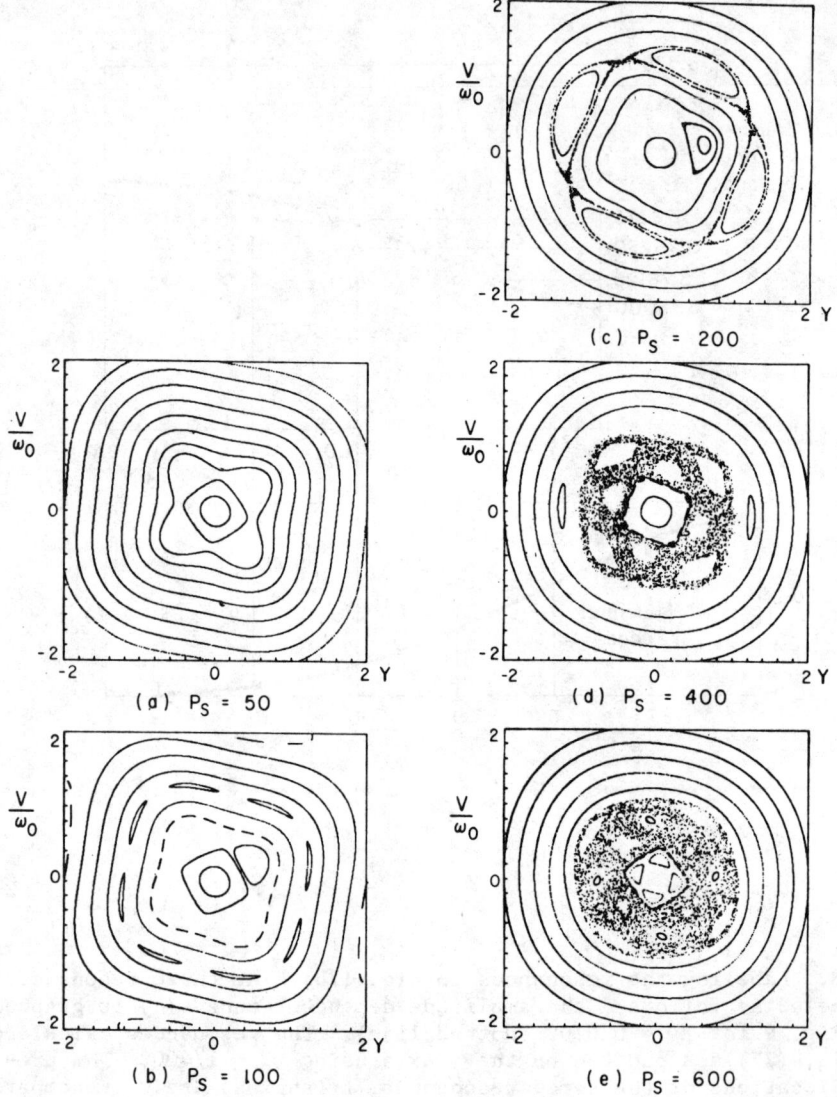

Fig. 19. Variations in synchrotron period for bunched beams.

(from Fig. 9, it should appear at about A = 0.45 r_o). As P_s increases, the tails of the Bessel functions fall off rapidly (see Fig. 11) and the multiplet edge becomes more distinct. In plot (b), the single island resonance near the center corresponds to the k, ℓ, m = 4, 2, 15 and the four island resonance on the edge to the k, ℓ, m = 4, 3, 15*. The eight island string between them is a "half integer" resonance 4, 2.5, 15 (or 8, 5, 30).

Tori Sections at Varying Modulation Phase: Figure 20

A single set of nested tori are shown here at four different cross sections. The points in each plot are still plotted once every P_s = 400 iterations but each plot shows the cross section at a different phase of the modulation oscillation (one fourth of a wavelength apart). The sequence shows that the actions, the cross sectional areas of the tori, are at least approximately conserved throughout the modulation cycle. Particles are not transported radially by the modulation. This sequence also illustrates the twisting of the island tori.

The Disintegration of a Torus: Figure 21

This is a series of four single particle plots showing the breakdown of radial confinement for a single particle. The tune shift and synchrotron period are both held constant here. The variation is in the modulation depth. Previous plots have shown a full modulation of the tune shift, i.e. $\Delta\nu$ varies from zero to $2\overline{\Delta\nu}$. Here the beam strength parameter for the unbunched beam is replaced by

$$\xi = \xi_o(1 + M\cos 2\omega_s n) , \qquad (78)$$

where D < M < 1 is the modulation depth. The multiplet depth is proportional to M. When M > 0.65, the multiplet is wide enough to reach and destroy the torus shown in this plot.

SUMMARY AND COMMENTS

Non-collisional diffusion of the betatron amplitude in an ideal storage ring beam is due to the overlap of nonlinear resonance and the accompanying disintegration of the tori of regular motion. By calculating the resonance widths and spacings, it is possible to make crude estimates of the stability border in the amplitude and parameter spaces. For proton rings with crossing beams, the principal source of nonlinear resonance is the beam-beam interaction. At ISABELLE, the beam-beam interaction has a negligible effect on the dynamical stability of unbunched beams, although for large tune shifts $0.02 < \Delta\nu < 0.06$, the 15, 4 resonance creates a significant oscillation of the betatron amplitude. When the ISABELLE beams are bunched, the beam-beam force

---
*This is a "tail" sideband because it exists outside the formal multiplet width. When ($k\overline{\Delta\nu}/2\nu_s$) approaches one, these tail resonances can be substantial.

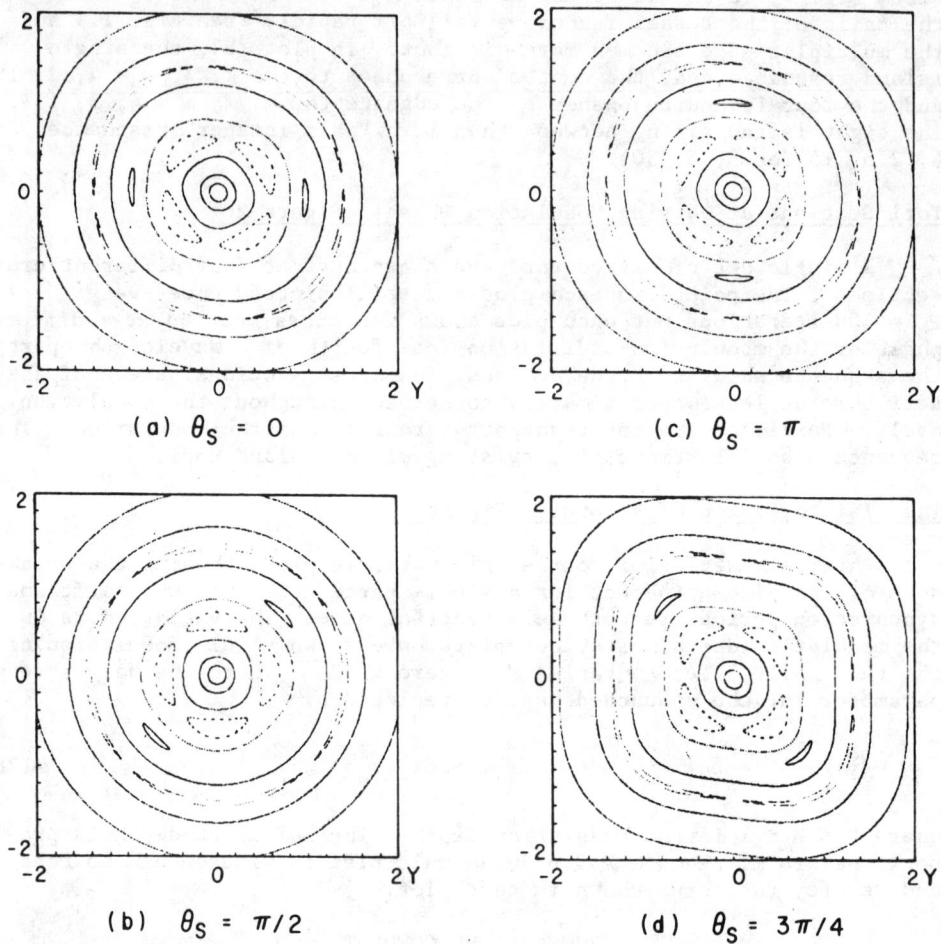

Fig. 20. Section plots at different synchrotron phase.

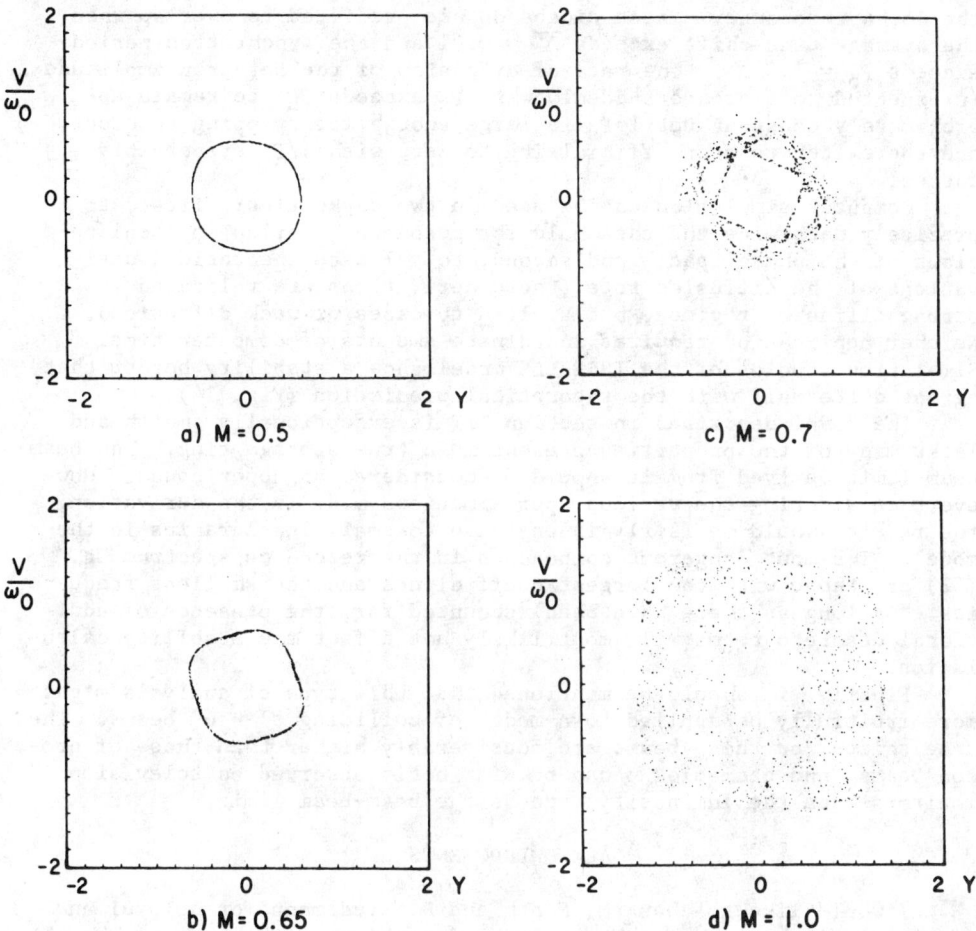

Fig. 21. The breakup of regular motion.

is modulated by the synchrotron oscillation, producing sidebands of the 15, 4 resonance. These sidebands are predicted to overlap when the average tune shift exceeds $\overline{\Delta \nu} = 0.01$ and the synchrotron period exceeds $P_b = 1.2/\overline{\Delta \nu}$. The rate of diffusion of the betatron amplitude is expected to increase suddenly when P_s exceeds P_b, to remain approximately constant until P_s is large enough for trapping to occur, and thereafter to drop off, falling to zero with $1/P_s$ or possibly faster.

Computer simulation can be used in two capacities: first, to precisely determine the threshold for resonance overlap in local regions of the phase space, and second, to validate theoretical derivations of the diffusion rate (these derivations are validated in strong diffusion regimes, but applied to cases of weak diffusion). Neither application requires inordinate amounts of computer time. Simulation studies of the ISABELLE model show a stability border that agrees quite well with the theoretical prediction (Fig. 14).

The model described in section III is exceptionally smooth and lacks many of the properties present in a true storage ring. The beam-beam limit derived from it should be considered an upper bound. However, considering the various approximations made in the derivation, the result should be fairly insensitive to small inaccuracies in the model. The most dangerous components in the resonance spectrum Eq. (52) are those with the largest coefficients and the smallest frequencies. As long as these have been accounted for, the presence of additional descrete terms will most likely not effect the stability calculation.

Finally, it should be mentioned that this type of analysis might more fruitfully be applied to a model of colliding $e^+ - e^-$ beams. The tune shifts for these beams are considerably higher than those of proton beams, and beam blowup can be distinctly observed on television monitors when the luminosity exceeds the beam-beam limit.

ACKNOWLEDGMENTS

I would like to thank M. Month and H. Wiedemann for helpful and inspiring discussions on the nature and problems of storage rings. I am especially in debt to B. V. Chirikov, from whose work most of the techniques in this paper have been derived. This work was supported under National Science Foundation Grant ENG 78-09424 and Department of Energy Contract EY-76-s-03-0034-PA215.

REFERENCES

1. M. Month, "Nature of the Beam-Beam Limit in Storage Rings", IEEE Trans. on Nucl. Scien., Vol. NS-22, 1376 (1976).
2. G. M. Zaslavskii and B. V. Chirikov, "Stochastic Instability of Non-linear Oscillations," Sov. Phys. Uspekhi 14, 549 (1972).
3. M. Berry, in Topics in Nonlinear Dynamics, edited by S. Jorno, A.I.P. Conf. Proc. (A.I.P., New York, 1978), vol. 46.
4. V. I. Arnold, "Instability of Dynamical Systems with Several Degrees of Freedom," Dokl. Akad. Nauk. SSSR 156, 9 (1964).
5. B. V. Chirikov, "Universal Instability of Many Dimensional Oscillator Systems," Physics Reports 52, No. 5, 263-379 (1979).
6. B. V. Chirikov, "Stability of the Motion of a Charged Partical in a Magnetic Confinement System," Sov. J. Plasma Phys. 4(3), May-June 1978.
7. J. Tennyson, M. A. Lieberman, and A. J. Lichtenberg, "Diffusion in Near-Integrable Hamiltonian Systems with Three Degrees of Freedom," (This Publication).
8. E. D. Courant and H. S. Snyder, "Theory of Alternating Gradient Synchrotrons," Ann. of Phys. 3, 1 (1958).
9. J. Green, "A Method for Determining a Stochastic Transition," J. Math. Phys., 20 June (1979).
10. I. Percival, "Variational Principles for Invariant Tori and Cantori (This Publication).
11. V. I. Arnold, Mathematical Methods of Classical Mechanics, Springer-Verlag, New York 1978.
12. F. M. Izrailev, S. I. Misnev, G. M. Tumaikin, "Numerical Studies of Stochasticity Limit in Colliding Beams (One Dimensional Model)," preprint no. 43, Institute of Nuclear Physics, Siberian Branch, Academy of Sciences of the USSR, Novosibirsk (1977).
13. A. W. Chao, M. Month, "Particle Trapping During Passage Through a High-Order Nonlinear Resonance," Nuc. Instr. and Meth. 121 129-138 (1974).

THE BEAM-BEAM INTERACTION AS A NON-LINEAR PROBLEM: A NUMERICAL STUDY

E.D. Courant
Brookhaven National Laboratory, Upton, New York 11973

ABSTRACT

The radial diffusion of single particle trajectories in intersecting storage rings is simulated on a computer. The motion is assumed to be linear except for a small nonlinear component resulting from the intersection of the particle trajectory with the opposing beam. The diffusion is examined for both bunched and unbunched opposing beams. The bunched model shows a substantial diffusion in contrast to the unbunched model which shows little or none (insignificant on the time scale considered). The difference is thought to result from an additional degree of freedom in the bunched system, which results in an increase in the rate of the Arnold diffusion.

INTRODUCTION

The motion of a particle in a storage ring may be approximately reduced to two components. The first is the betatron oscillation. It is the dominant component and is considered to be linear. The second is the so-called "beam-beam interaction", a small periodic kick that results when the particle crosses the beam of the neighboring ring. The beam-beam interaction is non-linear and is treated as a small perturbation to the linear betatron oscillation. The properties of linear motion are reviewed below, followed by a discussion of the effects of a small nonlinear perturbation.

Consider a system of 2N dynamical variables (coordinates and momenta) in the vicinity of an equilibrium point. If the equations of motion are linear, with coefficients that may be periodic in the time variable θ, then by Floquet's theorem there exist N normal modes where the motion goes with θ as

$$e^{\pm i\omega_r \theta} \qquad r = 1,\ldots N \quad . \tag{1}$$

If the N eigenfrequencies ω_r are all real and distinct, the motion is stable for all time. Every actual motion is a linear combination of these normal modes and is characterized by N independent bi-linear invariants (or integrals)

$$W_i = \sum_{rs} \omega_{rs}^{(i)} x_r x_s \qquad i = 1,\ldots N \quad . \tag{2}$$

Each of these is a positive definite quadratic form in the dynamical variables $x_r = (q_r, p_r)$. A given motion must then take place on the intersections of the N ellipsoids defined by the invariants W_i. This intersection defines an N-dimensional "ellipse" in the 2-N dimensional phase space. Close to a stable equilibrium, the ellipse is closed and bounded; a particle stays on this closed trajectory manifold forever.

Now suppose the motion becomes non-linear. The linear problem was characterized by a Hamiltonian

$$H_o = \sum_{rs} H_{rs} x_r x_s \quad . \tag{3}$$

A small non-linear part is now added:

$$H = H_o + G(x,\theta) \quad . \tag{4}$$

The perturbation $G(x,\theta)$ has terms of third and higher degrees in the x's. (The origin remains an equilibrium point.) How does this non-linear term effect the stability?

This problem has been extensively studied; there is no simple answer to the question. A naive examination would suggest three possibilities:
  a) The closed trajectory manifolds are simply distorted a bit. They remain closed and smooth.
  b) Stability limits are created. For sufficiently small amplitudes, a) applies. For larger amplitudes (above some "threshold"), the trajectories change their character and go out to infinity.
  c) Some or all of the closed manifolds break up completely. A given trajectory may no longer be confined to an invariant manifold.

The actual situation is most often described by possibility (c). However, in the region of phase space close to equilibrium, many of the invariant manifolds may remain intact. The results of Kolmogorov, Arnold and Moser can (very crudely) be described as follows.

If the frequencies of the linear problem are mutually incommensurable to a "sufficient degree", and the amplitude is small enough, then, in the vicinity of an invariant manifold of the linear problem there exists an invariant manifold of the nonlinear problem. These "KAM" trajectories are divided by thin regions of stochasticity. For small enough amplitudes, most trajectories are of the KAM type. A few, however, are not, and the behavior of these stochastic trajectories is dependent upon the dimensionality of the phase space:

1) If n = 1, the trajectories are one dimensional curves in the two dimensional phase space. Therefore, if there is just <u>one</u> closed KAM curve, everything inside it must stay inside for all time. The existence of KAM curves guarantees eternal stability for motion starting within them. Therefore there exists a stable region. Many numerical studies have shown that this region is often quite large and has a definite boundary called the "stochasticity limit". Far inside this boundary, the KAM curves seem so close that a computer (with finite resolution) cannot find the stochastic regions between them. The motion then proceeds practically as if case (a) applied.

2) For n ≥ 2, the situation is different. For n = 2, KAM trajectories lie on two dimensional manifolds in a four dimensional phase space. They cannot, therefore, enclose finite volume of the

phase space (just as a one dimensional string cannot enclose a volume of three-dimensional space). A system which is not on a KAM manifold is consequently not bound to a particular region of the phase space. It is free to travel arbitrarily far from its initial conditions. This process is called <u>Arnold diffusion</u> because V.I. Arnold explicitly exhibited at least one system in which it took place.

There are at least two problems in which Arnold diffusion may play a vital role:
   a) The dynamics of the solar system. The earth's orbit is significantly perturbed by Jupiter and other planets. Yet, there have been no major changes in this orbit over approximatly 10^{10} periods of revolution (the age of the solar system).
   b) Beam stability in storage rings. The lifetime of the beam must be of the order of 10^{10} periods of revolution, despite small nonlinearities in the motion due to magnet imperfections and the beam-beam interaction.

THE ISABELLE MODEL

For ISABELLE, let us consider the following model. The particles in a beam oscillate horizontally and vertically with similar (but not identical) frequencies. These betatron oscillations are assumed to be strictly linear except at the point where the particles cross the second beam. At these points, the particles experience an impulse force

$$\Delta \vec{p} = \vec{F}(x,y) \quad ,$$

where $\vec{F}(x,y)$ is nonlinear vector function.

The linear part of $\vec{F}(x,y)$ results in the so-called "tune shift", a change in the normal mode frequencies of the betatron oscillation. The nonlinear part creates a coupling between the x and y motion and a dependence of the betatron frequencies on amplitude. Since the linear part of $\vec{F}(x,y)$ is proportional to the nonlinear part, the tune shift may be used as a convenient measure of the coupling-nonlinearity strength. With the coupling, the system has two degrees of freedom with a periodic dependence on time. Therefore the existence of KAM curves does not ensure stability and Arnold diffusion is an a priori possibility.

The motion of the model system may be simulated on a computer using a pair of difference equations. The linear motion, expressed in terms of the normal coordinates, is just the motion of two uncoupled harmonic oscillators. The change in the normal coordinates and momenta over one revolution is a simple rotation.

$$\begin{pmatrix} x \\ \bar{p}_x \end{pmatrix}_{n+1} = \begin{pmatrix} \cos 2\pi\nu_x & \sin 2\pi\nu_x \\ -\sin 2\pi\nu_x & \cos 2\pi\nu_x \end{pmatrix} \begin{pmatrix} x \\ p_x \end{pmatrix}_n \quad (5)$$

$$\begin{pmatrix} y \\ p_y \end{pmatrix}_{n+1} = \begin{pmatrix} \cos 2\pi\nu_y & \sin 2\pi\nu_y \\ -\sin 2\pi\nu_y & \cos 2\pi\nu_y \end{pmatrix} \begin{pmatrix} y \\ p_y \end{pmatrix}_n \tag{6}$$

The beam-beam interaction results in an instantaneous change in the momenta at the crossing point

$$p_x(n+1) = \bar{p}_x(n+1) + \frac{\xi x}{4\pi} f(r^2) \tag{7}$$

$$p_y(n+1) = \bar{p}_y(n+1) + \frac{\xi y}{4\pi} f(r^2) , \tag{8}$$

where $r^2 = x^2 + y^2$. The functions $xf(r^2)$ and $yf(r^2)$ are proportional to the x and y components of the electric field at r. The r^2 dependence implies that the beams are round and that they collide head-on. These assumptions are not valid for either ISABELLE or the electron-positron rings, but it is hoped that the qualitative nature of the motion is fairly independent of these particulars.

The parameter ξ is the tune shift at small amplitudes. More generally it reflects the strength (density) of the "crossed" beam. If the density of the beam is not uniform (i.e. if it is bunched), ξ may vary from one crossing to the next. If this is the case, Eqs. (5)-(8) need to be supplemented by an additional difference equation for ξ_n. Allowing ξ to vary increases the effective dimensionality of the problem and is consequently expected to reduce the stability.

The full motion is described by the product of the two pairs of mappings, Eqs. (5)-(8), plus a possible implicit variation in ξ. For the computations that follow, two different beam profiles $f(r^2)$ are studied. The motion is computed for both constant values of ξ and periodically varying values of ξ. The x and y tunes are chosen to be $\nu_x = 0.641$ and $\nu_y = 0.611$, avoiding the major lower order resonances. Lastly, it should be noted that in the actual computations, the linear tune shift is subtracted from the perturbation terms on the right hand side of Eqs. (7)-(8). This is done by substituting a new function, $f(r^2) \equiv f(r^2) - 1$, in place of $f(r^2)$. This substitution should not effect the qualitative features of the motion.

The beam profiles that are considered are:
a) A uniform beam with a sharp edge at $r = 1$,

$$f(r^2) = \begin{cases} 1, & r < 1 \\ \frac{1}{r^2}, & r > 1 \end{cases} \tag{9}$$

b) A Gaussian shaped beam with a soft edge at $r \sim 1$,

$$f(r^2) = \frac{1-e^{-r^2}}{r^2} . \tag{10}$$

The major distinction between these two profiles is that case (b) is analytic while case (a) is not (the first derivative is discontinuous at r=1). The KAM theory is not necessarily applicable to problems with fewer than five continuous derivatives. Thus, it is quite possible that no KAM invariant surfaces exist at all for case (a). The motion would then be expected to exhibit radial diffusion regardless of the initial conditions.

It is generally believed that the beam-beam limit is defined approximately by $\xi = 5 \times 10^{-3}$. This value marks the beam density at which the diffusion rate becomes unacceptable for storage rings. For those runs in which ξ varies periodically, it increases linearly from zero to ξ_{max} in 200 turns, remains constant for 600 turns, and then ramps back to zero in another 200 turns. The total period is 1000 revolutions. The discontinuities in the time derivations at the tops and bottoms of the ramps do not invalidate the KAM theorem (since they occur in θ, which is normal to the mapping section, $\theta = 0$).

SIMULATION RESULTS

Three experiments are discussed:
1) Case (a) (uniform beam) with fixed ξ,
2) Case (b) (analytic beam) with fixed ξ, and
3) Case (c) (analytic beam) with period variations in ξ.

The initial conditions are chosen close to $x = y = 1$ and $p_x = p_y = 0$. For each experiment, the results are found to be qualitatively independent of small changes in the inital conditions.

Experiment 1:

The beam profile is given by Eq. (9) and the beam density by $\xi = -0.0095$ (about twice the conventional beam-beam threshold). The amplitudes of the x and y oscillations are plotted against the number of revolutions in Fig. (1). Both amplitudes exhibit diffusive type behavior. This may be due to Arnold diffusion, but it might also result from a global stochasticity created by the discontinuity in the first derivative of the beam profile. A third possibility is that machine roundoff causes the diffusion. The diffusion rate is approximately 1-2% per million revolutions.

Experiment 2:

The beam profile is given by Eq.(10), and the beam density by $\xi = -0.0095$ (as in experiment 1). The amplitude diffusion is shown in Fig. (2). For this case, the amplitudes do not appear to diffuse on the time scale considered. It is possible, therefore, that the trajectory is confined to an invariant manifold in the phase space. This result supports the conjecture that the diffusion in experiment 1 was due to the discontinuity in the first derivative of the beam profile.

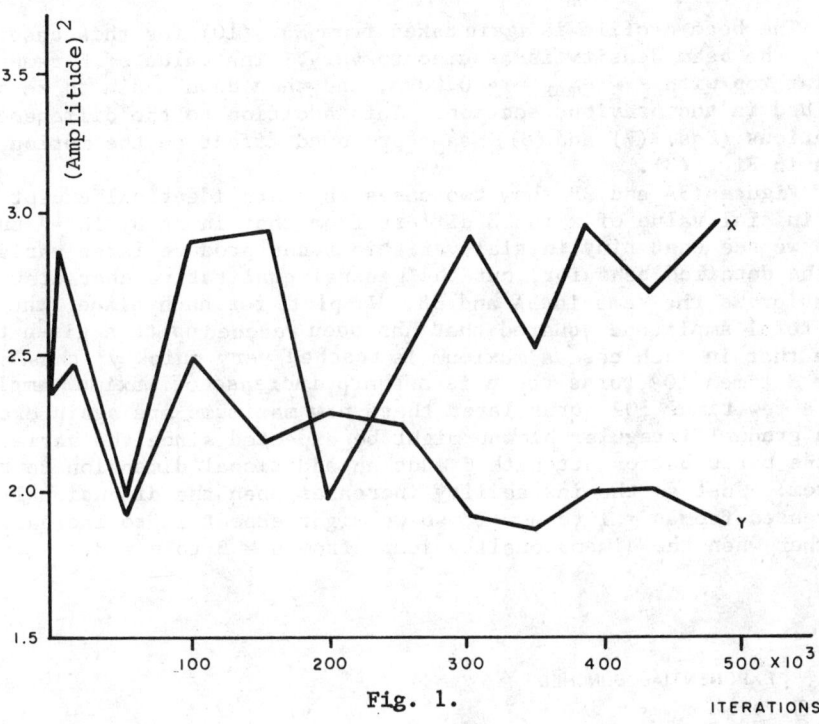

Fig. 1.

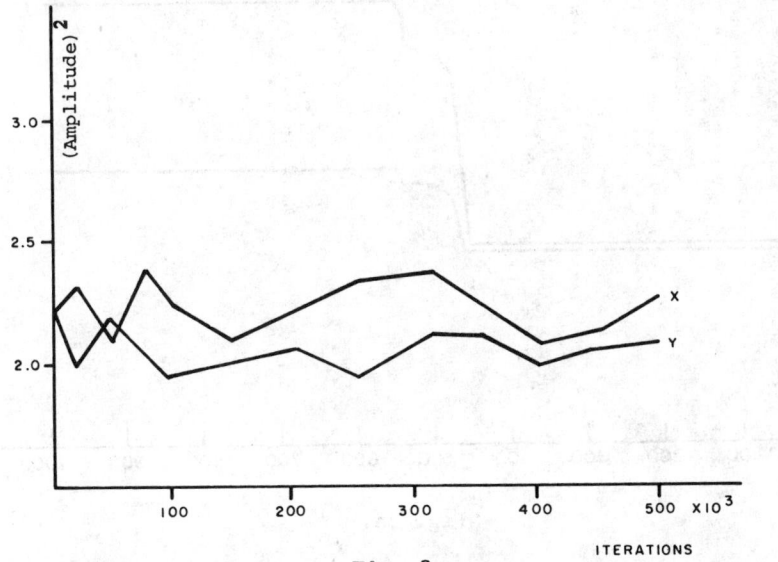

Fig. 2.

Experiment 3:

The beam profile is again taken from Eq. (10) for this case: however, the beam density is assumed to vary. The value of ξ ramps up to a flat top with $\xi = \xi_{max} = -0.0095$, and then down again to zero as described in the previous section. This addition to the difference equations (Eqs. (7) and (8)) has a profound effect on the motion, as seen in Fig. (3).

Figures 3A and 3B show two cases that are identical except that the initial value of x in 3B differs from that in 3A by 10^{-13} units. Thus we see that tiny initial variations can produce large variations in the detailed behavior, but the general qualitative character of the behavior is the same in 3A and 3B. We plot, for each plane, the maximum total amplitude squared that has been reached up to a given time. Note that in each case a maximum is reached very quickly; thus after a 2 or 3 times 10^5 turns there is a sharp increase of maximum amplitudes, and a few times 10^5 turns later these new maximums are again exceeded. This gradual irregular blowup might be expected since the variation in the perturbation strength ξ adds an additional dimension to the system. Just as the instability increases when the dimensionality is increased from n = 1 to n = 2, so we might expect it to increase still further when the dimensionality jumps from n = 2 to n = 3.

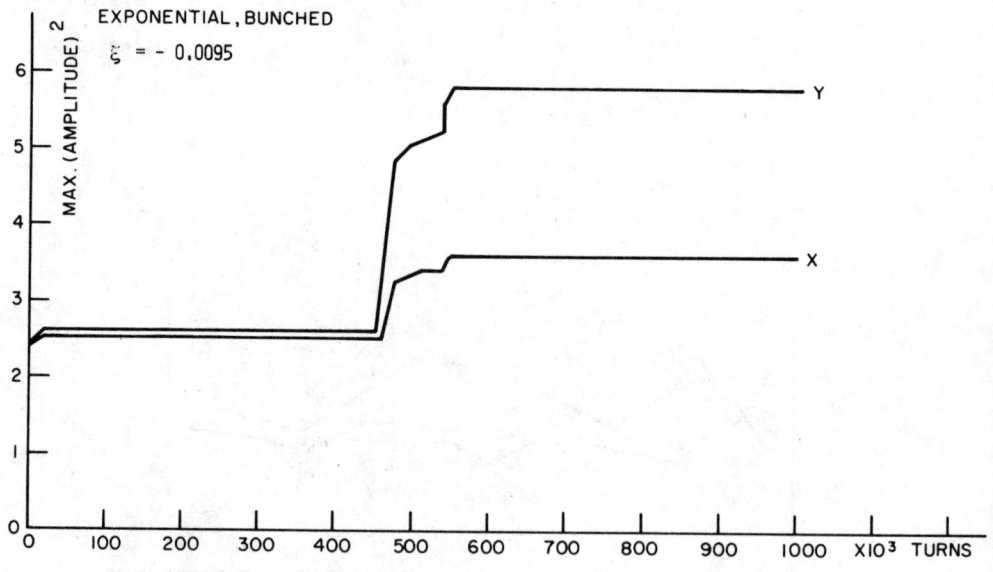

Fig. 3a.

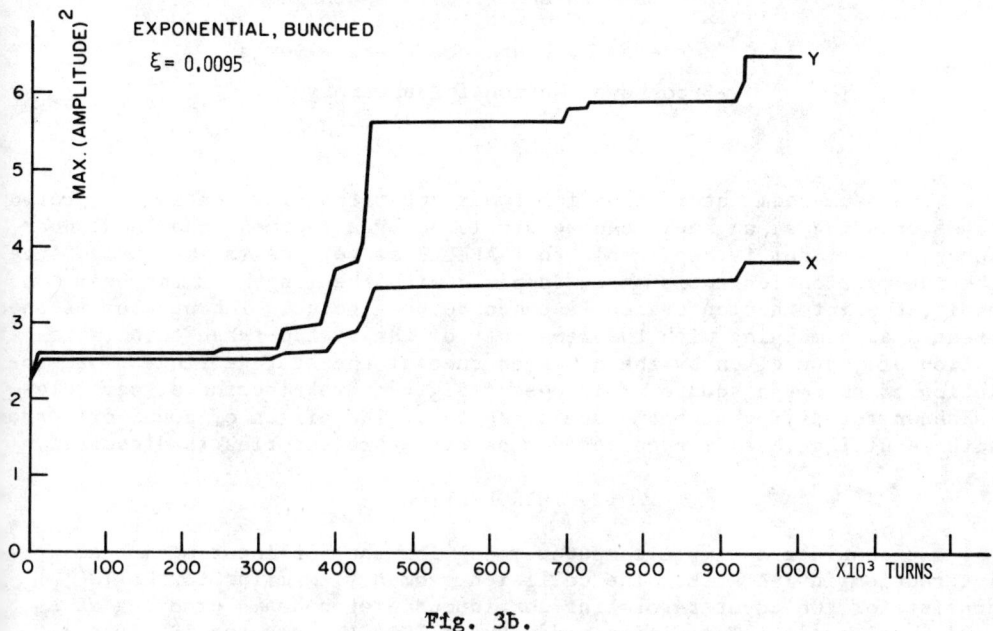

Fig. 3b.

CONCLUSION

The results of experiments 2 and 3 above, for the unbunched and bunched beam cases respectively, suggest that bunching can seriously reduce the life-time of a beam. The precise cause of the diffusion is not understood. It is thought, however, that beam non-uniformity by increasing the dimensionality of the problem to n = 3, either destablizes the motion entirely or adds substantially to the rate of Arnold Diffusion.

SIMPLE COMPUTER MODEL FOR THE NONLINEAR BEAM-BEAM INTERACTION IN ISABELLE

J. C. Herrera, M. Month and R. F. Peierls
Brookhaven National Laboratory

SUMMARY

The beam-beam interaction for two counter-rotating continuous proton beams crossing at an angle can be simulated by a 1-dimensional nonlinear force. The model is applicable to ISABELLE as well as to the ISR. Since the interaction length is short compared with the length of the beam orbit, the interaction region is taken to be a point. The problem is then treated as a mapping with the remainder of the system taken to be a rotation of phase given by the betatron tune of the storage ring. The evolution of the mean square amplitude of a given distribution of particles is shown for different beam-beam strengths. The effect of round-off error with resulting loss of accuracy for particle trajectories is discussed.

1. INTRODUCTION

We consider a simple computer model for the nonlinear beam-beam interaction in ISABELLE. The collision geometry, similar to the ISR, consists of two counter-rotating continuous proton beams crossing at a horizontal angle. The difference between ISABELLE and the ISR is essentially the magnitude of the crossing angle, the former being about 11 mrad, with the ISR having a crossing angle of about 25 times this value.

A simple discussion of the beam-beam force and the equation for the motion of a particle about its central orbit around the storage ring is presented in Section 2. Our model is a weak-strong one, meaning that one beam, the "strong" beam, is taken to be fixed while we study the motion of the particles in the weak beam passing through the strong one. The proper self consistent problem is not treated. Because a given particle passes through the strong beam from one side to the other, the net force experienced by a "weak" particle crossing the "strong" beam is only in the vertical direction. The horizontal force averages to zero. In other words, the beam-beam interaction for beams crossing at a non-zero angle is 1-dimensional.

Since the interaction length is short compared with the length of the beam orbit, the interaction region is taken to be a point. The problem can then be treated as a mapping, with the remainder of the system taken to be a rotation of phase given by the betatron tune of the storage ring.

Some simulation results are given in Section 3. We show in particular the evolution of the mean square amplitude of a distribution of particles for different beam-beam strengths. We observe that the mean square amplitude tends to grow if the beam-beam strength is sufficiently large. This growth is associated with the presence of round-off error, with resulting loss of accuracy for particle trajectories. We give a brief discussion of this loss of accuracy problem and we consider whether by taking averages over a sufficiently large particle distribution in the phase space we can obtain meaningful results in spite of the errors in the individual particle trajectories.

II. MOTION OF A PARTICLE CROSSING AN UNBUNCHED BEAM

Throughout the following discussion we shall consider the collisions in the weak-strong case, that is, we shall formulate the problem of a single proton crossing a continuous proton beam in the horizontal plane at an angle (α_0) and then making one pass through a perfectly linear machine before colliding again with the continuous beam. Under these circumstances the particle trajectory will suffer a net angular deflection only in the vertical plane and this change in slope per intersection is given by

$$\Delta \left(\frac{dy}{ds}\right) = \frac{2\pi r_p I}{ec\gamma\beta^2 \tan\frac{\alpha_0}{2}} \, \Phi\left(\frac{y}{\sqrt{2}\sigma_V}\right) , \qquad (1)$$

where Φ is the error function defined by

$$\Phi(x) = \frac{2}{\sqrt{\pi}} \int_0^x dt\, e^{-t^2} . \qquad (2)$$

[The significance of each of the symbols in these equations is presented in Ref. 1].

In discussing the change in particle coordinates for one revolution through such a system, it is convenient to change variables: from the transverse vertical coordinates $y(s)$ and the slope $y'(s)[\equiv dy/ds]$, dependent on the azimuthal distance s, to the normalized coordinates $\eta(\Psi)$ and $\dot{\eta}(\Psi)[\equiv d\eta/d\Psi]$, dependent on the betatron phase coordinate. The matrix equation yielding such a transformation is

$$\begin{bmatrix} y(s) \\ y'(s) \end{bmatrix} = \begin{bmatrix} \beta^{-\frac{1}{2}}(s) & 0 \\ \beta^{-\frac{1}{2}}(s)\alpha(s) & \beta^{\frac{1}{2}}(s) \end{bmatrix} \begin{bmatrix} \eta(\Psi) \\ \dot{\eta}(\Psi) \end{bmatrix} , \qquad (3)$$

where $\alpha(s)$ and $\beta(s)$, which appear in the transformation matrix are the conventional Twiss parameters for the linear machine. If we introduce the complex vaiable, $Z = \eta(\Psi) + i\dot{\eta}(\Psi)$, to describe the proton position and angle at any azimuth (see Fig. 1), we can then write the variable Z_2 after the intersection in terms of its initial value Z_0 and the increment $\Delta\dot{\eta}$ experienced at the collision point. Thus, we have

$$Z_2 = Z_0 e^{-i\Psi_1} + i\Delta\dot{\eta} , \qquad (4)$$

or, using Eqs. (1) and (3),

$$Z_2 = Z_0 e^{-i\Psi_1} + i\, \frac{(2\pi)^{3/2} \sigma_V^* |\Delta\nu_V|}{\beta_V^{*\frac{1}{2}}} \, \Phi\left(\frac{\beta_V^{*\frac{1}{2}} \eta_1}{\sqrt{2}\sigma_V^*}\right) . \qquad (5)$$

Here σ_V^* is the rms size of the strong beam at the intersection point and

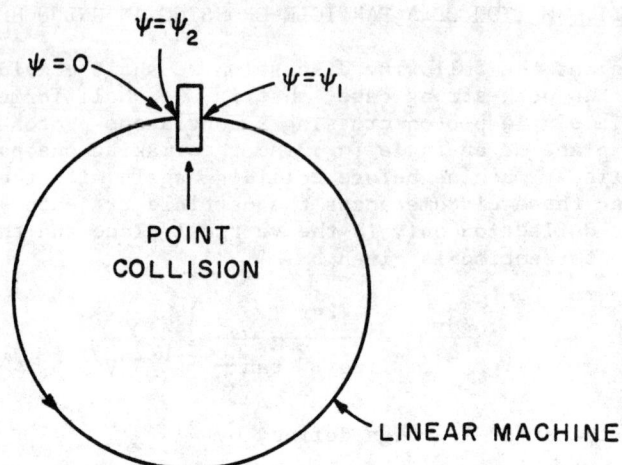

Fig. 1. Pictorial representation of a particle trajectory in a linear machine with a nonlinear point interaction.

β_V^* is the beta function at the same point for the linear machine. The quantity $|\Delta\nu_V|$ is the beam-beam strength parameter, which is written as

$$|\Delta\nu_V| = \frac{\beta_V^* r_p I}{\sqrt{2\pi} ec\gamma\beta^2 \sigma_V^* \tan\frac{\alpha_0}{2}} . \qquad (6)$$

Since we are interested in the excursions of the intersecting particle trajectory on successive revolutions with respect to the fixed strong beam, it is best to normalize the particle phase space variable Z to the square root of the linear invariant W_s of the strong beam. We choose W_s such that the corresponding boundary curve contains 95% of the particles for a beam which has a Gaussian distribution in phase space, that is to say,

$$W_s = \frac{6\sigma_V^{*2}}{\beta_V^*} . \qquad (7)$$

Using such a normalization, we can then express Eq. (5) as

$$\hat{Z}_2 = \hat{Z}_0 e^{-i\Psi_1} + i4\pi|\Delta\nu_V|\Theta(\hat{\eta}_1) . \qquad (8)$$

We have transformed the Z variables appearing in Eq. (5) according to

$$Z \to \hat{Z} = \frac{\eta + i\dot{\eta}}{\sqrt{W_s}} = \hat{\eta} + i\dot{\hat{\eta}} \qquad (9)$$

and have introduced the Θ function

$$\Theta(\hat{\eta}) = \frac{\sqrt{\pi}}{2\sqrt{3}} \Phi(\sqrt{3}\hat{\eta}_1) \ . \tag{10}$$

If we multiply the phase space variable $\hat{Z}_2$ by its conjugate, $\hat{Z}_2^*$, we obtain the invariant associated with the <u>unperturbed</u> linear system. Thus,

$$W = \hat{Z}_2 \hat{Z}_2^* \ . \tag{11}$$

We emphasize that this quantity is not an invariant of the <u>perturbed</u> system, but it does equal the square of the amplitude of the motion executed by the particle on successive revolutions. It is the mean value of W, over a distribution of N particles, which is investigated numerically in the next section. In general it can be written at any step, i, in the calculation as

$$\langle W \rangle_i = \frac{1}{N} \sum_{n=1}^{N} \hat{Z}_i(n) \hat{Z}_i^*(n) \ . \tag{12}$$

III. SOME SIMULATION RESULTS

The behavior of a distribution of particles in phase space under repeated application of the mapping, given in Eq. (8), was studied numerically. The mean square amplitude of the particles for the distribution, $\langle W \rangle$, defined in (12), was computed initially and at regular intervals during the sequences of iterated mappings.

Using a phase space normalized to the dimensions of the "strong" beam, the initial distribution was chosen to be uniform within the unit circle, corresponding to an actual radius, $\sqrt{6}\ \sigma_V$. The expectation value of the initial mean square amplitude for this distribution is, $\langle W_0 \rangle = 0.5$. The actual value of $\langle W_0 \rangle$ differed from 0.5 due to fluctuations associated with the finite size of the sample. Several different sample sizes were used. For purposes of comparison and in order to investigate growth rates, the ratio $\langle W \rangle / \langle W_0 \rangle$ is plotted in the figures.

There is concern about the effects of rounding errors in computations involving many iterations of a nonlinear mapping. To observe the cumulative effect of these rounding errors, runs were performed with and without roundoff, all using single precision arithmatic (16 digits). We find that except for very low strength values $[|\Delta v|_V < 0.01]$, precision is completely lost in a small number of runs, sometimes in as few as 100 interactions. However, the averaging approach used here tends to mask this effect and we find that although specific point accuracy is lost, the trends are essentially independent of the point precision.

The two parameters are the betatron tune (the phase advance per collision/2π) and the beam-beam strength parameter, $|\Delta v_V|$. A tune of 3.77 (as in ISABELLE) was chosen, and a range of values of $|\Delta v_V|$ from 0.01 to 0.3 was tried. Figure 2 shows the results for the mean square amplitude for $|\Delta v_V| = 0.01$, 0.03 and 0.1, plotted at every 5000 iterations up to maximum of 50,000 iterations. Figure 3 shows the effect of the use of a finite sample, specifically the reduction in the finite sample fluctuations as the sample size is increased from 50 to 1000 particles.

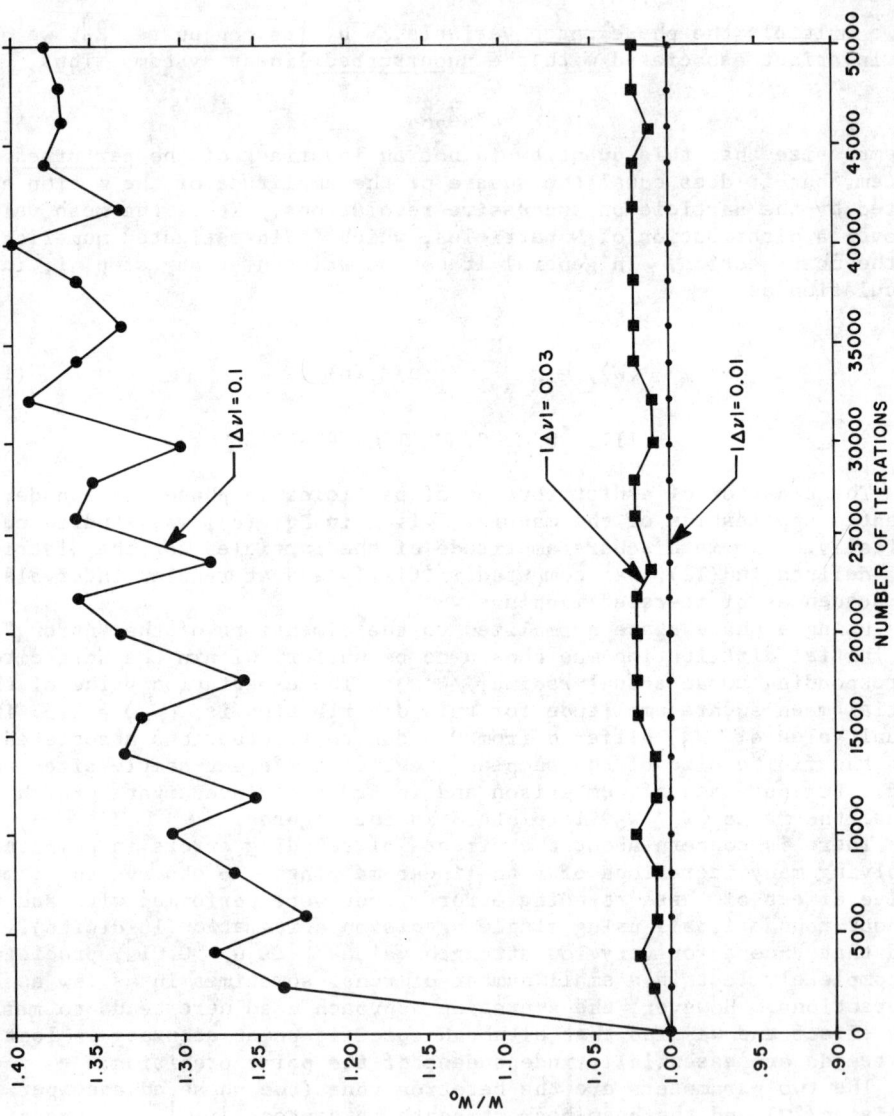

Fig. 2. Evolution of mean square amplitude for a sample of 1000 particles and for Beam-Beam strengths $|\Delta\nu| = 0.01, 0.03, 0.1$ units. Tune of linear machine is $\nu = 3.77$.

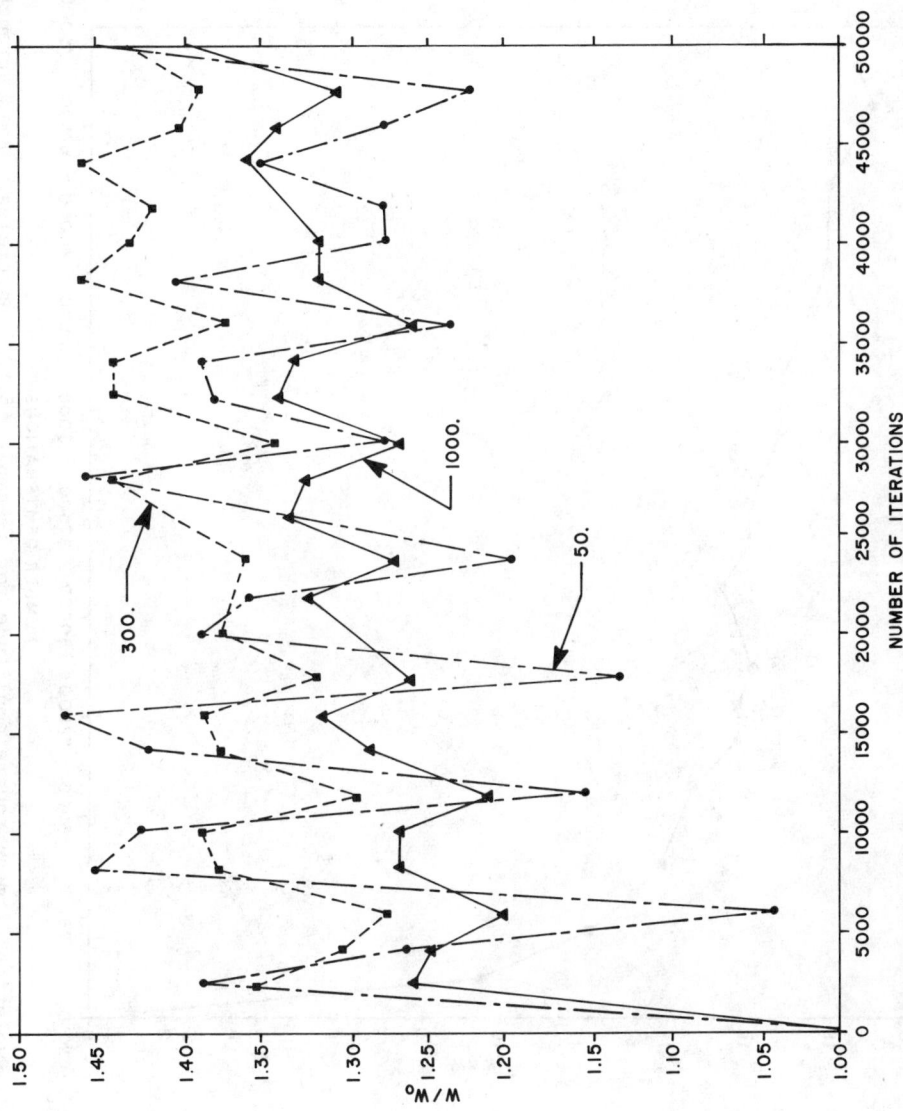

Fig. 3. Evolution of mean square amplitude for different sample numbers: 50, 300, and 1000 particles. Beam strength parameter $|\Delta\nu| = 0.1$. Tune of linear machine is $\nu = 3.77$.

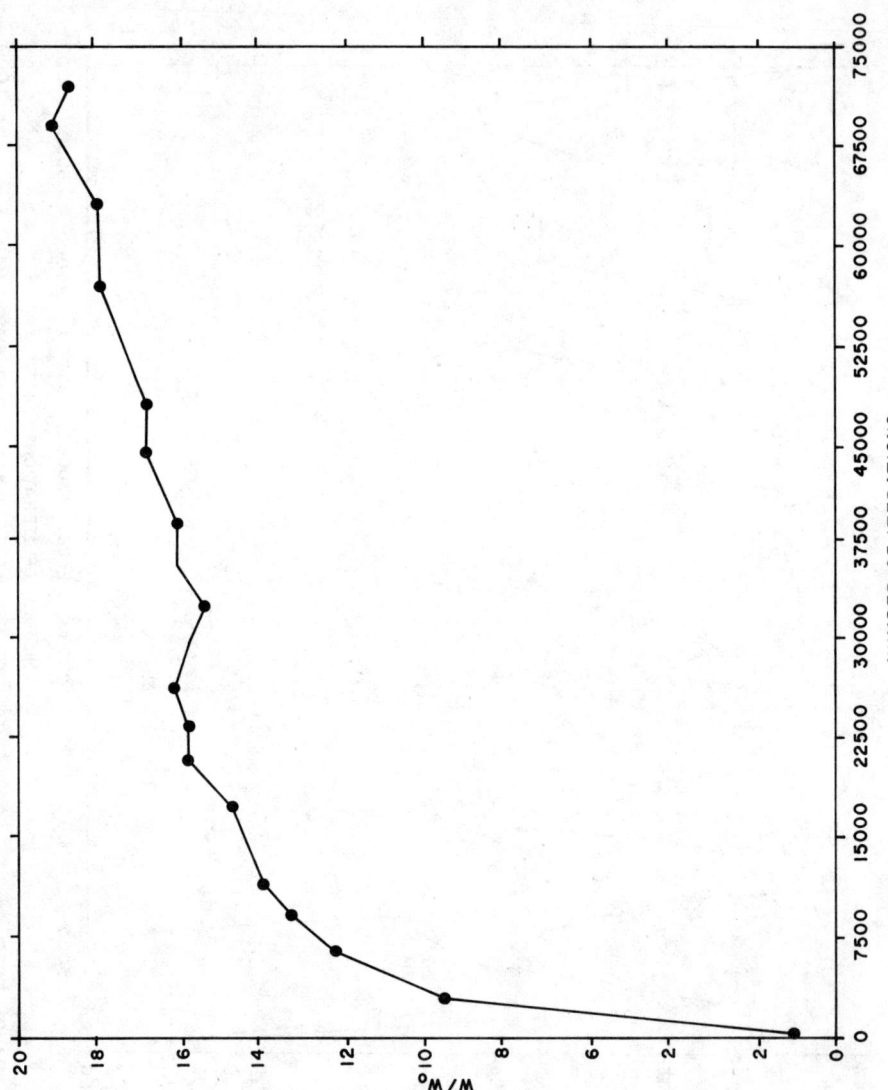

Fig. 4. Evolution of mean square amplitude for a sample of 1000 particles. Beam strength parameter $|\Delta\nu| = 0.3$. Tune of linear machines is $\nu = 3.77$.

Finally, Fig. 4 shows the results for $|\Delta v|_V = 0.3$ using 1000 particles in the sample.

The following conclusions can be drawn:

(1) The mapping appears to be stable for $|\Delta v_V| = 0.01$, although it is not obvious whether the stability is genuinely due to a threshold for growth (at some value of $|\Delta v_V|$ above 0.01) or just that the growth rate is too slow to be observable in a reasonable number of iterations. It is interesting that the model being used is a good approximation of the beam-beam interaction at the ISR and that beam observation at the ISR shows no observable beam-beam effects under normal operating conditions, i.e. $|\Delta v_V| \approx 0.001$, for up to $\sim 10^{11}$ interactions.

(2) For $|\Delta v_V| > 0.03$, the mean square amplitude shows growth. In the case of $|\Delta v_V| = 0.1$, there is a 40% increase of $\langle W \rangle$ in about 50,000 iterations. The growth rate is a strong function of $|\Delta v_V|$. Probably these effects are due to the nearby 4th order resonance at 3.75.

(3) For $|\Delta v|_V = 0.3$, a very large value indeed, the growth is extremely rapid initially, with $\langle W \rangle$ increasing by a factor of ~ 10 in the first 3000 iterations. Subsequently the growth rate falls so that when $\langle W \rangle / \langle W_0 \rangle$ is in the 18-20 range, the growth rate has fallen to about 10% growth in 30,000 iterations. Although not shown on this graph, the run was taken out to 300,000 iterations. Towards the end of the run, the slow growth rate observed at about the 100,000th run had not significantly diminished and measured about 5% in 30,000 iterations. We believe that the initial fast growth is due to the nearby 4th order resonance, but the apparent persisting slow growth is unexplained at the present state of our numerical analysis.

REFERENCES

1. J. C. Herrera, these preceedings.

BEAM-BEAM LIMIT SIMULATION OF SPEAR I*

A SUMMARY OF RESULTS

E. Close

Lawrence Berkerley Laboratory

ABSTRACT

We discuss here an attempt to simulate the beam-beam limit effect in the SPEAR I 1.5 GeV storage ring located at SLAC. A summary discussion is given of the models used and the results obtained. Remarks are made concerning the difficulties encountered in the simulation problem.

INTRODUCTION

One of the principal problems present in colliding beam storage rings is the limit on luminosity (counting rates) caused by what is commonly called the "beam-beam" limit. At this limit an increase in beam strength leads to a decrease in luminosity due to excessive growth in the size of one of the beams. Also, the stored beam is eventually lost due to a shortening of the beam lifetime. It was to investigate this beam-beam limit that the computer program PEP was developed.[1]

In this paper results from that study are summarized. The study was carried out over a period of time and it is not possible to cover in detail all the lessons learned and results obtained. A somewhat more complete survey and references to background material can be found in Ref. (2) and some recent detailed results in Ref. (3). In this paper are discussed the basic, original model and results obtained from it (Model I), an updated model that lead to better agreement with measured data (Model II), and some more recent results based on a variant of the second model (Model III). In conclusion a number of remarks are made that are derived from the experience gained during the course of this study.

It should be emphasized that these calculations were meant, insofar as possible, to simulate a real storage ring. This is in contrast to other types of studies when a specific nonlinear equation is studied with the object of understanding the properties of its solution space. This simulation attitude strongly influenced the model development and the presentation and interpretation of results.

Model I

Since the exact modeling of a colliding beam storage ring is not possible, a simplified model was developed. This simulation model, called Model I here, concentrated on the collective effect of one beam upon the particles in the other beam as the two beams intersected in the field-free interaction regions of an e^+e^- storage ring. Thus,

---

*Prepared for the U. S. Department of Energy under Contract W-7405-ENG-48

the simulation consisted of two bunched beams; a weak beam, beam 1, and a strong beam, beam 2.

The strong beam was defined analytically by a charge distribution. Initially this distribution was Gaussian in the variables x,y,z with rms beam widths $\sigma_x, \sigma_y, \sigma_z$. The storage beam was assumed to be highly relativistic, traveling parallel to the z-axis, and stationary in the sense that its charge distribution was not a function of the number of collisions it experienced with the weak beam. The three-dimensional character of the strong beam and the rapid change of the beta functions in the interaction region were included in the calculation by discretizing the interaction region along the z-axis. See Fig. 1B, 1C, also Table I.

(a) LINEAR TRANSFER MATRIX

$$M = \begin{bmatrix} \cos \mu & \beta \sin \mu \\ -\frac{1}{\beta} \sin \mu & \cos \mu \end{bmatrix} ; \quad \mu = \frac{2\pi \nu}{N}$$

$$\nu = \frac{1}{2\pi} \int_0^C \frac{ds}{\beta} \qquad X_i = M_{-L/2} \, M \, M_{-L/2} X_f$$

(b) BEAM IN INTERACTION REGION

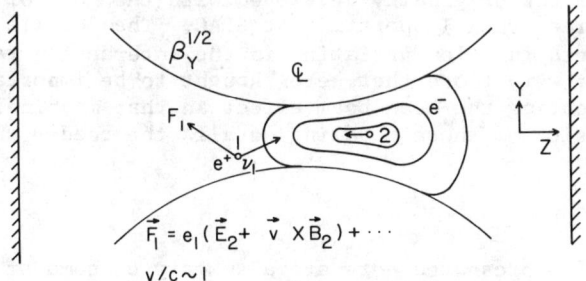

(c) DISCRETIZED INTERACTION REGION

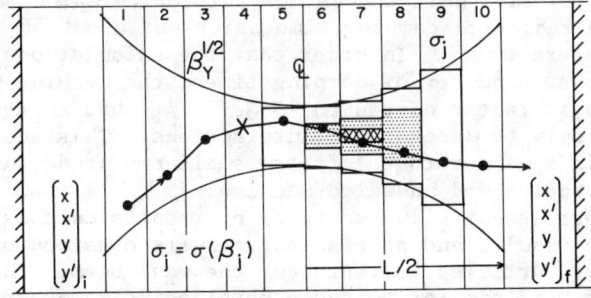

FIG. 1. Beam-Beam simulation model.

The weak beam, also assumed to be highly relativistic, was initially defined by drawing samples of test particles from a Gaussian distribution. For the weak beam the effects of beam growth and damping in the transverse x,y plane due to quantum radiation were included. Also, the x-motion was coupled into the y-motion in a manner that represented the natural coupling present in the ring.

The basic ring structure between interaction regions was represented by a linear transformation. See Fig. 1A. All energy loss effects, other than quantum noise perturbations, were ignored. Thus, all collisions for each test particle were time independent in the sense that a test particle was either always early, on time, or late when it arrived at the interaction region. The only forces considered were the basic Lorentz force of the strong beam bunch acting on the weak beam test particle, the random noise due to quantum radiation, and a balancing damping effect. The latter two effects were adjusted to yield the correct (experimentally measured) beam size. In order to compare the calculated results with experimentally measured results, a luminosity calculation was included.

The simulation using Model I was done by setting the model parameters such as the machine tunes (ν_x, ν_y), the beam distribution sizes ($\sigma_x, \sigma_y, \sigma_z$), the beam strengths and quantum noise parameters to values that represented the machine being simulated and then following the evolution of the weak beam charge distribution as a function of the number of interactions with the strong beam.

The model was originally developed with the goal of simulating the e^+e^- machine SPEAR I operating at SLAC. The details included, such as the beta function variation in the interaction region and the quantum noise, were those that were thought to be important for correctly representing the beam-beam effect in that machine. For a fuller discussion of these modeling details the reader should consult Ref. (2).

Results I

The results presented here are a summary of some of the main conclusions that were derived from computer runs made using the CDC program PEP based on Model I. A somewhat fuller discussion can be found in Ref. (2).

The natural time unit to use for this calculation was the transverse quantum radiation damping time shich was about 66×10^3 μsec, or 170,000 interactions. In order that the calculations could be carried out over a number of damping times, the machine parameters were scaled by a factor of ten to 15 GeV. At that energy the damping time was only 66 μsec or 170 interactions. This scaling was carried out in a manner that left the small amplitude tune shift the same for the scaled and unscaled machine.[2]

There were essentially two types of results calculated; single test particle results and statistical results obtained by choosing a sample of test particles to represent the weak beam. The single particle results were similar to those obtained using a sample.

Our basic conclusions from runs using Model I can be summarized as:
1. the beam strengths needed to exhibit beam blow up were unreasonably high and did not correlate with experimentally obtained results;
2. quantum noise and damping, although contributing adversely to a beam that showed growth, did not in themselves seem to trigger or cause the observed growth;
3. Model I must be missing some necessary details since it tended to show only stable results for beam strengths that were known to cause growth in SPEAR I.

These calculations also showed that it was necessary to calculate fields in a smooth accurate manner in order to eliminate the introduction of numerically spurious results that caused artificial beam growth. They also showed that beam shape was a factor that influenced the results. Flat beams grew at tune shifts that were different from those of round beams. For example, the small amplitude linear force tune shift for the separation of stable and unstable beams was about $\Delta\nu_x = 0.136$ for a round beam but appeared more like $\Delta\nu_y = 0.302$ for a flat beam. Thus, an understanding of the beam-beam limit in a particular machine would require that the beam be correctly modeled.

The results also showed that when strengths were such that beams were stable, they would exhibit stable behavior over a large number of interactions. We had single particle runs on the order of 10^6 interactions that tended to support this. When they were unstable they exhibited fast growth.

In Figs. (2) and (3) are shown the type of results obtained from Model I calculations. In each case luminosity is plotted versus the number of interactions. Figure 2 is for a flat beam with parameters chosen to represent the scaled SPEAR I machine, whereas a round beam was used in the calculations shown in Fig. (3). The fluctuations were attributed to the small sample size and smoothed for larger samples. As noted above, a rough or inaccurate impulse (field) calculation shows numerical blow up in Fig. (2) where in fact the more accurate, smooth calculation shows none. Both Figs. (2) and (3) show stable beam, no luminosity loss, at beam strengths that were known experimentally to cause beam loss in SPEAR I. The small amplitude linear force tune shifts are a measure of the interaction force. The blowup exhibited in Fig. (3) is at an unreasonably high interaction force level.

Model II

The original model had failed to produce results that correlated with experimental measurements. In general it had shown no beam growth when in fact there actually was. The basic model was, therefore, expanded to include the effect of momentum errors $\Delta p/p$.[4] This was done by including an early/late timing effect and also a phase modulation in the linear transfer matrices. Thus, as the test particles arrived at the interaction region, those for which $\Delta p/p \neq 0$ were sometimes early, on time, or late as their position in the weak beam bunch changed. Also they were transferred between interaction regions using a transfer matrix with appropriately updated the times.

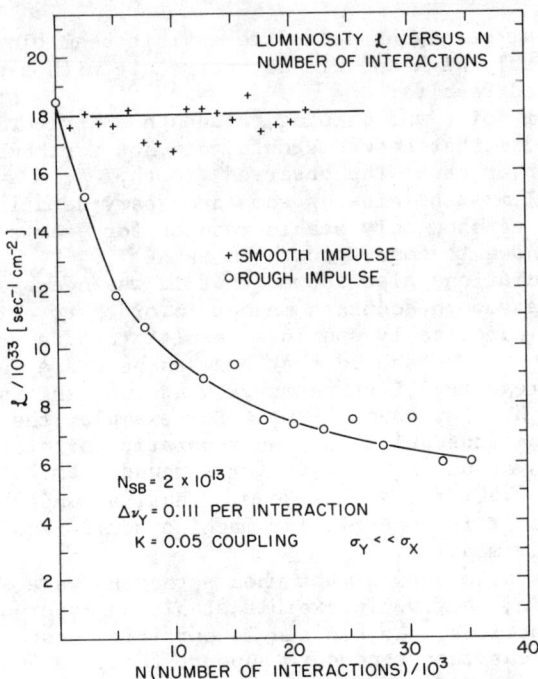

FIG. 2. SPEAR I scaled to 15 GeV flat beam luminosity calculations.

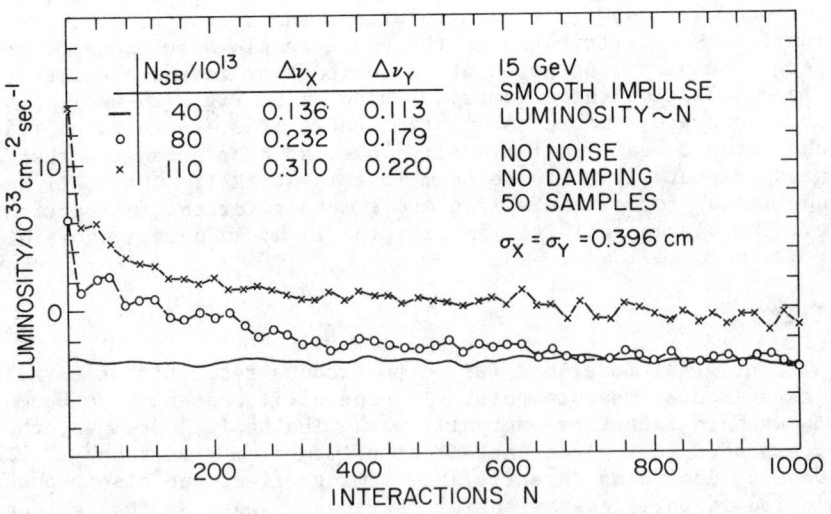

FIG. 3. SPEAR I scaled to 15 GeV round beam luminosity calculations.

At the same time a new charge distribution was used for the collective effect of the strong beam on the weak beam. See Table I. The closed form solution for this distribution as derived by Dr. Smith[4,5] was about a factor of 5 faster than the evalution of the bi-Gaussian by numerical quadrative.

Table I. Charge distributions for nonlinear fields.

1. Bi-Gaussian

$$\rho = \frac{\lambda}{2\pi\sigma_x\sigma_y} \exp\left[-\frac{1}{2}\left(\frac{x^2}{\sigma_x^2} + \frac{y^2}{\sigma_y^2}\right)\right]$$

2. L. Smith

$$\rho = \frac{\lambda}{2\pi\sigma_x\sigma_y}\left[1 + \frac{1}{2}\left(\frac{x^2}{\sigma_y^2} + \frac{y^2}{\sigma_y^2}\right)\right]^{-2}$$

$$\lambda = \pi/2$$

The model was also refined to include the fact that the beta functions which determined the strong beam shape in the interaction region depended on the beam strength.

Results II

Runs were made on a parameter set that corresponded as closely as possible to the 1.5 GeV SPEAR I machine running with about 8.2 mA of current. This beam intensity corresponded to beams with 4.0×10^{10} particles per bunch, linear tune shifts of $\Delta\nu_x = 0.059$ and $\Delta\nu_y = 0.072$ and was one of the higher intensity runs for which SPEAR I results were available

With the inclusion of longitudinal motion, the relevant damping time was the longitudinal quantum radiation damping time which was about 85,000 interactions. Single test particle runs were made over this length of time using the unscaled machine parameter set. Exploratory runs showed that off-momentum particles oscillating on the order of $3\sigma_z$ along the z-axis showed significant amplitude growth within 85,000 interactions, that is, within one longitudinal damping time τ

Because of these preliminary results, a series of runs that covered a $5\sigma_x$ by $10\sigma_y$ area in the initial value space were made. All test particles were started with zero slope. These runs were made with $\Delta p/p = 0$ and $\Delta p/p$ about $3\sigma_z$. Results of these runs were saved for later analysis and plotting.

The results that are shown in Fig. (4) are plots of the maximum excursion A_x that was achieved by the test particle that started initially at the value (x^o, y^o). The abscissa is the initial vertical displacement y^o, the ordinate the initial horizontal displacement x^o, the number plotted at any coordinate is the maximum amplitude A_x that the test particle would have in the ring at the time of the plot. All values have been divided by a measure of the beam width, σ_x for x^o and A_x, and σ_y for y^o.

The plots are given for a number of interactions N that correspond to 0.001, 0.5, and 1.0 damping times. Those points that do not have an amplitude number had no test particle tracked for that initial value. Due to the expense of generating these results, not all points in the $5\sigma_x$ by $5\sigma_y$ space shown were selected for tracking.

The interpretation of these plots is limited to rather elementary conclusions. The plot on the left is for $\Delta p/p = 0$ and the one on the right for $\Delta p/p \sim 3\sigma_z$. The cross-hatched area has particles that do not exceed $1\sigma_x$ in amplitude, the next area is for those that do not exceed $3\sigma_x$, and the darker region for those that do not exceed $6\sigma_x$ in amplitude. The evolution of these areas is shown for the damping times $t/\tau_z = 0.0, 0.5, 1.0$.

The same type of results are shown in Fig. (5) for the maximum amplitude A_y in the y plane. For $\Delta p/p = 0$ the results, although not as clear cut, tend to show that the beam is relatively stationary when synchrotron oscillations are not included. Again when $\Delta p/p \sim 3\sigma_z$, the 3 to $6\sigma_y$ region shrinks, and outside that region particles increase their amplitude of oscillation and there is at least one point out to $28\sigma_y$.

These results tend to indicate that not including synchrotron oscillations in the original model was a serious omission since their inclusion leads to beam growths of significant magnitude. Previous results obtained from Model I suggest that the inclusion of quantum noise and damping would have made these growths greater. Luminosity calculations might, however, not show much beam loss since not much growth occured within 3σ and most of the beam is contained within that region. Aperture restrictions for SPEAR I were about $47\sigma_x$ and $30\sigma_y$, so clipping was not experienced. However, from a lifetime or diffusion point of view there are points in the initial distribution space, $5\sigma_x$ by $5\sigma_y$ by $3\sigma_z$, that are getting close to the y-plane wall. Long tails grow for particles that are at $3\sigma_z$ and for these particles the growth of the tails of the distribution is rather fast and large. Thus, it might very well be that a lifetime calculation, if one could perform it, would show a short beam lifetime. Obviously, the results given here are not complete enough to draw any definite conclusions.

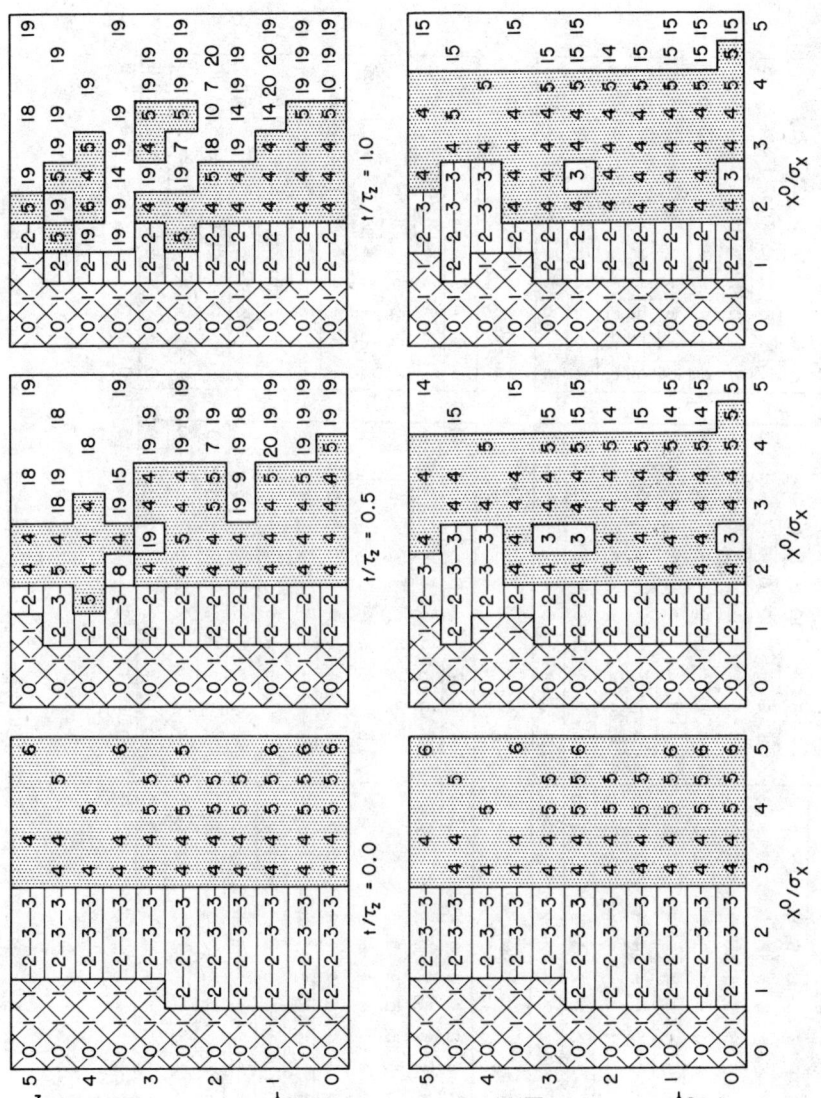

FIG. 4. SPEAR I at 1.5 GeV flat beam maximum x-plane amplitude plots.

218

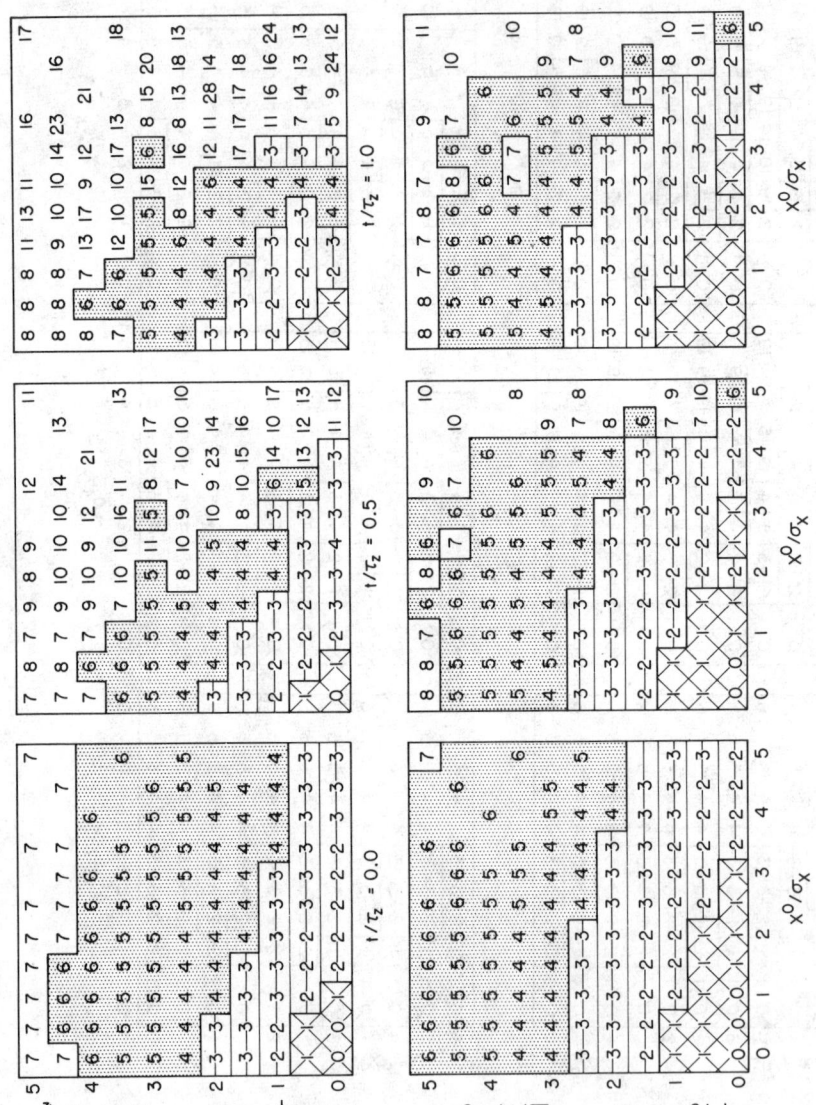

FIG. 5. SPEAR I at 1.5 GeV flat beam maximum y-plane amplitude plots.

Model III

In response to a renewed interest in the original beam-beam calculations, runs were made using a program WEA10 developed by Dr. Laslett of LBL.[6] This program is based on a model, called Model III here, that is similar to Model II. It does a somewhat more restricted simulation than was done using the program PEP and Model II. In particular Model III is an impulse calculation and does not take into account the longitudinal beam length or the variation of beta along the longitudinal axis. Also $\Delta p/p \neq 0$ appears only in the transfer matrix elements, no timing effects are included, all interactions take place at the symmetry point of the interaction region. It does, however, have the same coupling of the x motion into the y motion that was used in Model II. No quantum noise is considered. Since no use was made of this effect in Model II calculations, its omission in Model III is not important.

Results III

Many single test particle runs were made using Model III. These results are described in detail in Ref. (3). The main interest here is to note how these results relate to the original runs.

As clearly as possible the original SPEAR I parameter set was used as input. Since only a strength parameter is used in WEA10, the small amplitude tunes were used as a measure of the beam strength and they were set so that $\Delta \nu_x$ was the same as previously. Because no longitudinal beam length was included, $\Delta \nu_y$ did not turn out to be the same as before. There was in fact no way that this model would give the same tune shifts as Model II.

An initial test particle at $4\sigma_x$, $4\sigma_y$ was chosen since this test particle had previously exhibited large growth in amplitude within 85,000 interactions. The particle was run to 84,000 interactions and no noticeable growth in amplitude was observed. For all practical purposes the selected particle showed regular behavior with no growth.

By sampling the initial value space it was possible to find an initial value that did exhibit large growth and studies were made on its behavior as $\Delta p/p$ was varied and also as the synchrotron oscillation frequency νs was varied. These results are also reported on in Ref. (3). However, the interesting result obtained from Model III was that a change in the model which on the surface looked rather slight caused a completely different behavior of a particular initial value test particle. Thus, it would be necessary to redo the $5\sigma_x$ by $5\sigma_y$ sample set to see if the results were qualitatively the same with Model III. A lack of time and the cost of such runs prevented that from being done.

GENERAL REMARKS

The results summarized in this report were obtained trying to numerically simulate the beam-beam effect as it actually occured in an e^+e^- storage ring. The first attempt, Model I, although it contained many details of the interaction process, did not produce results that correlated with experimentally measured values. It also appeared that the inclusion of some details like quantum radiation/ damping were not necessary and in fact only made understanding the results more difficult.

The inclusion of synchrotron oscillations and tune variation due to momentum errors, Model II, gave beam growths that were large at beam strengths that were close to the experimentally determined limits. However, the results were limited to a $5\sigma_x$ by $5\sigma_y$ region evaluated at $0\sigma_z$ and $3\sigma_z$. The calculations needed to adequately populate a full three-dimensional space, much less the full six-dimensional phase space, are time consuming. It would be a very large task to do a complete set of runs over a range of parameters, such as beam strengths and synchrotron frequencies, and then analyze the results.

The third set of runs, derived from Model III and only mentioned here,[3] point out the perplexing fact the behavior of a specific orbit is i) very sensitive to slight parameter changes and ii) model dependent.

Thus, it seems that before conclusions can be reached a rather dense sampling of initial value space must be done and the behavior of the sample investigated. Also, since different models give different orbits, it would appear that only the total sample behavior can have any real validity and that two models would be judged equivlent if the total sample behavior were equivalent, regardless of what individual orbits did.

One of the guidelines that should be followed is to remember that the storage rings are analogue models and results must always be checked against them or there is no way of knowing whether the numerical calculations reflect anything that relates to a real machine.

It also appears important to correctly model the beam shapes that are in the machine being simulated. Although this is difficult, idealized beam shapes can lead to erronous results. In particular sharp edges on charge distributions and doubly valued charge distributions are to be avoided unless they really exist. Round beams are not like flat beams. The distribution suggested by D. Smith[4,5] is both numerically fast and physically reasonable.

The change in the beta function across the interaction region and the finite bunch length must be considered. A check that results are correct must be made when a simple impulse calculation is used. If the changes in the beta function and bunch shape are significant, a simple impulse calculation will not suffice. At least for the simulation of SPEAR I, a longitudinal discretization was necessary.

The problem of trying to simulate and understand the beam-beam effect is a rather perplexing one. From a theoretical viewpoint it is interesting to construct a "model", choose a set of parameters and then explore the properties of specific orbits, either numerically, analytically, or with a combination of both. However, from a practical point of view none of the parameters are known exactly and, further, what the model should be is not obvious. The calculated results are not only model dependent, but small parameter changes can lead to qualitatively different behavior of individual orbits.[3] Thus, it may make little sense to refer results from individual orbits, and only the collective sample behavior may be meaningful. But the sample size is of necessity small compared to beam bunches which, for the unscaled SPEAR I machine, were on the order of 10^{10} particles per bunch. So gross qualitative conclusions are eventually arrived at from results obtained using a very small sample.

This state of affairs necessitates the checking of calculations against experimental results and trying to build a computational model that accurately reflects a real machine. It was this approach that was taken in developing Model I and II. The results obtained with Model II are a somewhat encouraging indication that such a model can be built. Unfortunately a simulation calculation is expensive in time, effort, and money. Also it is not obvious that the model extrapolates to another, different machine.

What is really needed is a theory that would exhibit, as of a function of machine parameters, what happens to all orbits. This is what the usual linear orbit theory does. It would be nice if somehow in the present nonlinear problem equations could be obtained that show globaly how all solutions behave with respect to some of the relevant machine parameters. Just what these "solutions" represent is left open. However, they should be relatable to experimentally measured beam quantities.

ACKNOWLEDGMENTS

I would like to thank Dr. L. Smith of LBL for his guidance and active contribution to these studies, Dr. J. Laslett of LBL for the many hours spent in independent verification of calculations and his many helpful suggestions, the SPEAR group at SLAC for furnishing machine parameters and measured results, and Dr. M. Month of BNL for his persistent encouragement that has resulted in this paper.

REFERENCES

1. PEP a CDC7600 program, unpublished, initially developed by J. E. Augustin and E. Close in a joint BLB/SLAC collaboration.
2. E. Close, SPEAR I at 1.5 GeV, UCID-8055, Lawrence Berkely Laboratory, Berkeley, California.
3. E. Close, General Properties of Beam-Beam Interaction Orbits, LBID-006, Lawrence Berkely Laboratory, Berkeley, California.
4. L. Smith, private communication, LBL.
5. E. Close, Electric Fields for Beam-Beam Force Calculations, LBL-8903, Lawrence Berkeley Laboratory, Berkeley, California.
6. J. Laslett, WEA10 an interactive CDC6600 program, unpublished, Lawrence Berkeley Laboratory, Berkeley, California.

"STABLE AND UNSTABLE MOTION IN DYNAMICAL SYSTEMS"

J. Moser*

Courant Institute of Mathematical
Sciences, N. Y., N. Y. 10012

I. STABLE AND UNSTABLE MOTION IN DYNAMICAL SYSTEMS

I must confess to you that since this Conference started two days ago I have felt a rather embarassing as well as gratifying experience; most of the topics that I was going to discuss in my lecture have already been presented in one form or another by the speakers before me: M. Lieberman gave a fascinating talk on Arnol'd diffusion, John Greene impressed us all with his pictures of the regions of instability and Joe Ford described stable behavior in a comprehensive way. Furthermore, not being an expert on storage rings I fear that I have little to add that is new and unfamilar to most of the participants of this Conference.

Nonetheless, I would like to present here a number of general mathematical statements that can rigorously be proven, in the hope that they will provide a firm ground for future theoretical as well as practical investigations in the field of Accelerator Dynamics.

But there is yet another-perhaps more instructive-reason why I wish to give this overview of the rigorous stability results known to date: It is important to recall that the mathematical development of this subject, which long preceded the age of computers, did in fact anticipate most of the qualitative features of dynamical systems which numerical experiments nowadays give so much evidence for.

You have heard already many times in this Conference the term "stochasticity". Well, the phenomenon, which currently dynamicists refer to by that word, was not called that way originally. The so called "regions of stochasticity" were in fact termed "regions of instability" nearly fifty years ago by G. D. Birkhoff.[1-3] And this interesting so called "stochastic" behavior is observed in the neighborhood of homoclinic points a term introduced last century by H. Poincare.[2,]

So, the concepts that we are all using routinely when discussing the behavior of dynamical systems, were developed a long time ago. From theoretical considerations their quantitative aspects on the other hand, which most members of this audience are interested in, are only now beginning to be explored.

It is also important to note that the "abstract" and "nonpragmatic" approach which I will adopt in this lecture affords us the advantage of a wider scope and applicability. After all the study of stability and instability of dynamical systems was not prompted by the necessities of accelerator design! The original incentive was provided by Celestial Mechanics but the results are sufficiently general to be applicable to other fields of physics including the stability problem of particle beams in the storage rings of accelerators.

---

*I am indebted to Tassos Bountis for putting an informal lecture into presentable form, as well as for supplying references and figures.

I.A. Rigorous Stability Results

Let me begin by describing a phenomenon which can be handled essentially by the tools of Linear Algebra and which refers to the role of <u>sum</u> and <u>difference-resonances</u>. Assume that our system can be described by the Hamiltonian

$$H(x,y,p_x,p_y,t) = H^{(2)} + H^{(3)} + \ldots + H^{(s)} + \ldots \qquad (1)$$

where x,y are the position and p_x,p_y the momentum variables and $H^{(s)}$ denotes a polynomial in these variables of degrees s. We also take the quadratic part $H^{(2)}$ in (1) to have the "diagonalized" form

$$H^{(2)} = \tfrac{1}{2}(p_x^2 + p_y^2 + \omega_x^2 x^2 + \omega_y^2 y^2) =$$
$$= \omega_1 I_1 + \omega_2 I_2 \qquad (2)$$

where ω_1,ω_2 are the two normal mode frequencies of the system when $H^{(s)} = 0$, $s = 3,4\ldots$, and where we have transformed to Action-Angle variables (I_k, θ_k) with

$$\theta_k = \omega_k t + \delta_k, \quad k = 1,2.$$

Note that the absence of <u>linear</u> terms in (1) and the fact that $H^{(2)}$ has the positive definite form (2) makes the origin a <u>stable</u> equilibrium point of the system. Thus, when we leave out all the $H^{(s)} (s \geq 3)$ in (1), the motion is always stable. We want, however, to ask what happens to the solutions of all "nearby" quadratic Hamiltonians of the form

$$H_2 = H^{(2)} + \epsilon Q, \quad \epsilon \text{ "small"}, \qquad (3)$$

where Q is also a <u>quadratic</u> function of x,y,p_x,p_y and has in addition a 2π-periodic dependence on t i.e. $Q(t) = Q(t + 2\pi)$. We are thus seeking to determine the "<u>strong linear stability</u>" properties of our system.

Floquet Theory[4,5] tells us that the frequencies of the solutions of H_2 will be somewhat different than ω_x, ω_y but that the system will be stable provided ω_x, ω_y are not simple multiples of one another. The precise statement[27] is that stability is ensured as long as

$$j_1 \omega_x + j_2 \omega_y \neq j_0, \quad j_1, j_2 \geq 0, \quad j_1 + j_2 = 2, \qquad (4)$$

for integers j_1, j_2, j_0.

Let us observe that requiring j_1 and j_2 to be positive, cf. (4), implies that we know the <u>sign</u> of the frequencies ω_x, ω_y! However, the fact that ω_x, ω_y enter in $\overline{H(2)}$ as ω_x^2, ω_y^2 introduces a sign ambiguity at the outset. Thus the-admittedly-mathematical question clearly presents itself: How can we define frequencies with a sign?

This problem was first studied by M. G. Krein.[7] It was subsequently treated by Gel'fand and Lidskii[8], Jacoubovich[9] and others[27,28] in the fifties, and by now there have been books written on the subject. I will only give the main results here.

Since our system is <u>linear</u> we can write down its so called Floquet solutions in the form

$$X(t) = \begin{bmatrix} x(t) \\ y(t) \\ p_x(t) \\ p_y(t) \end{bmatrix} = e^{i\omega t} P(t), \qquad (5)$$

where $P(t)$ is a column four-vector whose complex components are 2π-periodic functions of t, i.e.: $P(t) = P(t+2\pi)$, and ω is a characteristic frequency of the system.

Now the real (or imaginary) part of $P(t)$ defines in the 4-dimensional phase space a "curve" to which one can assign a definite orientation. This orientation is due to the canonical nature of the equations of motion and can be defined by the requirement[27]

$$\text{Imag. Part } (JP, \bar{P}) > 0. \qquad (6)$$

In (6), (,) denotes the inner product operation, $\bar{P}(t)$ corresponds to the negative frequency $(-\omega)$ solution (5) and

$$J \triangleq \begin{bmatrix} 0 & I \\ -I & 0 \end{bmatrix},$$

where I is the 2×2 identity matrix.

Interchanging P and $\bar{P}$ in (6), or, equivalently, ω and $-\omega$ in (5), reverses the inequality sign in (6). Therefore, by insisting that the quantity in (6) be positive, we specify a definite orientation for the curve $P(t)$, and hence a definite sign (positive) for the frequency ω.

This argument carries over to the <u>nonlinear</u> case as well, i.e. the case with $H^{(2)}$, $H^{(3)}$,... not all zero, where its implications are less widely known. The statement is now one of <u>formal</u> stability[10,27] since it tells us that if ω_x, ω_y satisfy the conditions

$$j_1 \omega_x + j_2 \omega_y \neq j_0, \text{ for all integers } j_1, j_2 \geq 0, \qquad (7)$$

there exists a <u>formal</u> integral G, given in the form of a formal series-expansion

$$G = G^{(2)} + G^{(3)} + \ldots + G^{(s)} + \ldots \qquad (8)$$

This series is a formal integral in the sense that its Poisson bracket with H vanishes to all orders, i.e.

$$\{G, H\} = 0 \qquad (9)$$

and its quadratic part, $G^{(2)}$, is positive definite. Alex Dragt showed us in his lecture one way of constructing such formal integrals (8).

While the existence of integrals is intimately connected to the question of (global) stability of the motion, it is important to bear in mind that integrals of the type (8) above are, in most cases, divergent[12-14] and can only serve as crude approximations in a study of the stability properties of nonlinear systems.

Let me end this part of my talk with an observation. Note that as a result of the above analysis it appears that it is the sum-resonances that one is ultimately interested in, in these problems. This means that in a resonance diagram like the one that we have been shown by other speakers (see Fig. 1 below)

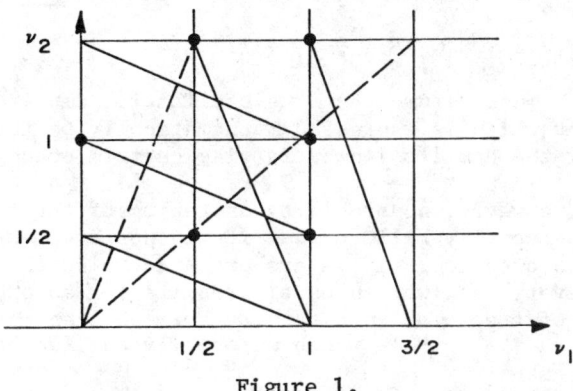

Figure 1.

the diagonal (solid) lines contribute very little to the overall behavior of the solutions. And yet we heard from A. Chao that difference-resonances after all do matter! So, how does one resolve this paradox?

I suggest that the mass reversal of which A. Chao spoke in his talk is actually equivalent to reversing the orientation of P(t) in (6) (or changing ω to $-\omega$ in (5)), thereby making what is a sum-difference appear to be a difference-resonance.

II. STABILITY FOR INFINITELY LONG TIMES

In the second part of this lecture I would like to turn my attention to the purely mathematical question of stability for all times for the nonlinear problem. I realize that my friends in the fields of Astronomy and Celestial Mechanics, who routinely compute orbits for hundreds and sometimes thousands of years, will find such an endeavor at best futile! Some members of this audience, however, who are interested in the stability of particle beams after some 10^{11} revolutions in the storage rings of their accelerators may be more understanding. Nevertheless, the fact remains that by posing the question of stability in this way we are in effect asking a purely mathematical question of apparently little if any practical significance.

Resolving this highly delicate problem of stability for infinite times requires the use of very sophisticated and subtle mathematical concepts, which I do not intend to go into here. I will only attempt to outline for you the main results as briefly as possible.

The kind of stability we are talking about can be defined as follows: We call a solution X(t) **stable** if for any $\epsilon > 0$, there exists $\delta > 0$ such that

$$|X(0)| < \delta \text{ implies } |X(t)| < \epsilon \text{ for all } t! \tag{10}$$

Take first n = 1, i.e. one degree of freedom (plus the 2π-periodic time dependence): is there such a stability criterion available? The answer is yes and the particular condition that must be met for stability is that low order resonances must **not** occur[15], i.e.

$$j_1 \omega_1 \neq j_0, \quad j_1 = 1,2,3,4. \tag{11}$$

In addition, one must assume that the oscillation amplitude is frequency dependent which is expressed mathematically by the time dependent part of the Hamiltonian satisfying certain nondegeneracy conditions.

For n > 1, however, no stability criterion of the type (10) is available. I personally believe that it may not even exist but this is truly an open question. If we are prepared to settle for a weaker criterion, however, then we can obtain results for an arbitrary number of degrees of freedom. Such results form the content of the so called KAM theory[16-18,3,6,11] which we briefly outline below starting with the n = 1 case:

Write the Hamiltonian as the sum of one part which depends solely on the Action variable I and one which depends on θ and t,

$$H = H_0(I) + \epsilon H_1(I,\theta,t,\epsilon H_1(I,\theta,t,\epsilon) \tag{12}$$

with $H_1(I,\theta,t) = H_1(I,\theta,t + 2\pi)$ and ϵ a sufficiently small parameter. The assertion now is that if the amplitude in the $\epsilon = 0$ case is frequency dependent, i.e.

$$\frac{\partial^2 H_0}{\partial I^2} \neq 0, \tag{13}$$

(cf. also comments below (11)), and if the **fifth** derivatives of H_1 are bounded, say by,

$$\sup_{k+\ell \leq m} \left| \frac{\partial^{k+\ell} H_1}{\partial I^k \partial \theta^\ell} \right| < M, \quad m = 5 \tag{14}$$

then for sufficiently small ϵ there exist **invariant tori**

$$I = I(\theta,t); \quad |I - I_0| \text{ "small"} \tag{15}$$

where I_0 is the action variable of the unperturbed system. These invariant tori, in fact, form a set of positive measure which implies that the system is **stable** since all solutions are confined to lie between such tori (15) for all time.

Note that condition (13) means that H_0 must be at least quadratic in I which clearly illustrates the stabilizing effect of nonlinearities on the motion of the system. Incidentally a condition like (14) is also necessary. There exists, in fact, a counterexample due to F. Takens[19] with m = 2 in (14), for which one can prove that KAM theory is not applicable.

The above statements can be extended to n(> 1) degrees of freedom. One starts again with a Hamiltonian expressed in the n action variables alone, i.e.

$$H_0(1) = H_0(I_1, I_2, \ldots, I_n). \qquad (16)$$

These action variables are n analytic constants of the motion, which for a system of n harmonic oscillators take the familiar form

$$I_j = \tfrac{1}{2} (p_j^2 + \omega_j^2 x_j^2)\omega_j^{-1}$$

They are independent of each other in the sense that their gradients are linearly independent. Moreover, mutual Poisson brackets vanish[11]

$$\{I_i, I_j\} = 0, \quad i,j = 1,2,\ldots,n$$

i.e. they are said to be in "involution".

A <u>near integrable</u> system, therefore, is defined as one which is sufficiently close to the integrable system (16), and is described by the Hamiltonain

$$H = H_0(I) + \epsilon H_1(I, \theta, t, \epsilon)$$

where (17)

$$I \equiv (I_1, I_2, \ldots, I_n); \quad \theta \equiv (\theta_1, \theta_2, \ldots, \theta_n)$$

The existence of invariant tori for such a system can be established provided that the (2n+3)-derivatives of H_1 are small enough, i.e. m = 2n+3 in (14), and H_0 satisfies the nondegeneracy condition

$$\det \left[\frac{\partial^2 H_0}{\partial I_k \partial I_j} \right] \neq 0, \qquad (18)$$

cf. (13).

The important difference concerning stability between the n=1 and n>1 cases is that for more than one degree of freedom the invariant tori no longer bound the orbits lying between them. It is thus possible for solutions to "leak out" along resonance lines of the "Arnol'd web" schematically represented below (Fig. 2). Lieberman showed us in his talk some very beautiful pictures of this phenomenon. It is precisely this "leaking out" of solutions to distant parts of phase space, which forms the basis of the so called Arnol'd Diffusion.

In conclusion we can say that for systems with n(>1) degrees of freedom the KAM tori cannot guarantee stability for all solutions of (17) since resonance lines (such as the ones of Fig. 2) form a

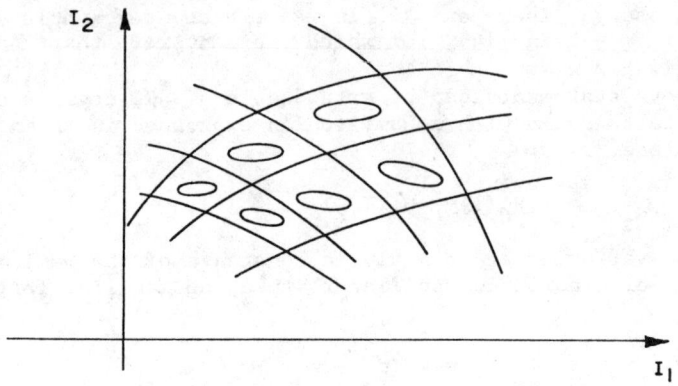

Figure 2.

network extending over most of the constant energy surface.

Finally, I would like to remark that the proof of the above mathematical statements is done by <u>construction</u> and not by indirect arguments. It is based on an iterative procedure which is very rapidly convergent and could be implemented on the computer without great difficulty. It is surprising, therefore, that up to now no such numerical scheme has been developed to actually calculate the invariant tori predicted by the KAM theory.

II. STABILITY IN DYNAMICAL SYSTEMS

After these remarks on stability I would like to discuss certain aspects of the <u>unstable</u> behavior that one encounters even in the simplest non-integrable dynamical systems. Consider, for example, the one-dimensional nonlinear pendulum driven parametrically by a periodic term,

$$\ddot{\theta} + (\omega^2 + \epsilon \cos t) \sin \theta = 0 \qquad (19)$$

The macroscopic (i.e. large scale) pictures that one obtains in the $\theta, \dot{\theta}$ plane for such a system are shown in Fig. 3.

Notice the invariant curves as well as the sequence of six islands around the central fixed point 0. In fact, what you see here is an awfully incomplete picture. The actual situation is infinitely more intricate and involved. The centers of the islands (which are the intersection points of a low order periodic orbit) are themselves topologically equivalent to 0: Thus there are islands about each one of them and about the centers of those islands there are other islands forming an exceedingly complicated pattern of "boxes within boxes" ad infinitum.

And this is still not the complete story! For example, the separatrices that you see in Fig. 3 around the stable (elliptic) fixed points of order 6, do not join smoothly in the region

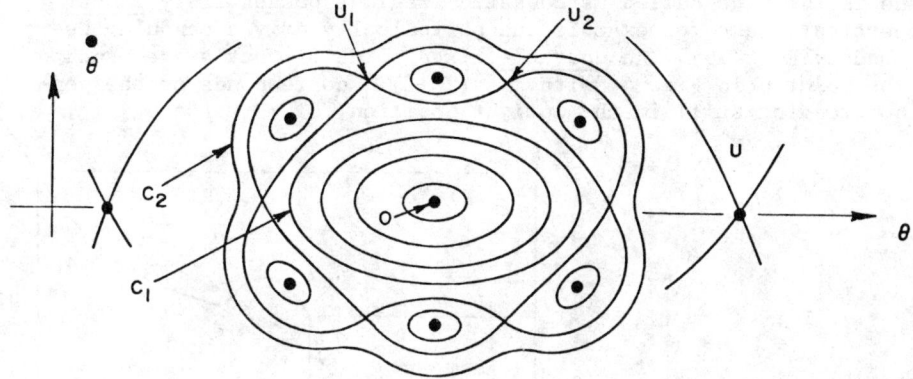

Figure 3.

between two successive unstable points U_1 and U_2, say. Instead they intersect each other in a highly complex way, shown schematically in magnification below

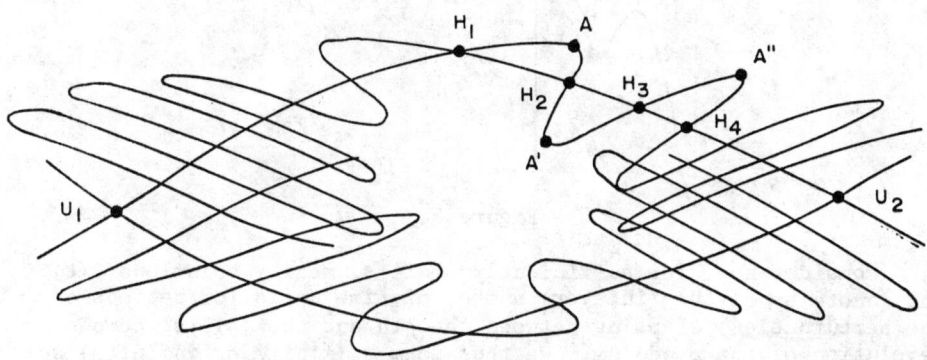

Figure 4.

The points of intersection H_1, H_2,... in Fig. 4 are the so called Homoclinic points and were known to Poincaré nearly 80 years ago[2]. What is amazing about these points is that if there is one there must be an infinite number of them! This is a consequence of the area preserving property of the "mapping"

$$(\theta, \dot{\theta})_{t=0} \longrightarrow (\theta, \dot{\theta})_{t=2\pi} \qquad (20)$$

defined by the solution of Eq. (19). Thus the shaded region H_1AH_2, say, in Fig. 4 is subsequently mapped on regions $H_2A'H_3$, $H_3A''H_4$,... etc. all having the same area as H_1AH_2. Moreover, all this must happen in a <u>finite</u> part of the plane since these unstable regions are themselves enclosed within invariant curves such as C_1 and C_2, see Fig. 3.

It is in these regions of infinitely often intersecting curves that iterates of the "mapping" (20) behave in a highly irregular and chaotic fashion. In fact, it is possible to prove that certain solutions in these so called "stochastic" regions posses truly <u>random</u> properties! Take for example the periodically driven pendulum described by Eq. (19). An <u>unstable</u> fixed point of that system such as the point U in Fig. 4 [with $\theta = (2k+1)\pi$] corresponds to the pendulum standing still in the upright position, cf. Fig. 5a.

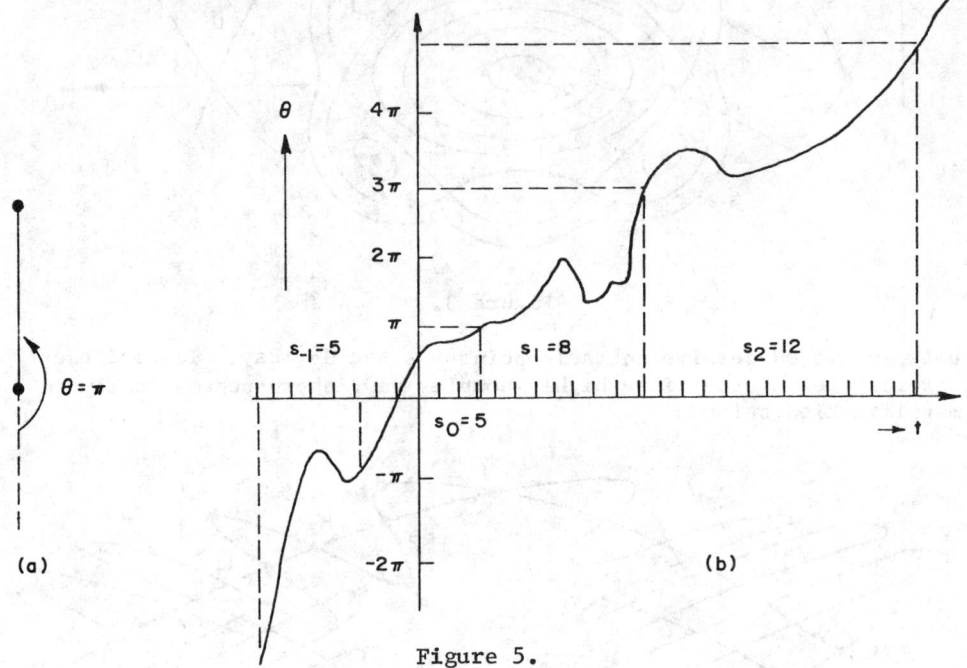

Figure 5.

Consider now, for sufficiently small ϵ, nearby solutions $\theta(t)$ and denote by s_j the (integer) number of time units (or periods of the perturbation) elapsing between the jth and the (j+1)st complete revolution of the pendulum. We thus form a (finite or infinite) sequence of integers

$$S \equiv (\ldots,s_{-2},s_{-1},s_0,s_1,s_2,\ldots), \tag{21}$$

cf. Fig. 5b. The assertion is that given <u>any</u> such sequence of sufficiently long integers (21), there exists an exact solution of (19) which behaves that way!

These ideas were originally discussed by G. D. Birkhoff[1]. They were rigorously established and formulated "geometrically" much later mainly due to work by S. Smale[20,21] and C. C. Conley[29], see also E. Zehnder[22]. The important thing to remember here is that such results are <u>generic</u> in nature and pictures of the type seen in Figs. 3,4 above are typical also in non-integrable Hamiltonain systems.

I also wanted to say something about Arnol'd Diffusion but Lieberman already gave us in his talk a much fuller and pictorial presentation than I had in mind. Let me only make one comment here, namely that Arnol'd has so far discussed this phenomenon only with regard to a carefully chosen example[23-26]

$$H = \tfrac{1}{2}(I_1^2 + I_2^2) + \epsilon(\cos\theta_1 - 1)[1 + \mu(\cos\theta_2 + \cos t)], \quad (22)$$

where ϵ,μ are small parameters. For $\mu = 0$, $\epsilon \neq 0$ this system has invariant tori

$$I_2 = \text{const}; \quad I_1 = \theta_1 = 0 \quad (23)$$

where I_1 and θ_1 are zero on the separatrix between bounded and unbounded motion. What Arnol'd shows for this particular example is that no matter how small $\epsilon,\mu(\neq 0)$ are, there exist solutions with I_2 <u>varying appreciably</u> over a finite range of values.

One must keep in mind, however, that in (22) the perturbation is of a very special type and does not represent the typical situation. While we may all believe with Arnol'd that the so called Arnol'd Diffusion is generic as an instability mechanism, this has not been demonstrated for Hamiltonians with a more general perturbation term.[26]

Furthermore the possibility that a single orbit can wander throughout all of the "Arnol'd web" has not yet been proven. One may well imagine that to prove it- if at all possible- will require a mathematical tour de force since the analytical difficulties one encounters when playing around with sets of homoclinic points are considerable. The best one has been able to do so far is to obtain lower estimates of the diffusion times,[24]

$$|I(t) - I_0| < \mu^b \text{ for } t \leq \exp(1/\mu^a)$$

which become exceedingly long for small values of the perturbation parameter μ (a,b are two positive constants).

In conclusion, I would like to stress two points which, in my opinion, are of fundamental importance: The first one is that the instability (or, if you prefer, the "stochasticity") of which I spoke here, is not a mathematical pathology that occurs in judicisouly selected models. It is a very real phenomenon present in all nonintegrable systems, which are, in a well defind sense, the majority among all dynamical systems.[25] Moreover, "stochasticity" is not a numerical artifact produced by the inherent errors of computer calculations. To all remaining skeptics I would like to show some pictures obtained by F. Rannou[25] (see Figs. 6-8). She had the clever idea to use an algebraic mapping (analogous to the "standard" mapping discussed by Greene in his talk) and operate only on the <u>integers</u>! Thus instead of points she permuted little squares in the plane, completely avoiding in this way the effect of machine errors. Her results, which vividly portray all the interesting features of stable and unstable regions of dynamical systems, are therefore free of numerical uncertainties. Compare the qualitative simplicity of the pictures with those of 31.

The second and final point I would like to make is that all we know to date about stability concerns systems which are "near"

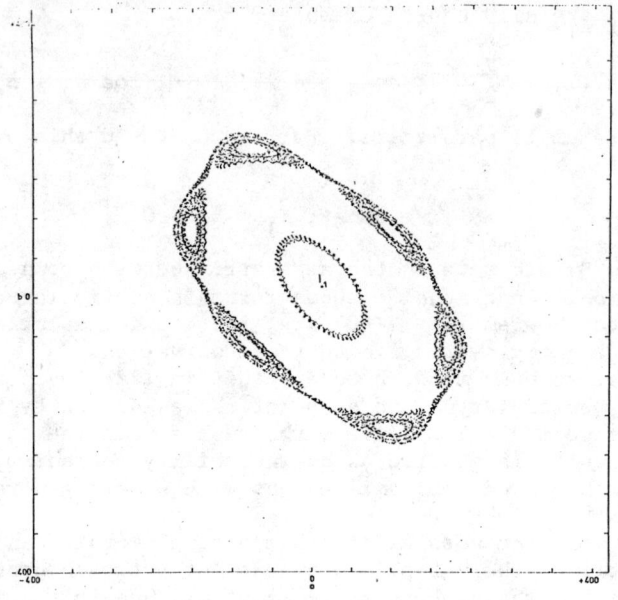

Fig. 6.

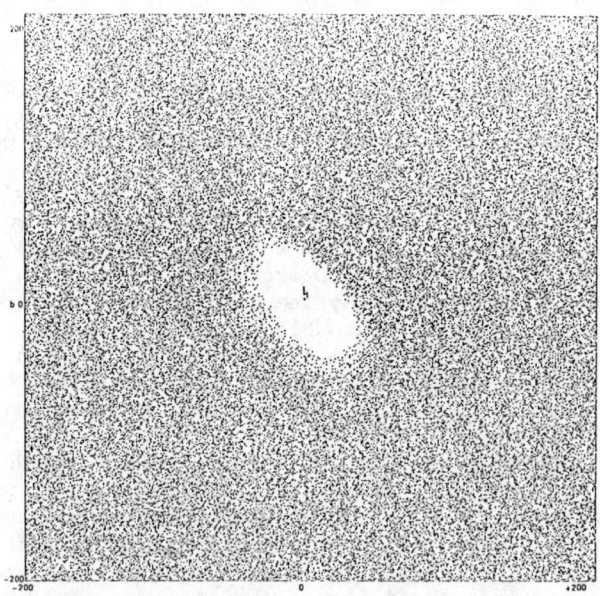

Fig. 7.

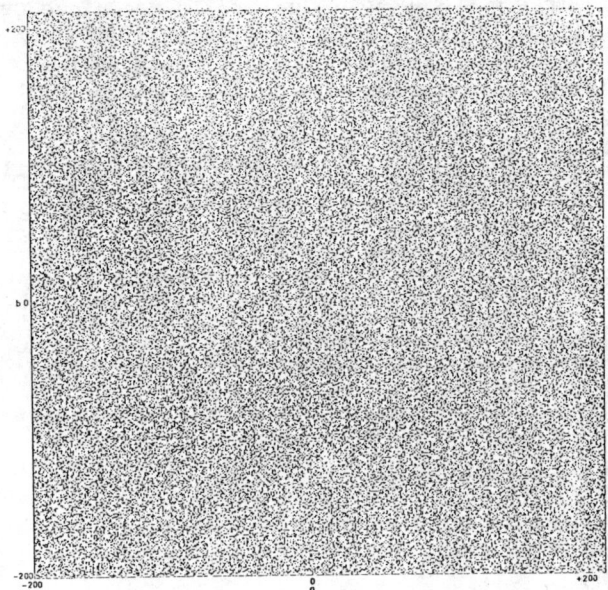

Fig. 8.

integrable ones. These integrable systems are rare and often hard to recognize, yet they are highly desirable: not only do we know a lot about them, but as it turns out they determine to a considerable degree the behavior of the nonintegrable systems close by. To illustrate that an integrable is not easily recognizable, we mention the example

$$H = \frac{1}{2} \sum_{k=1}^{m} (p_k^2 + \omega_k^2 x_k^2) + \frac{1}{2} \sum_{k<j} \frac{\omega_k + \omega_j}{\omega_k - \omega_j} (x_k p_j - x_j p_k)^2$$

which represents an integrable system with integrals

$$I_k = p_k^2 + \omega_k^2 x_k^2 + \omega_k \sum_j {}' \frac{(x_k p_j - x_j p_k)^2}{\omega_k - \omega_j}, \quad 0 < \omega_1 < \omega_2 \ll \omega_k$$

in volution, which is certainly not apparent!

REFERENCES

1. G. D. Birkhoff, "Dynamical Systems," Am. Math. Soc., 1927; revised ed. 1966. Also see: "Nouvelles Recherches sur les systemes dynamiques," Mem. Pont. Acad. Sci. Novi Lyncaei, 1, 85-216 (1935), Chapter IV.

2. H. Poincaré, "Les Methodes Nouvelles de la Mécanique Céleste III," Gauthiers-Villars, Paris (1892); Dover Press (1957) NASA Translation TT F-450, Washington (1976) esp. paragraphs 395-404.
3. J. Moser, "Stable and Random Motions in Dynamical Systems," Princeton University Press, Princeton, J. J. (1973).
4. G. Floquet, Ann. Ec. Norm. Supér (2), 12, 1883.
5. E. Coddington and N. Levinson, "Theory of Ordinary Differential Equations," Gordon & Breach, New York (1965), Chapter 3.
6. V. I. Arnol'd and A. Avez, "Ergodic Problems in Classical Mechanics," Benjamin Inc., New York (1968).
7. M. G. Krein, Pamyati, A. A. Andronova, Izvestia Akad. Nauk (1955) pp. 413-498; Math. Review 17, No. 738.
8. I. M. Gel'fand and V. B. Lidskii, Usp. Math. Nauk. 10, No. 1 (63) (1955) pp. 3-40. [Transl. Amer. Math. Soc. 2(8)(1958) pp. 143-181].
9. V. I. Jacoubovich, Mat. Sbornik. 37, 79(1955) pp. 21-68 [Transl. Amer. Math. Soc. 2(10)(1958) pp. 125-175].
10. V. I. Arnol'd, Russian Math. Surveys 18, 85 (1963); Soviet Math. Surveys 18, 85 (1963); Soviet Math. 2, 247 (1961).
11. M. V. Berry, in "Topics in Nonlinear Dynamics," ed. S. Jorna, Am. Inst. of Phys. Conf. Proc. Series Vol. 46, A.I.P., New York (1978).
12. J. Moser, "Lectures on Hamiltonian Systems," Mem. Am. Math. Soc. 81, 1-60 (1968).
13. C. L. Siegel, "On the Integrals of Canonical Systems," Ann. Math. 42, pp. 806-822 (1941).
14. F. Gustavson, The Astron. Journal, 71, pp. 670-686 (1966).
15. J. Moser, The Astron. Journal, 63, pp. 439-443 (1958).
16. A. N. Kolmogorov, Proceedings of the International Conference of Mathematicians, North Holland, Amsterdam Vol. 1, 315 (1957); reprinted in "Stochastic Behavior in Classical and Quantum Systems," Volta Memorial Conference, Como (1977), Editors: G. Casati, J. Ford Springer Verlag (1979).
17. V. I. Arnol'd, Russ. Math. Surveys, 18, 9 (1963).
18. J. Moser, Nachr. Akad. Wiss. Gottingen, II Math. Phys. Kl. 1 (1962); see also: C. Siegel and J. Moser, "Lectures on Celestial Mechanics," Springer, New York (1971).
19. F. Takens, "A C^1-Counter-example to Moser's Twist Theorem," Indag. Math. 33, 379 (1971).
20. S. Smale in "Differentiable Dynamical Systems," Bull. A.M.S. 73, pp. 747-817 (1967).
21. S. Smale, "Differential and Combinatorial Topology," ed. S. S. Cairns, Princeton Univ. Press. pp. 63-80 (1965).
22. See Chapter III, Section 6 of Reference 3.
23. V. I. Arnol'd, Dokl. Akad. Nauk USSR 156, 9 (1964).
24. N. N. Nekhoroshev, Funct. Anal. Pril. 5, 82 (1971). (Translated in Funct. Analysis).
25. F. Rannou, Astron. and Astrophys. 31, pp. 289-301 (1974).
26. Recent work towards such generalizations is reported in B. V. Chirikov, "Universal Instability of Many-Dimensional Oscillator systems" Physics Reports (Phys. Lett. Sect. C) to appear, 1979.

27. J. Moser, Comm. of Pure and Appl. Math. 11, pp. 81-114 (1958).
28. M. Levi, Stability of linear Hamiltonian Systems with periodic coefficients IBM Research Report RC6610(#28482) Yorktown Heights, N. Y., June 29, 1977.
29. C. C. Conley, Low energy transit orbits in the restricted three body problem, SIAN. J. Appl. Math; 16, 732-746 (1968).
30. R. W. Easton and R. McGehee, Homoclinic Phenomenon of Orbits doubly asymptotic to an invariant three-sphere, Ind. Univ. Math. J. 28, 211-240, 1979.
31. F. M. Izreilov, S. I. Misnev, G. M. Tumavkim, Numerical Studies of Stochasticity limit in colliding beams (one-dimensional model) Preprint of the Institute for Reactor-Physics INP77-43, Novosibirsk, 1977.

EXACT RESULTS FOR SOME LINEAR AND
NONLINEAR BEAM-BEAM EFFECTS

Robert H. G. Helleman
The La Jolla Institute
P. O. Box 1434
La Jolla, California 92038
and
Twente University of Technolody
Department of Theoretical Physics
Enschede, The Netherlands

ABSTRACT

The nonlinear equation of a motion for a particle's displacement from its ideal orbit in a storage ring, when "kicked" by a second (periodically) colliding beam, can be translated into a nonlinear difference equation. Near the center of the beam this equation is virtually linear and we obtain the (Mathieu-like) regions of (in)stability, and all other properties there, analytically and in closed form. Adding in the slow (synchrotron) oscillations along the (now) 'bunched' beam, as a square wave modulation, another analytic expression is found which simply shows that if the peak beam-beam force, during a synchrotron oscillation, exceeds the previous (in)stability border, instability arises in this case. The order periodic solutions (resonances) are obtained for any, odd, nonlinear b-b force, indicating that the planned tune for "ISABELLE" may be too close to a major (period 4) resonance. Approximating the actual (error function) b-b force by a (3-) piece-wise-linear force (itself the exact b-b force for a rectangular beam) a region of nonlinear stability for all time is found (in ISABELLE as "$\Delta\nu$"≈ 0.03) which is about 50% wider than the width of the beam. This sizable region of nonlinear stability is bordered by an invariant curve, constructed with the aid of a technique employed earlier by McMillan and by Laslett.

I. INTRODUCTION

Each particle in a storage ring oscillates transversally about its ideal "circular" path, the so called betatron oscillations[1]. Employing just the first linear terms in a Taylor expansion of the magnetic field about this ideal path one has in the past obtained a great deal of insight and success in calculating these oscillations and building accelerators[1]. This indicates that (bending) magnets can apparently be made without appreciable nonlinearities. A different situation arises however when two such rings intersect and the beams collide periodically. Even though most of the particles just miss each other they do feel a strong nonlinear electromagnetic force, exerted periodically during the short time when the beams cross. Yet the particles are required (/requested) to perform $\approx 10^{11}$ ring revolutions and participate in $\approx 6 \times 10^{11}$ "near collisions" without exhibiting large amplitude (betatron) oscillations (exceeding the aperture of the vacuum chamber) and be lost. Note that this number of revolutions is roughly the same as the number of revolutions the earth and moon have ever made about the sun. Since the

long term stability of that classic nonlinear (three body) problem has not yet been decided after 300 years of study[2-5] the stability of colliding beams might conceivably be a difficult problem as well. Elsewhere we have reported on our ongoing variational studies of this nonlinear stability problem[6,7]. Here I wish to report some exact stability results, for a few linear as well as nonlinear beam-beam effects. My conclusions, from them, are:
1. That it is very useful to start with the first order, linear, part of the beam-beam effect since:
   a) The beam-beam force is virtually linear over a considerable region about the origin of the phase plane, where one hopes to focus the beam.
   b) Stability results can be derived analytically and in closed form, thus providing a global overview of the major effects.
   c) It seems to predict conclusion no. 3, below, cf. Fig. 2.
2. That "synchrotron oscillations" with a period much, much longer than the period of revolution do not materially change the above first order effects, i.e. solutions near the origin become unstable when the peak beam-beam effect, over a few revolutions, exceeds the stability border obtained without the synchrotron oscillations;
3. That the design value of the tune Q, in (1.1), for the planned ISABELLE rings (at $Q \approx 22.6$), is too close to a period 4 resonance of the nonlinear equations.

The increase in the width of the beam, observed in numerical calculations for ISABELLE[12], appears to be mostly due to patametric amplification near this period 4 resonance (i.e. due to the large islands of period 4, of Figs. 4-11). That for a (3-) piece-wise-linear approximation of the (nonlinear) error-function beam-beam force[9] I can rigorously establish a region of stability for all time, at one value of the beam-beam strength, i.e. of the fore-factor of the force, (at $\Delta\nu \approx 0.03$, in ISABELLE); a region which is about 50% wider than the beam width (i.e. it does require all three peices of the present nonlinear force). No conclusions are drawn for any other b.b. strengths since the geometric technique used, cf. McMillan and Laslett[18,19], is applicable to one case only. Note that this (3-) piece-wide-linear (nonlinear) beam-beam force is the exact force for a beam whose particle density is rectangular, across the beam.

Since the particle displacement y, outside the collision regions, was successfully described[1,6] by a simple harmonic oscillator,

$$d^2y/d\theta^2 + Q^2y = 0, \tag{1.1}$$

where θ is the azimuthal angle (proportional to the time) about the center of the ring and Q is called the "tune" of the machine, we model the beam-beam effect by a periodically and nonlinearly "kicked" oscillator,

$$d^2y/d\theta^2 + Q^2y = P(\theta)F(y), \tag{1.2}$$

where $P(\theta)$ is a periodic "pulse" (of unit area) which is $\neq 0$ over the collision region (in θ) and ≈ 0 otherwise. The function $F(y)$ is the nonlinear function describing the electromagnetic force on a test particle at displacement y (from its ideal orbit).

Equation (1.2) is a nonlinear Hill's, or Mathieu-, Equation whose linear version is difficult enough to analyse and is still the subject of much research in applied mathematics[8]. Here I only consider the transverse oscillations in one y- direction, perpendicular to the plan of revolution. For the nearly flat, ribbon-like, beams of the proposed ISABELLE storage rings[9,10] this already is a good approximation. Actually, one should consider the influence of all the test particles, forming a beam themselves..., back on the particles in the first beam, whose integrated force caused the F(y) in (1.2). Not having adequately understood even the (uncoupled) effect modeled by (1.2) one clearly is not yet in a position to investigate the mutual coupling between the individual orbits of all particles in the two beams.

Since the two beams collide over a very short θ interval only, we model the periodic pulse $P(\theta)$ by a periodic δ - function and "normalize" F(y) by extracting a factor B (the "beam-beam length"):

$$d^2y/d\theta^2 = - Q^2 y + B\left[\sum_t \delta(\theta - t2\pi)F(y)\right], \qquad (1.3)$$

where t is the integer counting the number of passages through a collision region. B is proportional to the current in the strong beam[9].

This differential equation can be translated exactly into a difference equation (i.e. an area preserving mapping[3-5]):
Between pulses the equation is linear and can be solved exactly. During a pulse the momentum p ($\equiv dy/d\theta$) is changed by an amount $B \cdot F[y(t2\pi)]$ (integrate (1.3) once, over the width of the pulse). Whence:

$$y_{t+1} = y_t C + (P_t/Q)S \qquad (1.4)$$

$$P_{t+1} = - Q y_t S + P_t C + B F(y_{t+1}), \qquad (1.5)$$

with

$$C \equiv \cos(2\pi Q) \quad \text{and} \quad S \equiv \sin(2\pi Q), \qquad (1.6)$$

These equations were also used by others, for some numerical investigations[11,12]. The P_{t+1} is defined as the momentum p $[(t+1)2\pi+]$, just after the (t+1)st. pulse is completed[13]. Writing p_t explicitly from (1.4),

$$P_t = (y_{t+1} - y_t C)Q/S, \qquad (1.7)$$

we can eliminate the p's from (1.5) to obtain a single (second order) difference equation, which will be the basic equation for this paper,

$$y_{t+1} + y_{t-1} = 2C y_t + (BS/Q)F(y_t), \qquad (1.8)$$

(letting $t+1 \to t$), still equivalent with (1.3)-(1.5)[11,12,14]. Note that a "phase plane" plot of p_t versus y_t [i.e. the "surface of section" for (1.3)] is just a linear transformation of the new "phase plane": y_{t+1} versus y_t, according to (1.7).

Since the second (strong) beam, whose influence the test particle feels, usually has a Gaussian particle density (in y) one integrates out the total Lorentz force on the test particle to find that F(y) is the "Error Function" (times $\frac{1}{2}\sqrt{\pi}$) of y)[9], considered (among others) here in section 4.1.

The error function has a sizable (virtually) linear section near its center. Hence linear, or first order, beam-beams effects play an important roll near the origin of the y_t, y_{t+1} phase plane. This is also the first point where one would like to focus the beam (to increase the collision rates). If the origin were to become linearly unstable, i.e. hyperbolic (instead of being elliptic), one would have a very hard time focussing the beam even if the displacements would be stabilized elsewhere, in some nonlinear sense, cf. section 4.2. The orbits would drift out to large distances (since major islands[3-5,14-17] or resonances do then come close to the origin) and back, but spend only a short time near the origin, resulting in greatly reduced collision rates with the other beam. Hence it is of importance to study this linear effect first, i.e. study the equation

$$y_{t+1} + y_{t-1} = (2C + BS/Q) y_t. \qquad (1.9)$$

In section 2 we solve this equation exactly and find its regions, in the B versus Q plane, of bounded and unbounded solutions similar to the corresponding regions for a Mathieu Equation[7,8]. However here we obtain their boundaries analytically and in close form (2.4), cf. figures 1 and 2. Already from these results, it is apparent that the design-tune Q for the planned ISABELLE rings is very close to a periodic solution of (short) period 4. The maximum value of the beam-beam strength allowed by (1.9), $B_{\ell b}$, for ISABELLE at its planned tune, is shown to satisfy ("$\Delta\nu$"$_{\ell b}$ ≡) $B_{\ell b}/4\pi Q \approx 0.17676$, cf. (2.5). This is the point at which (even) the origin of the phase plane turns hyperbolic, as discussed above. However long before we reach this strength the "parametric amplification" of the displacements will exceed the available aperture of the vacuum chamber, in the linear equation (1.9) [as well as the nonlinear (1.8), cf. (2.9)].

In section 3 we consider the slow periodic oscillations of the factor B in (1.9), caused by the so called synchrotron oscillations"[9,10,1] which are particle oscillations along the direction of the beam. They are unimportant when one operates with a continuous beam but become noticeable when the beam is cut into discrete "bunches" as is often done for other reasons[1,9,10]. We model this effect in (3.1), with a square wave modulation of the B in (1,9). The stability and instability regions for such an ("alternating gradient") equation can be found analytically (in the same way as for the Kronig-Penny model of solid-state physics or the Courant-Snyder analysis of the A.G.S.). The result is a complicated formula (3.7), (3.4) which becomes easier to analyse when we know that, in ISABELLE, the period of the synchrotron oscillations is much larger than one revolution period ($\approx 10^6$x). In that case we find the total solution to be unstable when the peak of the square wave (modulation) exceeds the boundaries found before in (2.4), cf. figures 1,2, during a few revolutions, as one might expect intuitively, in that case.

In section 4.1 we find the locations of the four lowest and most important resonances as a function of all the parameters for any nonlinear odd beam-beam force F(y), in (1.3)-(1.8), i.e. we find the nonlinear periodic solutions of period 1, 2, 3 and 4, cf. Ref. 11 [in the linear case (1.9) the periodic solutions of period 1 and 2 are (on) the borders, between the regions of bounded and unbounded solutions, in the B versus Q plane, cf. figures 2 and 1]. We also display in figures 3-11, a number of phase plots for the nonlinear $F(y) = (\frac{1}{2}\sqrt{\pi})$ erf(y) obtained numerically at values of B (or $\Delta\nu$) ranging from 0.1 $B_{\ell b}$ to 0.9 $B_{\ell b}$. From these figures as well as the analytic formulae for the period 4 solutions

it appears that the planned tune, Q = 3.7666..., for ISABELLE is too
close to the period 4 solution (which bifurcates off the origin at
Q = 3.75 and B = 0). The numbers obtainable from our period 4 formulae and the size of the B islands in figures 7-11, "explain" the
numerical results reported in Ref. 12.

In section 4.2 we discuss how a region of nonlinear stability can
be obtained in those few cases when one is able to find and construct an
invariant curve, i.e. a continuous curve containing all orbits $\{y_t, y_{t+1}\}$
whose starting point may be any point on the curve. Simplifying the error
function to a (3-) piece-wise-linear function $\bar{F}(4.8)$ [i.e. $\bar{F}(y)$ rises linea
near the origin and becomes a constant c above some y (and -c below -y)]
we do construct, in section 4.3, such an invariant curve for this nonlinear force $\bar{F}$, with the aid of a geometric technique employed earlier
by McMillan and by Laslett[18,19]. This force $\bar{F}$ would even be the exact
one if the particle density across the beam were rectangular, rather
than (approximately) Gaussian. We find this invariant curve at $(\Delta \nu \equiv)$
$B/4\pi Q \approx 0.03$ (for ISABELLE's tune). It does not at all imply that other
beams (or the same beam at other values of B or $\Delta\nu$) are unstable but merely indicates the limited range of applicability of the simple geometric
techniques used. Yet, in general, it is difficult enough to construct
even one such stable region for a nonlinear (nonintegrable[20,2-4,5]) mapping.

The region of nonlinear stability enclosed by this invariant curve
is a slightly hexagonal "square", which is about 50% wider than the
width of the beam. This further explains the results of Ref. 12 and
the rigorously establishes the nonlinear stability for all time of some
rectangular beams. The virtues of being rigorous have been adequately
sung by the mathematician A. N. Komogorov:

"It is better to be right than to be rigorous"

A. N. Kolmogorov,
quoted by B. V. Chirikov.

2. LINEAR BEAM-BEAM EFFECTS

In this section we obtain analytically and in closed form the solutions, and stability regions, for the first order, linear, beam beam
effect, described in Refs. 11, 25. We rewrite (1.9) as

$$y_{t+1} + t_{y-1} = 2 \cos(2\pi\sigma) y_t, \text{ with} \quad (2.1)$$

$$\cos(2\pi\sigma) \equiv C + BS/2Q, \quad (2.2)$$

where σ is called the "winding number", (and $\omega \equiv 2\pi\sigma$, the revolution number) and is allowed to be imaginary if necessary. Note that in the ab-
sence of a beam-beam interaction the winding number is equal to Q. The
winding number σ actually is the "driven", or perturbed "tune". Since
(2.1) is still linear, in y, all solutions have the same winding number
σ. In the nonlinear case (1.8) the different solutions have different
winding numbers, in general. We obtain the general solution of (2.1) as

$$y_t = ae^{it2\pi\sigma} + a^* e^{-it2\pi\sigma} = \{y_1 \sin(t2\pi\sigma) - y_0 \sin[(t-1)2\pi\sigma]\}/\sin(2\pi\sigma)$$
$$(2.3)$$

[$\sigma = m/2$ may yield "secular" solutions, $\propto$ t, as well[8,7]].

The above solution (2.3) is bounded for all t when σ is real, i.e. when

$$-1 < C + BS/2Q < +1, \qquad (2.4)$$

cf. (2.2). Thus we obtain from (2.4) the limits on the beam-beam strength caused by the first order term of F(y) in (1.8),

$$-\cot(\pi Q) < B/2Q < \tan(\pi Q), \quad \text{for } Q \in (m, m+\tfrac{1}{2}) \qquad (2.5.a)$$

$$\tan(\pi Q) < B/2Q < -\cot(\pi Q), \quad \text{for } Q \in (m-\tfrac{1}{2}, m), \qquad (2.5.b)$$

resulting in the same boundary curves in each half-integer Q interval[11], see Fig. 1.

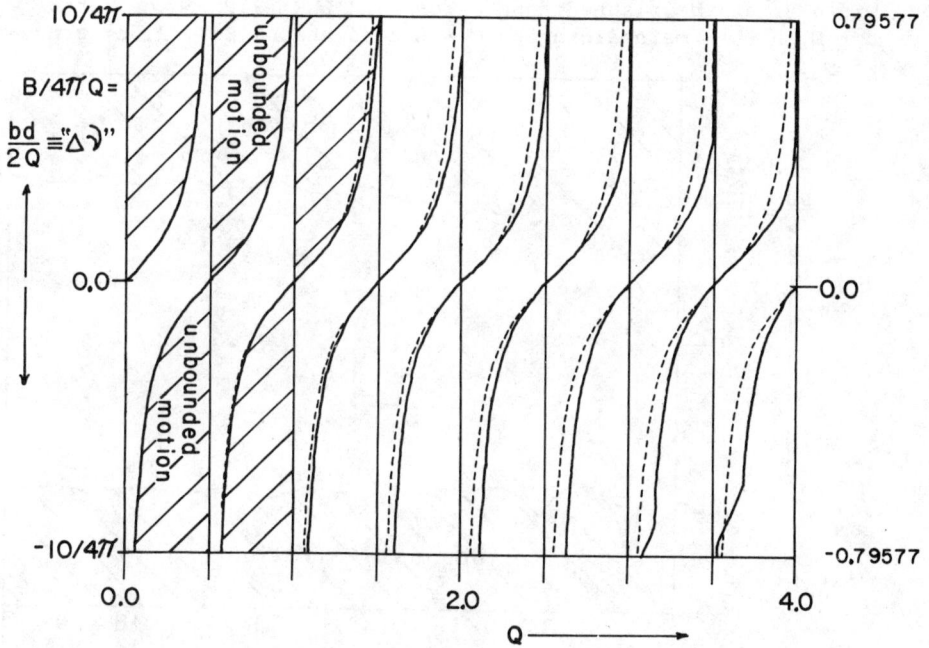

(at d=0.05; dotted: d⟶0)

Fig. 1a. Regions of bounded and unbounded solutions of (2.1) (dotted) in the B versus Q plane, for the first order, linear, beam-beam effect, as given by (2.5). The fully drawn curves separate the similar regions for a pulse of finite width (0.05×2π, here) obtained in Eq. (1.2), cf. (2.5.c).

It should be stressed that with the simple solution (2.3) of the linear equation (2.1), (2.2) we have in fact obtained the general solution of the corresponding linear Hill's (differential) equation (1.3) with a periodic δ - coefficient. The stability regions of this linear Hill's equation (1.2) for pulses of finite width, d2π (and strength b, such that $B = bd2\pi$.) are slightly different. They can be obtained directly from a linear differential equation (1.2) by the same method (of splitting a linear equation (1.2) into 2 equations valid for different time intervals) as used here in section 3.

Hence, we do not duplicate the derivation but merely state that those "finite pulse" solutions of a linear (1.2) are bounded for all time when

$$-1 < \cos[(1-d)2\pi Q] \cos\left(d2\pi \sqrt{Q^2-b}\right) + \frac{\frac{1}{2}\,b-Q^2}{Q\sqrt{Q^2-b}} \sin[(1-d)2\pi Q] \sin\left(d2\pi \sqrt{Q^2-b}\right) < +1 \quad (2.5.c)$$

cf. Ref. 8, Eq. (8.2) [note the ½ is left out there]. The fully drawn curves in Fig. 1a are the (+1 and -1) boundaries defined by (2.5c.). In Fig. 1b the (2.5.c) boundaries are graphed as a function of the pulse width d, at the planned tune Q for ISABELLE (9). Note that as we let $d \to 0$, $b \to \infty$ (maintaining $bd2\pi = B$ a constant) Eq. (2.5c) reduces to (2.4).

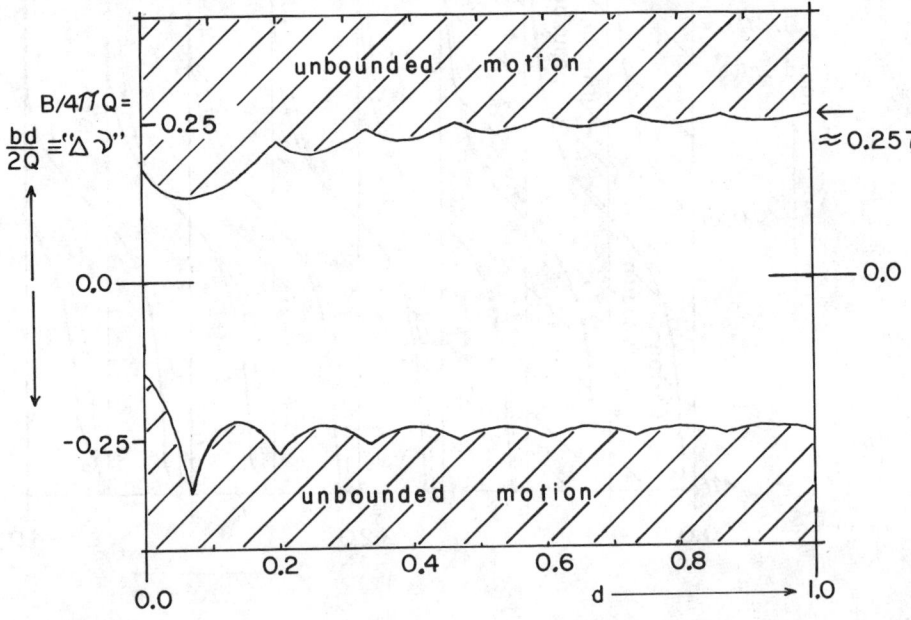

(at Q=22.6/6)

Fig. 1.b. Regions of bounded and unbounded solutions of a linear beam-beam differential equation (1-2), at fixed tune Q. The b.b strength B is plotted versus the pulse width $d(2\pi)$.

At the proposed value of Q for ISABELLE, i.e. at Q = 22.6/6, determined by many additional considerations[9,10,12], this "linear boundary" (2.56) becomes

$$0 < \text{"}\Delta\nu\text{"}_{\ell b} \equiv \frac{B_{\ell b}}{4\pi Q} < -\frac{\cot(\pi Q)}{2\pi} \approx 0.17676.. \quad (2.6)$$

The highest value of $\Delta\nu \approx 0.0064$, stays well below this limit (2.6). From (2.6) and (2.5) we see that it would be advantageous to choose the highest possible (half-integer) Q interval as well as the highest possible Q value within this interval, cf. Fig. 1, as high as in compatible with other design criteria. Note that at high Q values the finite pulse width becomes increasingly important; compare the fully drawn and dotted curves in Fig. 1.

From (2.2) we immediately find all solutions with the same winding number σ, in the B versus Q plane,

$$\frac{B}{2Q} = \frac{\cos(2\pi\sigma) - \cos(2\pi Q)}{\sin(2\pi Q)} = \frac{2\sin[\pi(Q+\sigma)]\sin[\pi(Q-\sigma)]}{\sin(\pi Q)}, \quad (2.7)$$

graphed in Fig. 2 (thus providing details of a part of Fig. 1).

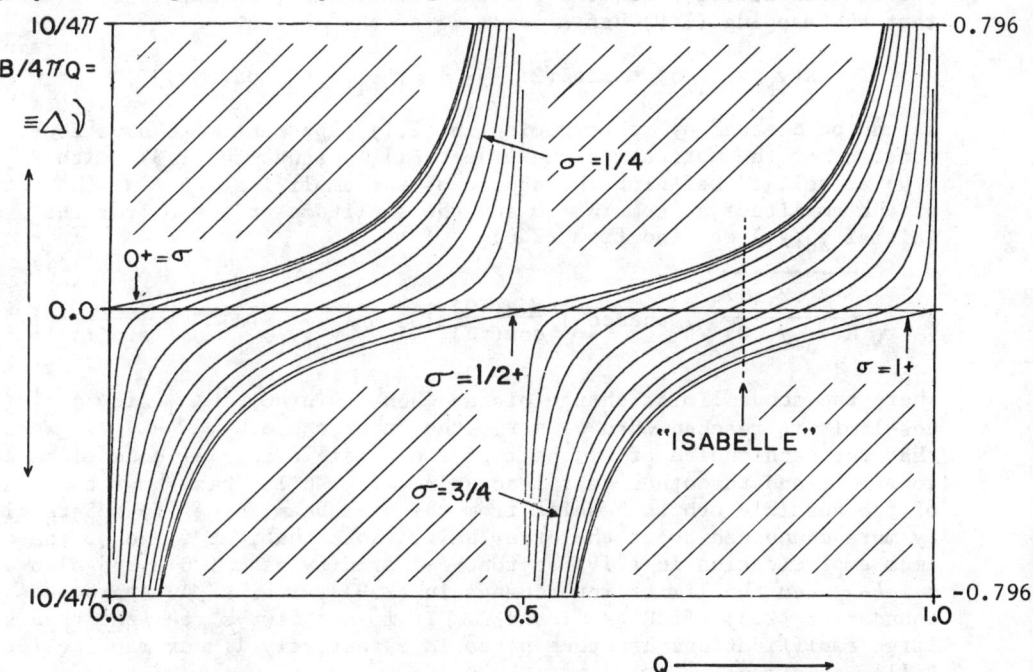

Fig. 2. B versus Q curves, at constant winding number σ (at σ intervals of 0.05), as given by (2.7). Note how close the operating point for ISABELLE (transposed from 2.7666.. to 0.7666..) is to a '4-th order resonance', at $\sigma = 3/4$.

Figure 2 is important when surveying the major resonances ($\sigma \equiv m_1/m_2$ with small integers m_1 and m_2) one faces when choosing any particular tune Q. It is apparent from Fig. 2 that the presently planned tune for ISABELLE leaves no room for an increase in $\Delta\nu$ by an order of magnitude (e.g. by increasing each beam current 3-fold) before one hits the 4th order resonance ($\sigma = 3/4$) at $\Delta\nu \approx 0.0167..$ (cf. Fig. 4 in section 4). In the nonlinear cases (1.2), (1.8) similar results can be obtained locally, using our variational techniques, described elsewhere[21-24,16,7,14]. For

For each different set (or class) of initial conditions one obtains a more or less deformed version of Fig. 2,[7,16,21-23], which reduces to Fig. 2 itself as we let the initial conditions come closer and closer to the central region, where F(y) is nearly linear. The main novelty of those local variational methods is that they converge (quadratically) at all,[24,21,7] a highly unusual feature when applied to a nonlinear (non-integrable[20,13]) system,[4,5]. Essential use is made of the fact that, in general, the winding number σ is different for any two solutions.[24]

The δ-"kicks" in (1.3) are equally likely to decrease, as well as increase, the instantaneious displacement y (θ). Hence, higher amplitudes (as well as lower ones) can occur with them, than without them: "parametric amplification." Even if the orbit remains bounded this amplification may lead to considerable loss of particles, when the amplitude exceeds the aperture of the vacuum chamber. In order to calculate the amplification factor A for the present linear case, we make use of the fact that the mapping (1.9) has a constant of the motion,

$$K(y_t, y_{t+1}, \sigma) \equiv \sin^2(2\pi\sigma) y_t^2 + [y_{t+1} - y_t \cos(2\pi\sigma)]^2. \qquad (2.8)$$

as can be checked by substituting in (2.1) [apparently we have found a constant of the motion for the linear Hill's equation (1.3) (with F(y) = y) as well]. Defining the square of the amplification A as the ratio of the amplitude at some B $\neq$ 0 and the amplitude at B = 0 (for the same initial y_0, y_1) we find from (2.8)

$$A \equiv \sqrt{\frac{K(y_0, y_1, \sigma)}{K(y_0, y_1, Q)}} \quad , \text{ whence } \left| \frac{\cos(\pi Q)}{\cos(\pi\sigma)} \right| \leq \underset{y_0, y_1}{\text{MAX}} (A) \leq \left| \frac{\sin(\pi Q)}{\sin(\pi\sigma)} \right| , \qquad (2.9)$$

where the above limits change places when $|\sin(\pi Q)| < |\sin(\pi\sigma)|$ (the cos-limit is reached when $y_1 = y_0$, the other one at $y_1 = -y_0$). We see that for each choice of Q,σ half of the possible initial conditions lead to A > 1, and the other half lead to A < 1. While this shows that half of the possible orbits benefit from the beam beam force (7) we are clearly more concerned about the other half. Note that, at fixed Q, the maximum amplification in (2.9) approaches infinity as $\sigma \downarrow 0+$, and also as $\sigma \downarrow \frac{1}{2}-$ [when the limits interchange in (2.9)], i.e. as $\Delta\nu$ approaches the boundaries (2.5) cf. Figs. 1 and 2. It is not always realized that such large amplifications can take place in a perfectly linear mapping (or Hill's equation).

3. SYNCHROTRON OSCILLATIONS

In the case of a "bunched" beam [9-11,1] the slow particle oscillations along the direction of the beam, known as synchrotron oscillations, create a slow modulation of the beam-beam strength B, in (1.3)-(1.9), as well. This modulation may be sinusoidal, parabolic or otherwise (in time) depending on the charge distribution over the length of a bunch. Here we simplify it to a square wave. The main result derived for the linear case (1.9) is an intuitively obvious one: if the maximum value of B exceeds the linear boundaries (2.5) for anumber of revolutions the motion is unstable here, as well. If the synchrotron period were of the order of magnitude of the revolution period the result (3.7), (3.4) would be more complicated.

Hence our previous model (1.9) changes into

$$y_{t+1} + y_{t-1} = 2 C y_t + BS/Q[1 - \beta h (2\omega_s t)]y_t \qquad (3.1)$$

where $h(\varphi)$ is a square wave which switches back and forth between +1 and -1, every π. Equation (3.1) is the same as

$$\begin{cases} y_{t+1} + y_{t-1} = 2 \cos\omega_+ y_t, & \text{for } 0 < t \leq \mu \\ y_{t+1} + y_{t-1} = 2 \cos\omega_- y_t, & \text{for } \mu + 1 \leq t \leq 2\mu \end{cases}, \text{ etc. modulo } 2\mu \qquad \begin{array}{c}(3.2)\\(3.3)\end{array}$$

$$\text{with } \mu \equiv \pi/2\omega_s \text{ and } \cos\omega_\pm \equiv C + BS/2Q(1 \pm \beta), \qquad (3.4)$$

cf. (2.2) and below (imaginary ω are allowed). The exact solutions of (3.2) and (3.3) are two simple sines, cf. (2.3). Thus after one period of the synchrotron oscillations, i.e. after 2μ revolutions, the displacements are

$$\begin{pmatrix} y_{2\mu+1} \\ y_{2\mu} \end{pmatrix} = M_- M_+ \begin{pmatrix} y_1 \\ y_0 \end{pmatrix} \text{ with } M_\pm \equiv \begin{bmatrix} \sin\omega_\pm(\mu+1) & -\sin\omega_\pm \mu \\ \sin\omega_\pm \mu & -\sin\omega_\pm(\mu-1) \end{bmatrix} \qquad (3.5)$$

The propagator over 2μ of the exact solution of (3.1) is the 2 x 2 matrix $M \equiv M_- M_+$ with det $M_\pm = 1$ (an area preserving mapping). Thus the properties of its eigenvalues $\lambda_{1,2}$ are solely determined by tr(M), as usual[14,15]. Hence the solutions of (3.1) remain bounded for all time if:

$$|\lambda_1 + \lambda_2| = |tr(M)| < 2 \qquad (3.6)$$

which, after writing out (3-5), becomes

$$\left| \frac{\cos\mu(\omega_+ + \omega_-)[1-\cos(\omega_+ + \omega_-)] - \cos\mu(\omega_+ - \omega_-)[1-\cos(\omega_+ - \omega_-)]}{\cos(\omega_+ + \omega_-) - \cos(\omega_+ - \omega_-)} \right| < 1. \qquad (3.7)$$

For the planned ISABELLE rings, $\omega_s \approx 2.5 \; 10^{-6}$: therefore $\mu \approx 6 \; 10^5..$, in units in which the revolution period[10] is 1. Therefore, as we vary B [and with it $\Delta\nu$ and $\omega_\pm$ (3.4)] ever so slightly, if only in its 4th digit..., the μ-dependent cosines in (3.7) oscillate many time between -1 and +1 (when $\omega_\pm$ remain real) while the other terms change only slightly [such small changes may already accur due to charge density fluctuations in the strong beam]. So we replace (3.7) by its maximum under these small variations, i.e. the solutions of (3.1) remain bounded if,

$$\left| \frac{2-\cos(\omega_+ + \omega_-) - \cos(\omega_+ - \omega_-)}{\cos(\omega_+ + \omega_-) - \cos(\omega_+ - \omega_-)} \right| < 1, \qquad (3.8)$$

where I ignored the small "ripple" on B, induced by these exceedingly slow synchrotron oscillations and considered only the amplitude of the ripple. Note that (3.8) is always satisfied if the $\omega_\pm$ are real. Hence the solutions of (31.) become unbounded, according to (3.4), only if $B(1 \pm \beta)$ does not stay within the interval

$$-\cot(\pi Q) < B(1 \pm \beta)/2Q < \tan(\pi Q), \quad \text{for } Q \in (m, m+\tfrac{1}{2}), \qquad (3.9)$$

plus the analogue of (2.5.b) for the remaining intervals. Compared with (2.5.a) this result (3.9) is intuitively obvious. However, when ω_s is much higher [e.g. in electron rings[11] inequality (3.7) is less simple to analyze and could remain satisfied in some cases even if (only) one of the $\omega_\pm$ becomes complex. In that case the region of bounded solutions would be larger than (3.9) at some particular Q values.

4. NONLINEAR BEAM-BEAM EFFECTS

By most available indications the nonlinear equations (1.3), (1.5), (1.8) appear to be nonintegrable ($\approx$ "nonseparable")[20,3,18,19] cf. Figs. 7-11. There are no methods available which can provide a general solution, valid globally (or even "regionally"), of such nonintegrable equations[5,4]; one of the classic problems in nonlinear mechanics[3], yet usually not mentioned in the standard graduate courses in Classical Mechanics. Only the classic "Three Body Problem", the original example of a nonintegrable system[5,4,3], has acquired enough notoriety to be mentioned unfortunately most often as somewhat of a couriosity whereas we know that most Hamiltonian systems are nonintegrable (Poincaré-Siegel)[4,3,5,20]. This, and other theorems, also indicate that no global ("regional") solutions exist in terms of elementary analytic functions. Hence, it is not too surprising that no (convergent) methods have been found to construct them. Note that our variational solutions of such nonintegrable systems[24,21,22,7,14,23,16] are extremely local: each new solution is valid for one set of initial conditions only (and may change discontinuously when re-done for nearby initial conditions), and are currently restricted to the class of (long-period) recurrent solutions. Applications to the present equations (1.3), (1.5), (1.8)[7] will be discussed elsewhere.

Hence, our results in this nonlinear section must, of necessity, be "local" in one sense or another: In section 4.1 we find, by inspection, the individual periodic orbits of periods 1,2,3 and 4, for any odd nonlinear beam beam force, i.e. the main nonlinear resonances of the system. To illustrate the effect of the major resonance of period 4, at ISABELLE tune, pictures of the phase plane were generated numerically; displayed in Figs. 3-11. The regions of "stochastic" or "random" behavior which arise in nonintegrable systems seem to appear near the hyperbolic (unstable) points of period 4[20,3,14-19], cf. Figs. 7-10.

In section 4.2, I do establish global nonlinear stability for all time of a sizable region of the phase plane, but only for one particular value of B ($\Delta\nu$) and only for one particular approximation of the error function, needed in the beam-beam force.

4.1 Lower Order Resonances For Any Odd Force

In the linear case the border between the regions of bounded and unbounded solutions is formed by the solutions of period 1 or 2 [and some secular solutions[8], cf. Figs. 2, 1 and (2.5), (2.7)], here obtained for

247

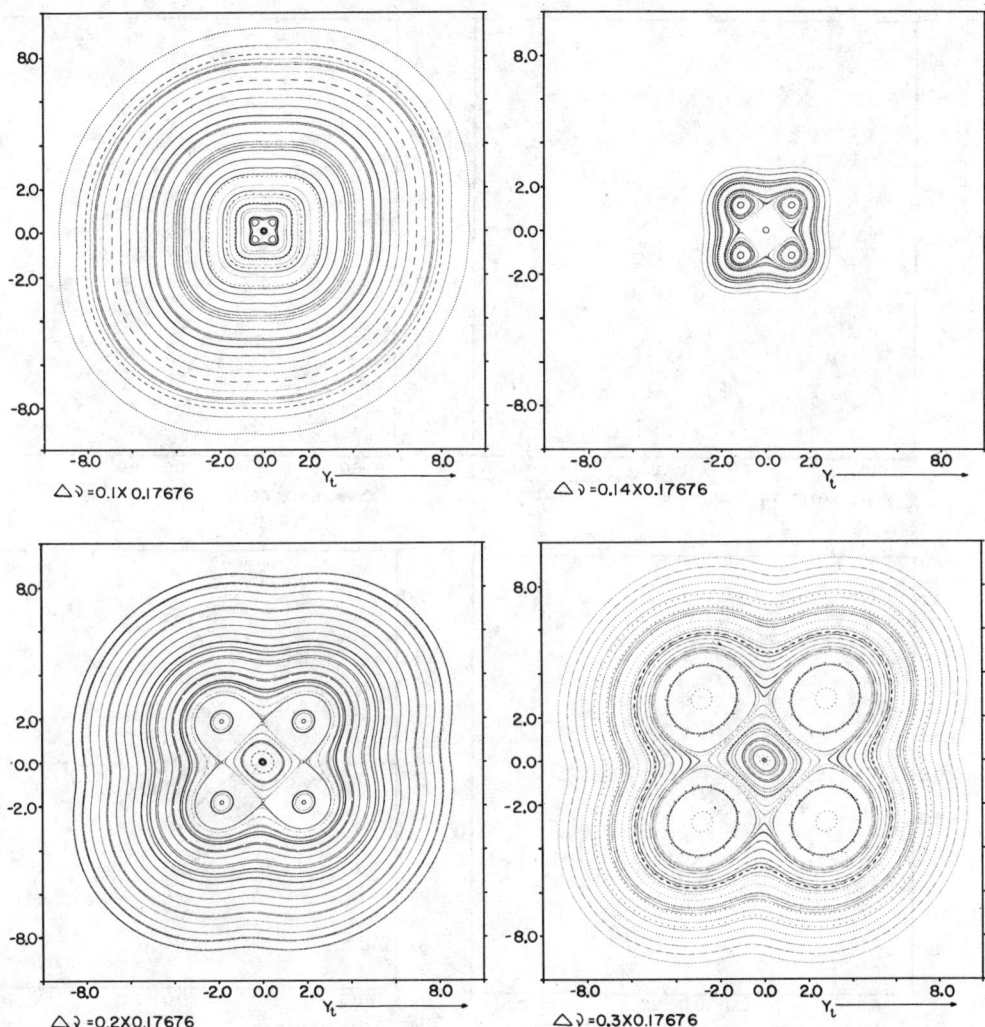

Figs. 3-6. y_t, y_{t+1} Phase Plane, orbits of the mapping (1.8), with $F(y) = (\frac{1}{2}\sqrt{\pi})$ erf(y), each Fig. with orbits for $\approx$ 40 different initial conditions, as $\Delta\nu = 0.1$ so 0.3 of $\Delta\nu_{\ell b}$, cf. Figs.

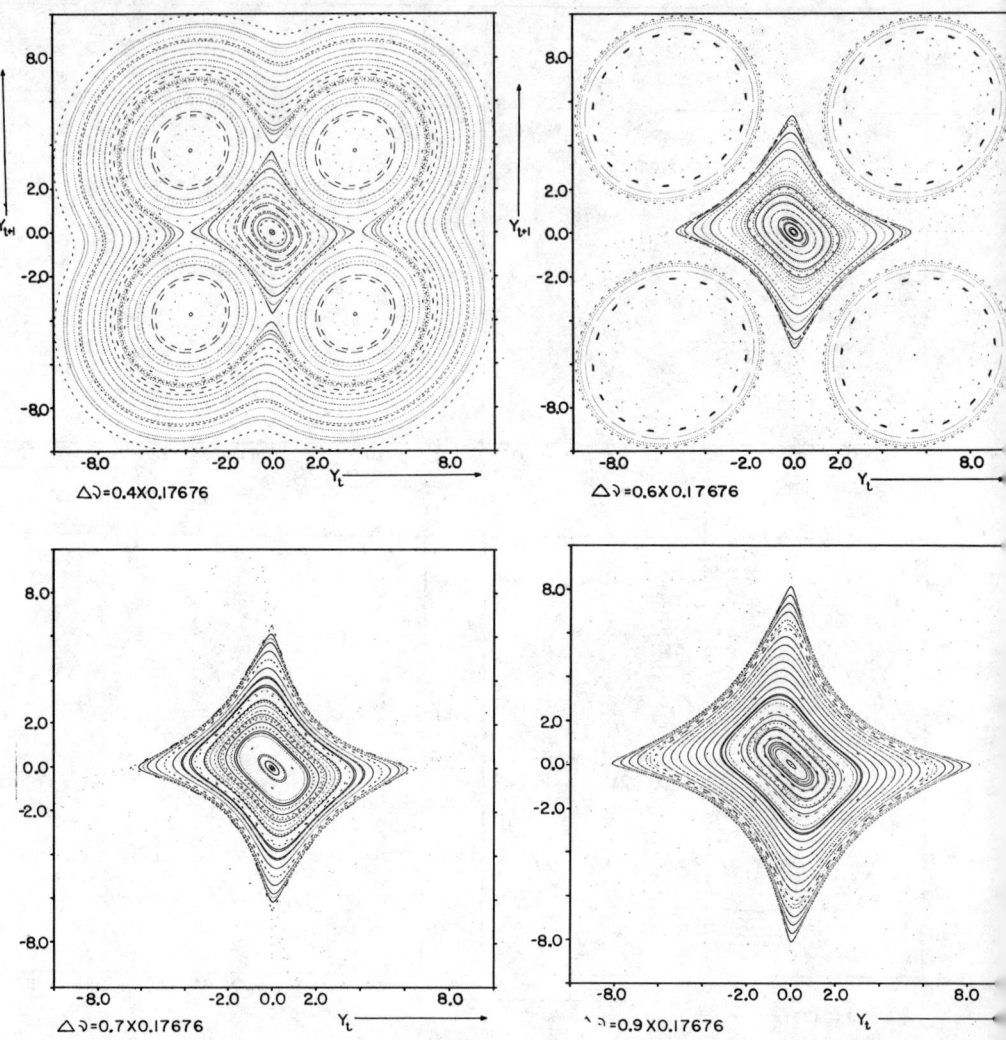

Figs. 7-10. y_t, y_{t+1} Phase Plane, orbits of the mapping (1-8), with $F(y) = (\frac{1}{2}\sqrt{\pi})$ erf(y), each Fig. with orbits for $\approx$ 40 different initial conditions, as $\Delta\nu = 0.4$ so 0.9 of $\Delta\nu_{\ell b}$, cf. Figs.

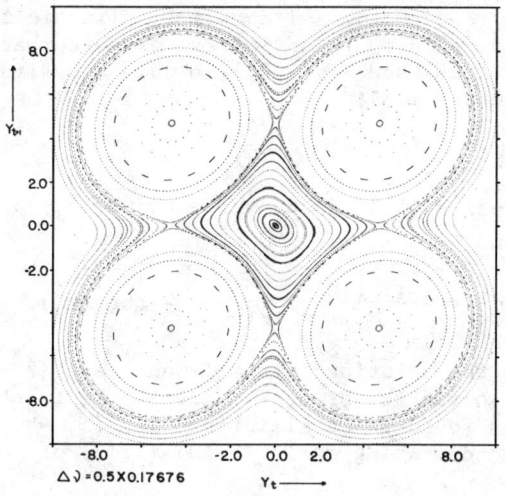

Fig. 11. y_t, y_{t+1} phase plane, orbits of the mapping (1.8), with $F(y) = (\frac{1}{2}\sqrt{\pi})$ erf(y), for ≈ 40 different initial conditions. See text.

any odd nonlinear force. The solutions of period 3, obtained in (4.3), only appear for high values of $\Delta\nu$ (i.e.B) at ISABELLE's tune, as we would suspect from Fig. 2, cf. Fig. 10. The solutions of period 4, obtained in (4.6), and the associated islands seem to set a practical limit ($\Delta\nu \lesssim 0.02$) to the beam beam strength at ISABELLE's present tune Q. Note that each result obtained is the corresponding linear result (2.7) multiplied by a simple factor, $y/F(y)$.

Period 1

Let the solution of the nonlinear mapping (1.8) be: y_t: = a,a,a,... etc., then, in addition to the **trivial** solution a = 0, we find from (1.8):

$$2\pi\Delta\nu \equiv \frac{B}{2Q} = \frac{(1-C)}{S} \frac{a}{F(a)} = \tan(\pi Q) \cdot \frac{a}{F(a)}. \qquad (4.1)$$

Note that tan (πQ) is the solution in the linear case, cf. (2.7), (2.5), when $F(a) = a$. Given some (e.g. error) function $F(y)$, Eq. (4.1) is easily solved for a (and -a). The stability types of these solutions have been calculated using our "determinant method"[16,17,15,14]. The above (two) period-1 solutions are found to be elliptic [i.e. linearly stable under infinitesimal perturbations[14,15]] and are therefore the centers of two islands in the y_t, y_{t+1} phase plane. These stability-type calculations can be easily reproduced using formulae (2.10)-(2.12) of Ref. 14, assuming $a/F(a) \geq 1$, as in any sensible approximation of the error function. In that case the period 1 solutions "split off" the origin of the phase plane only when $|B|2Q \geq \tan(\pi Q)$, cf. (4.1). It is also at this value that the origin, another solution of period 1 (but a desirable resonance), switches from elliptic to hyperbolic (i.e. linearly unstable under infinitesimal perturbations[14,15]), as we saw in (2.5), (2.7) and Figs. 2,1, a not so desirable event, discussed earlier above (1.8).

Period 2

Let the solution of (1.8) be y_t: = b, -b, b, -b,..., etc, then from (1.8),

$$2\pi\Delta\nu \equiv \frac{B}{2Q} = -\frac{(1+C)}{S} \cdot \frac{b}{F(b)} = -\cot(\pi Q) \cdot \frac{b}{F(b)}. \quad (4.2)$$

where $\cot(\pi Q)$ is the solution in the linear case (2.7), (2.5). Equation (4.2) is easily solved for b, or -b, given some function F. The above period 2 orbit is found to be elliptic, i.e. (b, -b) and (-b, b) are the centers of 2 islands, using again (2.10)-(2.12) of Ref. 14.[15-17]

Period 3

Let a solution of (1.8) be y_t: = c,o,-c,c,o,-c,...etc., then from (1.8),

$$2\pi\Delta\nu \equiv \frac{B}{2Q} = -\frac{(\frac{1}{2}+C)}{S} \cdot \frac{c}{F(c)}, \quad (4.3)$$

The first erm on the r.h.s. is again the solution in the linear case, cf. (2.7) for $\sigma = 1/3$ or $2/3$. Thus the c and -c solutions of (4.3) yield separate orbits here as well. Both orbits of period 3 are found to be elliptic when $BS \leq 0$ and hyperbolic otherwise. The two additional hyperbolic orbits or period 3, interlacing[3-5,14-19] the above elliptic ones, are found as follows:

Let a solution of (1.8) be y_t: = d,e,e,d,e,e,..., etc. (or -d,-e, -e,...,) then from (1.8)

$$e = Cd + \frac{BS}{2Q}(F(d)) \text{ and } d = (2C-1)e + \frac{BS}{2Q}(F(e)). \quad (4.4)$$

Plotting these two curves in a d-versus-e plane provides quick insight in the location of their intersection points which, given some F, are easily calculated. From the slopes of these curves (4.4) at the origin we see that they do have intersection points [other than (0,0)] as soon as the earlier c, in (4.3), differs from 0, i.e. when

$$|B/2Q| > |(\tfrac{1}{2}+C)/S|, \quad (4.5)$$

i.e. these two hyperbolic orbits of period 3 and the previous two elliptic ones bifurcate off the origin at the same value of B, as expected in general[3-5,14-19], but not always satisfied by such orbits of short period[14,5] One of the two orbits of period 3 is (barely) visible in Figs. 8,9, where $F(y) = (\tfrac{1}{2}\sqrt{\pi}) \text{ erf}(y)$.

Period 4

Let one solution of (1.8) be (I) y_t: = g, g, -g, -g,..., etc. or another one be (II) y_t: = g,0,-g,0 then, for both orbits, from (1.8),

$$2\pi\Delta\nu \equiv \frac{B}{2Q} = -\frac{C}{S} \cdot \frac{g}{F(g)} = -\cot(2\pi Q), \frac{g}{F(g)}, \qquad (4.6)$$

Where the cotangent is the solution in the linear case, cf. (2.7) for σ = 1/4 or 3/4. The above orbit (I) is found to be elliptic, after bifutcating off the origin, and the above orbit (II) to be hyperbolic. These major resonances are depicted on Figs. 3-11, using $F(y) = (\frac{1}{2}\sqrt{\pi})$ erf(y) planned tune of ISABELLE[9]. The four major islands about the above elliptic orbit (I) would lead to a high nonlinear "parametric amplification", at Δν (orB) values where the linear amplification (of the initial amplitude) would still be small, cf. (2.9). This period 4 orbit appears for the first time, at the origin, when

$$\Delta\nu \equiv \frac{B}{4\pi Q} = -\frac{\cos(2\pi Q)}{2\pi} \approx 0.01673....\bigg|_{ISABELLE} \qquad (4.7)$$

Inspecting Figs. 3-11 we see that the islands become sizable (≈ 1) at Δν ≈ 0.0247, eventually becoming enormous (≈ 10) as Δν increases to 60% of its (linear) boundary value ≈ 0.177, cf. (2.6) (the r.m.s.-width of the Gaussian beam is ≈ 0.71). Points inside the island "rotate" slowly about the centers of the islands (I) (every 4th mapping). Hence the particles could "rotate" out to large distances and Δν ≈ 0.025 - 0.03 would appear to be a practical upper limit, at ISABELLE's tune, just on the basis of the effects modeled by (1.8). The size of the islands is not only compatible with the growth of the width of the beam observed in the numerical simulations of Ref. 12 but even "explains" the considerable change in the growth rate going from Δν = 0.02 to 0.06[12], reported in Ref. 12. Formula (4.7) would seem to predict that if they had chosen Δν ≤ 0.0167 they would have observed no growth whatsoever, cf. Fig. 3. All this tends to show is that the mechanism underlying the growth reported might simply be a 4th order resonance, or parametric amplication in general rather than anything more fanciful (and certainly excluding Arnold Diffusion which cannot take place in a system with so few degrees of freedom[3-5,15]). Reference 12 reports the numerical facts only.

Finally we observe that these resonances start at the origin, at some particular B, "quickly" more away as we go to higher values of |B|, due to the factor y/(F(y) in all formulae, and eventually "slow down" to move, linearly with B, out to infinity.

4.2 Nonlinear Stability

Here we demonstrate the different stability arguments one is forced to resort to in the nonlinear case. Consider first a piece-wise-linear approximation of the error-function (times $\frac{1}{2}\sqrt{\pi}$) beam-beam force in (1.8)

$$\bar{F}(y) \equiv \begin{cases} y & \text{for } y < \tfrac{1}{2}\sqrt{\pi} \\ \tfrac{1}{2}\sqrt{\pi} & \text{for } y > \tfrac{1}{2}\sqrt{\pi} \\ -\tfrac{1}{2}\sqrt{\pi} & \text{for } y < -\tfrac{1}{2}\sqrt{\pi} \end{cases}, \qquad (4.8)$$

This nonlinear function $\bar{F}$ has the same asymptotic behavior as the error function (times $\tfrac{1}{2}\sqrt{\pi}$) for $y \to 0$ and for $y \to \pm\infty$, does not differ much from it and is the exact one if the particle distribution in the strong beam is rectangular (constant within the beam and zero outside).

In this section we shall see that when the beam-beam strength B (or $\Delta\nu$), in front of $\bar{F}$, cf. (1.8), has a particular value all particle orbits starting inside some particular hexagonal region about the origin of the phase plane, will remain inside forever, i.e. we have stability for all time. Substituting the design parameters of ISABELLE[9] we find a hexagonal (nearly rectangular) region at $\Delta\nu \approx 0.03056...$ (or $B/B_{\ell b} \approx 0.173$), its border extending to ≈ 1.62 from the origin, at the furthest and to ≈ 1.145, at the closest [while $\tfrac{1}{2}\sqrt{\pi} \approx 0.886$, in (4.8)]. This "slightly hexagonal" region very nearly coincides with the "rectangular" region spanned by the 'hyperbolic' points of period 4 in Figs. 4 and 5 as $\Delta\nu \approx 0.0247$ and ≈ 0.03535. Comparing Figs. 4,5 with the others, Figs. 3-11, one expects that similar stable regions also exist for most other $\Delta\nu$ values. If so, their size cannot easily be established in those cases (the Kolmogorov-Arnold-Moser Theorem[3-5,15]) does establish the existance of some (infinitesimally?) small stable regions about the origin, in some cases, but it does not indicate their size or shape). Using a technique employed earlier by McMillan and Laslett[18,19], here, we merely look for regions bordered by very simple polygons. In all likelihood, it is due to this severe restriction that we find such a region at one value of $\Delta\nu$ (per Q) only. Yet to construct even one (large) region is not simple. It is a bit more surprising that the region obtained is nearly the best (largest) we could hope for in practice.

We saw above, and in section 4.1, that a larger region would have covered the hyperbolic points (of period 4) and parts of the island. Thus it would most likely not be as stable region (unless we enclosed the four islands as well cf. Figs. 4-7, 11). Secondly we discussed, also in section 4.1, that it would not be advisable to go beyond $\Delta\nu \approx 0.025$-0.03 where the area enclosing the four islands becomes very large.

The above sizable region of stability at $\Delta\nu \approx 0.03$ might further explain why no beam growth was observed, in Ref. 12, at $\Delta\nu \approx 0.02$. In addition it rigorously establishes the stability for all time of some rectangular beams.

This hexagonal curve actually is an "invariant curve", i.e. starting with any phase point on this curve all ensuring points of the orbit lie on this very same curve. In the linear (integrable[3,20]) case we have already found an infinitey of such invariant curves, since we obtained a constant of the motion K in (2.8) for any and all initial phase points: Each curve of constant K [in (2.8) an ellipse] is an invariant curve.

Finding such invariant curves for a nonlinear (nonintegrable[3,20] mapping is much more difficult and only occasionally can we construct a single (sizeable) one, long enough to be of practical use. Yet the K.A.M. theorem predicts the existence of many such[5,4,15] but the proofs do not guarantee any curves larger than "infinitesimally" short ones.

In an area preserving mapping of the phase plane on itself invariant curves are very important since they constrain all orbits which start on the inside to stay on the inside, <u>forever</u>[5,4,3]. This is virtually the only tool available to determine stability for all time, in a nonlinear mapping (or a 2 degree of freedom Hamiltonian system). In order to visualize why all points must remain inside an invariant curve it is perhaps easier to again consider our mapping points as the intersection points of a continuous orbit in a 3 dim. (reduced) phase space with a 2 dim. plane, as we did in changing from (1.3) to (1.8).

The (continuous) invariant curves turn out to be the intersections of (continuous) "invariant tori" ("inner-tubes") with that 2 dim-plane[5,4,3]. If our orbit points in the plane would appear both inside and outside the invariant curve, the continuous orbit (in 3 dim. space) would intersect the invariant torus somewhere. At that point there is a contradiction since any orbit starting at some point of the invariant torus must, by definition, remain on the torus for all time. Hence orbit points of an (area preserving) mapping of the plane are forced to remain inside an invariant curve if the orbits starts inside.

4.3 An Invariant Curve

The second-difference mapping (1.8) is equivalent to the two first-difference mappings,

$$x_{t+1} = y_t \tag{4.9}$$

$$y_{t+1} = -x_t + f(y_t), \text{ with } f(y_t) \equiv 2Cy_t + BS\ F\ (y_t)/Q, \tag{4.10}$$

cf. (1.4), (1.5). Thus the new x_{t+1}, y_{t+1} phase plane is identical with the previous y_t, y_{t+1} phase plane. Useful symmetries come to the fore if I rewrite (4.9), (4.10) as

$$x_{t+1} = y_t \text{ and } y_{t+1} = x_t - 2[x_t - \tfrac{1}{2}f(y_t)]. \tag{4.11}$$

The first two terms in each equation describe a reflection about the line $y = x$, while the bracket term, alone, makes a subsequent reflection of y_{t+1} about $[\tfrac{1}{2}f(y_t) =] \tfrac{1}{2}f(x_{t+1})$, parallel to the y_{t+1} axis. Hence any curve which is invariant under these two combined reflections is an "invariant curve" of the mapping (1.8) as well and contains a stable region as discussed before. These symmetry arguments were proposed earlier by McMillan, deVogelaere and employed by them as well as by Laslett and Greene[18,19,15].

Using the piece-wise linear $\overline{F}$ (4.8) we find from (4.10), the $\tfrac{1}{2}f(y)$ in (4.11)

$$\tfrac{1}{2}f(x) = \begin{bmatrix} ax & , \text{ for } |x| < \tfrac{1}{2}\sqrt{\pi} \\ bx + \tfrac{1}{2}\sqrt{\pi}(a-b), \text{ for } x > \tfrac{1}{2}\sqrt{\pi} \\ bx + \tfrac{1}{2}\sqrt{\pi}(b-a), \text{ for } x < -\tfrac{1}{2}\sqrt{\pi} \end{bmatrix}, \text{ with} \tag{4.12}$$

$$a \equiv C + BS/2Q, \text{ and } b \equiv C. \tag{4.13}$$

So, when $|y_t| < \frac{1}{2}\sqrt{\pi}$, the mapping (4.9)-(4.11) is perfectly linear and possesses an infinitey of invariant curves, the ellipses we found in (2.8) (parametrized by K). Consider the "last" of these ellipses, before $\frac{1}{2}f(x)$ switches its slope, as depicted in Fig. 12. In view of the above-mentioned reflection symmetry (parallel to the y-axis) about $\frac{1}{2}f(x)$ the ellipses exhibit this symmetry about y = ax. The "last" one is the one through the point F ($\frac{1}{2}\sqrt{\pi}$, $\frac{1}{2}\sqrt{\pi}$ a) cf. Fig. 12, where the slope of $\frac{1}{2}f(x)$ changes.

We first construct a hexagonal curve and subsequently derive the condition under which it is an invariant curve: Because of the above symmetry the tangent to the ellipse, at F, intersects the x-axis perpendicularly, at E(and the line y = - x at C). The intersection of the sec-one part of y = $\frac{1}{2}f(x)$ (4.12) with the x-axis is called B. From B we draw a line perpendicular to y = x (intersecting it at A and the line CE at D). The hexagon we shall consider is obtained by mirroring ABC about y = x and the total again, about y = - x.

This hexagon will be an invariant curve if it exhibits both symmetries discussed earlier. The symmetry about y = x is there by definition. The symmetry parallel to the y-axis about both parts of $\frac{1}{2}f(x)$ will be satisfied if DF = FC in Fig. 12 Using elementry geometry this yields the condition

$$a = -b/(1 + 2b), \tag{4.14}$$

[in the ISABELLE case we are mostly interested in b > 0 and a < 0, cf. (4.13) and Fig. 12]. This technique of constructing exceedingly simple invariant curves, in particular cases, is further elaborated in Ref. 18.

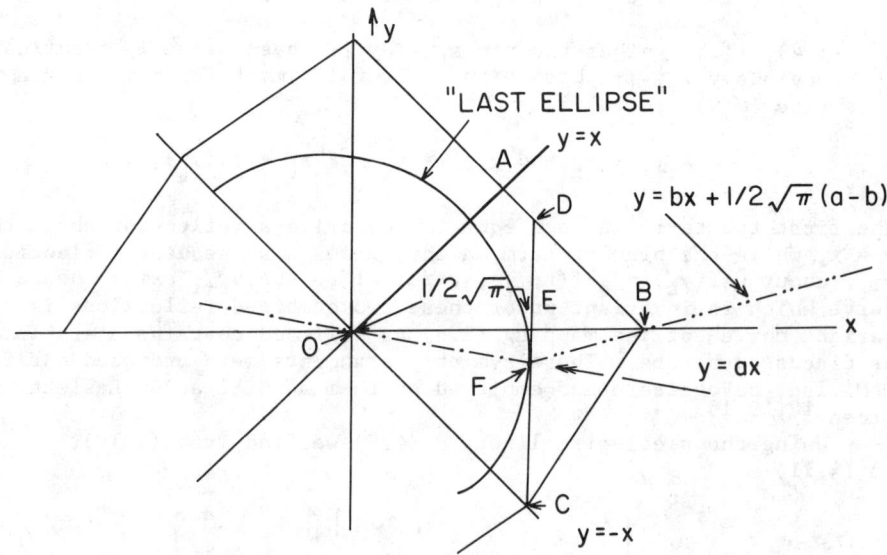

Fig. 12. Construction of the hexagonal 'Invariant Curve', consisting of ABC, mirrored about the line y = x and the total mirrored again about the line y = - x.

Upon substution of our parameters (4.13) the condition (4.14) becomes

$$(\Delta \nu \equiv) \frac{B}{4\pi Q} = \frac{C(1 + C)}{\pi(-S)(1 + 2C)} \quad .$$

For the ISABELLE values[9,10] this yields $\Delta\nu \approx 0.03056$ ($B/B_{\ell b} \approx 0.174$). In that case $\sigma B \approx 1.619$, $\sigma A \approx 1.253$. This completes the construction of our hexagon, in Fig. 12, invariant under the mapping (4.11), (4.12) or (1.8) with (4.8).

The stability types and other properties of the solutions of nonlinear mappings including (1.8), are the subject of a forthcoming article by T. Bountis and the author[17], parts of which are contained in Ref. 16. More recent developments in non-linear mappings, beam-beam effects and nonlinear dynamics in general will be found in Refs. (26, 27).

5. ACKNOWLEDGMENTS

This paper would never have been started without the long hours of work by Dr. Eminhizer, on projects leading up to it, or without the encouragement, help and advice of Drs. M. Month, D. Sutter and L. J. Laslett. The highly artistic pictures exhibited in Figs. 3-11 are the work of David Tucker (and the beam-beam force). Part of this work was supported by D.O.E. under contract EG-77-C-03-1538. I would like to compliment Drs. Month and Wiedemann for organizing this successful interdisciplinary symposium.

6. REFERENCES

1. H. Bruck, Circular Particle Accelerators, Los Alamos Translation LA-TR-72-10 Ref. (1972) or (in French) by Presses Universitaires de France (1966).
   J. D. Lawson, The Physics of Charged Particle Beams, Clarendon Press, Oxford (1977).
   A. A. Kolomensky and A. N. Lebedev, Theory of Cyclic Accelerators, Wiley, N. Y. (1966).
2. Topics in Nonlinear Dynamics, ed. S. Jorna, Am. Inst. Phys. Conf. Proc. Volume 46, A.I.P., N. Y. (1978).
3. M. V. Berry, "Regular and Irregular Motion", Ref. 2, p. 16; and his article in Ref. 27.
4. J. Moser, "Lectures on Hamiltonian System", Memoirs A.M.S. 81, 1-60 (1968); Ref. 2, p. 1 and his articles in this volume and in references 26 and 27.
5. J. Moser, Stable and Random Motions in Dynamical Systems, Princeton University Press (1973).
6. "A Review of Beam-Beam Phenomena", ed. M. Month, Brookhaven Report BNL 25703 (1979).
7. C. R. Eminhizer, P. A. Vuillermot, T. C. Bountis and R. H. G. Helleman, "Variational Studies of the Beam-Beam Interaction", Ref. 6, p. 25-59.
8. W. Magnus and S. Winkler, Hill's Equation, John Wiley Interscience, N. Y. (1966).
   H. P. McKean and E. Trubowitz, "Hill's Operator...", Pure Appl. Math. 24, 143-226 (1976).
   T. Bountis, in this volume.

9. J. C. Herrera, Ref. 6, p. 1-13, and his articles in this volume and in Ref. 27.
10. H. Hahn, M. Month and R. R. Rau, Rev. Mod. Phys. $\underline{49}$, 625-679 (1977) and the articles by M. Month in this volume and in Ref. 27.
11. F. M. Izraelev, S. I. Misnev and G. M. Tumaikin, "Numerical Studies of Stochasticity Limit in Colliding Beams (I-D. Model)", Preprint I. Ya. Ph. 77-43, Institute of Nuclear Physics Novosibirsk (1977); Note that the colliding particles here have opposite charges, hence, in that paper one is interested in the region of (our) negative B values and finds the instability regions to the left of the half integer Q values.
12. J. C. Herrera, M. Month and R. F. Peirels, "Simple Computer Model for the Nonlinear Beam-Beam Interaction in ISABELLE, Ref. 6, p. 14-24; all $\Delta \nu$ values printed there should be twice larger (private communication from the authors).
13. Since p_{t+1} is the momentum after the (t+1) st. pulse is completed, cf. (1.5), the value of $p[(t+1)2\pi]$, precisely in the middle of the pulse, for the differential equation (1.3), is the same as

$$p[(t+1)2\pi] = p_{t+1} - \tfrac{1}{2}BF)y_{t+1})$$
$$= -Qy_t S + P_t C + \tfrac{1}{2}BF(y_{t+1})$$

14. R. H. G. Helleman, "On the Iterative Solution of a 'Stochastic' Mapping", p. 343-370 in: 'Statistical Mechanics and Statistical Methods', ed. U. Landman, Plenum Publ. Co., N. Y. (1977).
15. J. M. Greene, J. Math Phys. $\underline{20}$, 1183 (1979); $\underline{9}$, 760 (1968) and his article in Ref. 27 and this volume. I. Percival, in this volume.
16. T. Bountis, Thesis, Physics Dept., Univ. of Rochester, N.Y. (1978); and University Microfilms, Inc.
17. T. Bountis and R. H. G. Helleman, to be submitted to J. Math, Phys. (1979).
18. E. M. McMillan, p. 219-244, in "Topics in Modern Physics, A Tribute to Edward A. Condon", eds. E. Britton and H. Adabasi, Colorado Assoc. Univ. Press, Boulder (1971).
19. L. J. Laslett, Ref. 2, p. 221.
20. R. H. G. Helleman, Ref. 2, p. 400.
21. R. H. G. Helleman and T. Bountis, p. 353-375, in <u>Stochastic Behavior in Classical and Quantum Hamiltonian Systems</u>, eds. G. Casati and J. Ford, Lecture Notes in Physics, Vol. 93; Springer Verlag (1979).
22. C. R. Eminhizer, R. H. G. Helleman and E. W. Montroll, J. Math. Phys., $\underline{17}$, 121-140 (1976); espec. sections 5 and 4.
23. C. R. Eminhizer, Thesis, Physics Dept. University of Rochester, N. Y. (1975); and University Microfilms, Inc.
24. R. H. G. Helleman, Ref. 2, p. 264-285.
25. A. W. Chao, Ref. 6, p. 60.
26. Proceedings of the (July) 1981, "Les Houches" Summerschool on <u>Chaotic Behavior in Deterministic Systems</u>, eds. R.H.G. Helleman and G. Jooss, North Holland Co. (1982).
27. Proceedings of the (December 17-21) 1979, '<u>International Conference on Nonlinear Dynamics</u>', ed, R.H.G. Helleman et al., Ann. N.Y. Ac. Sci. (1980).

KAM Surfaces Computed from the Hénon-Heiles Hamiltonian

J. M. Greene

Plasma Physics Laboratory, Princeton University
Princeton, New Jersey 08544

ABSTRACT

The Hénon-Heiles Hamiltonian is a well-known example of a nonintegrable system with two degrees of freedom. As is usual with such systems, the Hamiltonian generates three different kinds of orbits. Periodic orbits close on themselves. Other orbits cover two dimensional manifolds known as KAM surfaces after Kolmogorov, Arnol'd and Moser. Finally, some orbits fill out a finite portion of a three-dimensional constant-energy manifold in the phase space. In this paper the Hénon-Heiles Hamiltonian is treated by the method of residues, a method that has been developed previously. The purpose of this calculation is to find where KAM surfaces exist. The method depends on the numerical calculation of the stability of periodic orbits, together with heuristic schemes of extrapolation, to estimate the stability of other nearby periodic orbits. The crucial assumption is that the existence of KAM surfaces depends on the stability of nearby periodic orbits. This assumption appears to be verified for the system treated here, as it has been for previous cases that have been tested.

I INTRODUCTION

Some of the most fascinating problems of classical mechanics have yet to be solved completely. First among these is that of understanding how simple, deterministic equations can yield apparently random solutions. The simplest Hamiltonian system that exhibits this character is that of two coupled nonlinear oscillators. The orbits of such a system can be described in a four-dimensional phase space, but, since energy is conserved, individual orbits are confined to three-dimensional manifolds of constant energy.

These systems yield three different kinds of orbits. Periodic orbits are one-dimensional closed curves, and motion along these orbits is eminently predictable for all time. A triumph of our times was the proof of the existance of orbits that cover two-dimensional toroidal manifolds.[1] These manifolds are known as KAM surfaces after their discoverers, Kolmogorov, Arnol'd and Moser. Motion along these surfaces is well organized and essentially predictable. Finally, there are orbits that apparently fill a three-dimensional manifold of the phase space. Since unbelievable accuracy in the initial conditions is required to achieve even limited predictability for these orbits, they are called stochastic. Further, the latter type of orbit tends to wander more widely in phase space than do the other two types, with important physical consequences in a variety of different problems. Thus a study of these different types of orbit can be a matter of some significance.

The relation and interaction between these three types or orbits is the subject of this paper. One such interaction that arises in the system considered here is due to the fact that each KAM surface encloses a volume. In a two degree of freedom, energy conserving system, orbits inside a closed KAM surface remain inside for all time. This greatly restricts the wandering of some stochastic orbits. On the other hand, given KAM surfaces do not exist at all values of the energy in generic systems of the type considered here. Breakup of such surfaces may be accompanied by a considerable increase in the wanderlust of critical stochastic orbits, with various consequences in various problems. It can be physically quite important to know whether a given KAM surface exists or not.

Another connection between orbit types has been explored numerically in two previous papers.[2,3] These papers developed the concept that periodic orbits are stable when they are neighboring to KAM surfaces, and that, if these periodic orbits are rendered unstable by some large perturbation, the neighboring KAM surface disappears. This is useful because it is possible to estimate the stability of these orbits from extrapolation of calculations based on a few well chosen periodic orbits.

The earlier papers used simple mappings as abstract models of dynamical systems. The object of this paper is to extend the previous calculations to a Hamiltonian system. The Hénon-Heiles Hamiltonian[4] is an appropriate choice for such a purpose. It is simple, and has been studied by many previous authors over many years. A good review of this work has just been written.[5]

The Hénon-Heiles Hamiltonian, and the formalism needed for the computations, is given in Sec. II. Section III is devoted to a qualitative description of the orbits of such a system. This provides a framework for assessing the results of various calculations given in Sec. IV.

II FORMALISM

The Hénon-Heiles Hamiltonian is[4]

$$H = \tfrac{1}{2}(\xi^2 + \eta^2 + x^2 + y^2) + x^2 y - y^3/3 \tag{1}$$

where the momenta conjugate to x and y have been denoted ξ and η to minimize the number of subscripts. The equations of motion can be written in terms of the vectors

$$\underline{Z} \equiv \begin{bmatrix} x \\ \xi \\ y \\ \eta \end{bmatrix}, \qquad \underline{\nabla H} \equiv \begin{bmatrix} \partial H/\partial x \\ \partial H/\partial \xi \\ \partial H/\partial y \\ \partial H/\partial \eta \end{bmatrix}, \tag{2}$$

yielding

$$\frac{d}{dt}\underline{Z} = J\underline{\nabla H}, \tag{3}$$

where J is the matrix

$$J \equiv \begin{bmatrix} 0 & 1 & 0 & 0 \\ -1 & 0 & 0 & 0 \\ 0 & 0 & 0 & 1 \\ 0 & 0 & -1 & 0 \end{bmatrix}. \qquad (4)$$

The orbits lie on constant-energy surfaces in the phase space of the system,

$$H(\underline{Z}) - E = 0. \qquad (5)$$

When E is in the range $0 < E < 1/6$, one branch of the constant-energy surface is bounded. Only such orbits will be considered here.

It is convenient to deal with these orbits by introducing a Poincaré surface of section. This is a cross section through the orbits at the plane $x = 0$, $\xi > 0$, and is parameterized by y and η. The information $x = 0$, $\xi > 0$, y, η, and E completely characterizes an orbit. Following an orbit from one crossing of the surface of section to the next defines a mapping of the surface of section onto itself. This mapping will be the object of consideration in this paper, rather than the full orbit in phase space.

This mapping will be denoted by T. Thus,

$$(y_{n+1}, \eta_{n+1}) = T(y_n, \eta_n), \qquad (6)$$

where (y_{n+1}, η_{n+1}) and (y_n, η_n) are the coordinates of successive intersections of a given orbit with the surface of section.

The length of an orbit in units of the number of crossings of the surface of section is a useful quantity. In particular, a periodic orbit that passes through the surface of section Q times before closing on itself will be called a periodic orbit of length Q. Then,

$$T^Q(y_o, \eta_o) = (y_o, \eta_o). \qquad (7)$$

The curve corresponding to $\xi = 0$ in the surface of section is more or less egg shaped, depending on E, when $0 < E < 1/6$. This curve marks the boundary of the accessible region.

In addition to individual orbits, attention will be focused on bundles of nearby orbits. The difference between two orbits, $\underline{z} = \underline{Z}(t) - \underline{Z}_o(t)$, satisfies, in the linear approximation,

$$\frac{d\underline{z}}{dt} = J(\nabla\nabla H)\big|_{\underline{Z}=\underline{Z}_o} \underline{z}(t) \qquad (8)$$

where $\nabla\nabla H$ is the symmetric matrix of second derivatives of H and $\underline{Z}_o(t)$ is a solution of Eq. (3).

Since H has no explicit time dependence, one solution of Eq. (8) is

$$\underline{z}_1 = J\underline{\nabla}H\big|_{\underline{Z}=\underline{Z}_o}. \qquad (9)$$

The motions corresponding to $\underline{Z}_0$ and $\underline{Z}_0 + \underline{z}_1$ lie on the same orbit, but are displaced in phase.

The other useful piece of information about Eq. (8) is that conservation of energy yields

$$\nabla H^t \cdot \underline{z} - e = 0 \tag{10}$$

where ∇H^t is the transpose of ∇H and e is the perturbed energy.

The compact way of dealing with a variety of initial conditions on $\underline{z}$ is to introduce a time displacement operator U,

$$\underline{z}(t) = U(t)\underline{z}(0) . \tag{11}$$

Then U is a four by four matrix that satisfies

$$\frac{d}{dt} U = J(\nabla\nabla H)U \tag{12}$$

with initial conditions $U = 1$.

In fact, each column of U can be taken to be an independent solution of Eq. (8) since columns are not mixed in the evolution of U. When we wish to focus attention on a given energy surface it is convenient to deal with solutions confined to that surface. This can be accomplished by introducing a matrix V,

$$V \equiv U(t)S_1 \tag{13}$$

where S_1 is the constant matrix

$$S_1 \equiv \begin{bmatrix} \xi & 0 & 0 & 0 \\ 0 & 1/\xi & -y(1-y)/\xi & -\eta/\xi \\ \eta & 0 & 1 & 0 \\ -y(1-y) & 0 & 0 & 1 \end{bmatrix} \Bigg|_{\substack{Z = Z_0 \\ x = 0 \\ t = 0}} \tag{14}$$

evaluated at the initial position on the orbit, which is taken to lie in the surface of section. Then

$$(\nabla H)^t V = (0,1,0,0). \tag{15}$$

The first column of V corresponds to the solution $\underline{z}_1$ of Eq. (9), and the third and fourth columns are solutions constrained to the same energy surface as $\underline{Z}_0(t)$. Thus all information about nearby orbits lying in the energy surface can be found from $\underline{Z}_0(t)$ together with a numerical integration for the last two columns of V.

Nearby orbits may have different periods. Thus, at $t = T$, when $\underline{Z}_0(T)$ lands on the surface of section, the nearby orbits may already have passed through that surface. The correction of sliding along an orbit is equivalent to adding a multiple of the first column of V to one of the other columns. Further, only the components of V lying in the surface of section are needed to describe the mapping of the surface of section onto itself. Thus the desired information is contained in the two by two matrix M defined by

$$M \equiv S_3^\dagger S_2 V S_2 \tag{16}$$

where the sliding operator

$$S_2 \equiv \begin{bmatrix} 1 & 0 & 0 & 0 \\ 0 & 1 & 0 & 0 \\ -\eta/\xi & 0 & 1 & 0 \\ y(1-y)/\xi & 0 & 0 & 1 \end{bmatrix} \bigg|_{\substack{Z = Z_o \\ x = 0 \\ t = T}} \tag{17}$$

is a constant matrix with the components evaluated at a time when the central orbit crosses the surface of section, and

$$S_3 \equiv \begin{bmatrix} 0 & 0 \\ 0 & 0 \\ 1 & 0 \\ 0 & 1 \end{bmatrix} \tag{18}$$

is a projection operator.

Since attention in this paper will be focused on T, the mapping of the surface of section onto itself, rather than on the full orbits, the linearization of the orbits will be used only to obtain the matrix M. This matrix is the linearization around T.

In this paper particular attention will be focused on periodic orbits, that is, orbits that satisfy

$$\underline{Z}(T) = \underline{Z}(0). \tag{19}$$

The eigenvalues of $U(T)$ and $M(T)$ corresponding to these orbits are useful in understanding the mapping. The eigenvalues of these two matrices can be related as follows. It is straightforward, if tedious, to show that $U(T)$ has a doubly degenerate eigenvalue of unity, that

$$\text{Det } U = \text{Det } M = 1 \tag{20}$$

and that

$$\text{Trace } U(T) = 2 + \text{Trace } M(T). \tag{21}$$

Hence the eigenvalues of M are the other two eigenvalues of U.

A scheme for treating the eigenvalues of periodic orbits that enables one to estimate the eigenvalues of other nearby periodic orbits has been given in Refs. 2 and 3. A quantity known as the residue is introduced through

$$R \equiv (2 - \text{Trace } M)/4. \tag{22}$$

This contains information equivalent to the eigenvalues of M. Orbits are stable when $0 < R < 1$, and are unstable when $R < 0$ or $1 < R$. Further, it was shown in the above references that these residues tend to have an exponential dependence on orbit length. It was thus found convenient to define a mean residue

$$f \equiv (4R)^{1/Q} \tag{23}$$

where Q is the number of times the orbit intersects the surface of section before closing on itself. More extensive discussion of these definitions has been given in Refs. 2 and 3.

Note that, when Z_o is a periodic orbit, Eq. (8) is a linear differential equation with periodic coefficients. It is thus related to Hill's equation.[6] While Eq. (8) is fourth order, its essential information is carried by the two by two matrix M. This makes its relation to the second order Hill's equation even closer. It is thus useful to note that the residue, R, is essentially equivalent to the Hill determinant in its relation[6] to the eigenvalues of M.

The remaining computational problem to be solved before obtaining the results to be described is that of finding periodic orbits. In previous work with discrete mappings a simplification occurred when the mapping could be factored into two involutions. This simplification also occurs for the Hénon-Heiles Hamiltonian and will be described below. Without this simplification the calculations described in the succeeding sections would be exceedingly tedious or impossible.

The symmetry to be exploited is that which follows from the reversal of velocities. Namely, if $[x(t), \xi(t), y(t), \eta(t)]$ is an orbit, then $[x(-t), -\xi(-t), y(-t), -\eta(-t)]$ is also an orbit. From this it follows that if the velocity vanishes at a given point, $\xi = \eta = 0$, an orbit retraces its steps with the opposite velocity. Thus if the velocity vanishes at two points on an orbit, that orbit must be periodic with a period of twice (or perhaps some submultiple) the time taken to go from one stopping point to the other.

It is useful to translate this symmetry of orbits to a symmetry of the mapping T. This is consistent with the philosophy that this mapping is the primary object of consideration, so that all the properties of the orbits should be reduced to properties of T. We will now show that this symmetry takes the form of a particular factoring of the mapping.

Consider the three operators S, $\sqrt{T_+}$, and $\sqrt{T_-}$. The first of these inverts the sign of the momenta,

$$S \begin{bmatrix} x \\ \xi \\ y \\ \eta \end{bmatrix} = \begin{bmatrix} x \\ -\xi \\ y \\ -\eta \end{bmatrix}, \tag{24}$$

The second, $\sqrt{T_+}$, follows an orbit from an intersection of the surface of section to the next intersection of the surface $x = 0$, $\xi < 0$, and the third, $\sqrt{T_-}$ is the complementary operator that follows an orbit from the surface $x = 0$, $\xi < 0$, back around to the next intersection of the surface of section. It can be seen immediately that $S\sqrt{T_+}$ and $\sqrt{T_-}S$ are mappings of the surface of section onto itself, and that the mapping generated by the Hénon-Heiles Hamiltonian is the product of these two mappings,

$$T = (\sqrt{T_-}S)(S\sqrt{T_+}). \tag{25}$$

From the symmetry of the Hamiltonian under change of sign of ξ, η and t, it follows that

$$(S\sqrt{T_+})S\sqrt{T_+}) = 1 \tag{26}$$

and

$$(\sqrt{T_-}S)(\sqrt{T_-}S) = 1. \tag{27}$$

Mappings with this property are called involutions. Thus the Hénon-Heiles mapping has been factored into two involutions.

Consider an orbit segment that starts at the surface of section and passes through a point of vanishing velocity, $\xi = \eta = 0$, before its next intersection with the surface $x = 0$, $\xi < 0$. The starting point of this orbit is a fixed point of the mapping $S\sqrt{T_+}$, since the orbit retraces its steps after its velocity vanishes. Similarly, orbits whose velocities vanish just prior to crossing the surface of section define fixed points of $\sqrt{T_-}S$. Thus periodic orbits can be found equally well from finding two points where the velocity vanishes, or finding two points where the orbit crosses the surface of section at fixed points of $S\sqrt{T_+}$ or $\sqrt{T_-}S$.

Abstract generalizations of this concept have been treated in detail by deVogelaere[7] and some useful theorems concerning the fixed points of involutions have been given by Finn.[8] These ideas have been used previously[2,3] in the development of efficient schemes for finding periodic orbits. They can be taken over directly for use in the present problem.

There are other symmetries of the Hénon-Heiles Hamiltonian that yield independent factorizations of the mapping. In particular, the equations of motion are invariant under change of sign of x, η and t, since the Hamiltonian is even in x. Following the preceding arguments, orbits that twice pass through points where $x = 0$ are periodic, this symmetry can be translated into a factorization of the mapping of the surface of section, and one such factor is the reflection of the surface of section about the $\eta = 0$ axis. Further, from the threefold symmetry of the Hamiltonian, there are two more similar but independent factorizations of this mapping.

Another practical method for calculating periodic orbits has been developed by Helleman[9] and applied by Helleman and Bountis[10] to the Hénon-Heiles Hamiltonian. For historical reasons, their method has not been used here.

III PHENOMENA

In this section the basic phenomena of the Hénon-Heiles Hamiltonian, for energies less than 1/6, are sketched out, and a particular problem is identified for further study in the next section. The picture outlined in this section is based primarily on numerical computations, though it might be possible to confirm many of these results through perturbation theory.

We first consider the limit of very small energies, and notice that the volume of available phase space is very small. It is convenient to examine this with a microscope, and thus introduce the Hamiltonian,

$$H' = \tfrac{1}{2}(\xi'^2 + \eta'^2 + x'^2 + y'^2) + \sqrt{E}(x'^2 y' - y'^3/3). \tag{28}$$

The orbits derived from this Hamiltonian on the energy surface $H' = 1$ are equivalent to those derived from Eq. (1) on the energy surface $H = E$ with the scaling $\underline{Z}' \equiv \underline{Z}/\sqrt{E}$ with the notation of Eq. (2). However, the limit of vanishing E is well defined for the new Hamiltonian.

In the limit of vanishing energy the orbits of Eq. (28) are completely degenerate. Every orbit is periodic with the same frequency, and the surface of section mapping is the identity, $T = 1$.

For small but finite energy, eight of these simple periodic orbits survive the perturbation of the second term of Eq. (28). These have been denoted $\Pi_1, \ldots, \Pi_8$ by Churchill et al.[5] The positions of these orbits in the surface of section have been sketched in Fig. 1. The orbits that are stable, with small positive residues as defined in Eq. (22), are denoted with an o in Fig. 1, and the unstable orbits are denoted with an x. The orbit Π_1 parades up and down the $x = 0$ axis and so corresponds to the boundary of the object in Fig. 1. It is stable.

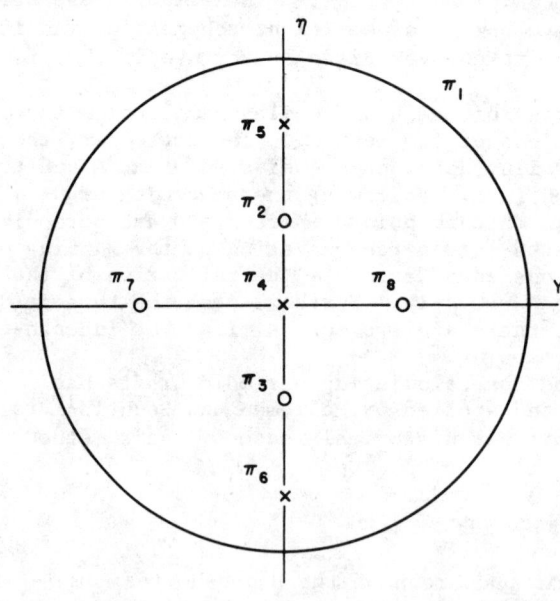

Fig. 1. Schematic representation of the periodic orbits in the surface of section for small energy.

The orbits Π_1, Π_5, Π_2, Π_4, Π_3, and Π_6 have a vanishing velocity at two points. Thus they are members of the first type of periodic orbit discussed in the previous section. Similarly, the orbits Π_1, Π_7, Π_4, and Π_8 have two points where x and η vanish simultaneously, and are members of the second type of periodic orbit.

A number of things happen when the energy is slowly increased. Some of these phenomena will now be sketched in, to the extent that they are needed in order to understand the significance of the calculations to be described in the next section.

First consider the immediate vicinity of Π_8. Linearization around this orbit, as described in the preceding section, yields the result that nearby orbits wind around Π_8 on concentric elliptical surfaces. The winding number for one of these orbits can be defined as the number of times the orbit encircles, or links, Π_8 divided by the number of times it passes through the surface of section, in the limit when both numbers are very large. This winding number will be denoted $\ell/2\pi$. It is related to the residue R, calculated for Π_8 from Eq. (22) by[2]

$$R = \sin^2 \ell/2. \qquad (29)$$

It can be determined numerically that for small, and moderate, values of the energy, R and ℓ are increasing functions of the energy, and that they vanish when the energy vanishes.

At each value of ℓ there is a bifurcation of Π_8. When $\ell/2\pi$ is irrational the product of the bifurcation is a KAM surface. This is an orbit that is not periodic, since it has an irrational winding number, but which ergodically covers a toroidal surface enclosing Π_8. Its intersection with the surface of section generates a closed curve surrounding the point corresponding to Π_8. Clearly, this closed curve is invariant under the mapping T. With increasing energy the KAM surface, identified from energy to energy by its winding number, expands away from Π_8.

For rational values, $\ell/2\pi = P/Q$ where P and Q are relatively prime integers, the products of the bifurcation are periodic orbits. Generally there are only two such orbits with a given winding number, but sometimes there are four or six, or even more. The symmetry of the Hénon-Heiles Hamiltonian may induce these higher multiplicity bifurcations, but this has not yet been explored. In any event these new orbits have vanishing residue at the point of bifurcation. For slightly larger energy half have small positive residue and are stable, and half have small negative residue and are unstable. The magnitudes of the residues of these orbits also increases with energy, similar to the behavior of Π_8. These orbits also bifurcate in the same way as Π_8, and so on in an infinite regression. In this paper only orbits that bifurcate directly out of Π_8 are considered, and the rest are ignored. Before leaving these secondary orbits however, note that the shapes of the secondary KAM surfaces can be calculated from a linearization around the periodic orbits. They are extraordinarily needle-shaped ellipses aligned with the neighboring KAM surfaces that bifurcated out of Π_8. Thus in the limit near Π_8 there is little practical difference between the neighborhood of a KAM surface and that of a periodic orbit.

At the energy for which $R = 1$, $\ell = \pi$, there are several possibilities. In some very symmetric cases ℓ may continue to increase with energy, with a corresponding decrease in the residue, R. More frequently, as with the well known special case of the Mathieu equation, the residue continues to increase, ℓ becomes complex, and the orbit becomes unstable. It has been shown that the residues of Π_1, Π_2 and Π_3 oscillate with increasing energy,[5] just as the discriminant[6] of the Mathieu equation is an oscillatory function of its argument, but a monotonic dependence on energy is more common for randomly chosen periodic orbits.

Parenthetically, it might be mentioned that there are certain pathological cases where there is an infinite multiplicity in the bifurcation of periodic orbits discussed above, and the bifurcated orbits have vanishing residues for all energies. These systems are called integrable. They are clearly of no interest for describing the real mixed up world, and the Hénon-Heiles system is not in that class.

Next, attention will be turned to the neighborhood of the unstable periodic orbits Π_4, Π_5, and Π_6. There is strong evidence for the existence of another kind of orbit here, one that chaotically fills a three-dimensional manifold of phase space, or a finite area of the surface of section. For small energies this type of orbit is confined to a small region near the x points, and along thin whiskers that extend between the x points and connect them.

As KAM orbits expand outward toward these chaotic orbits they become more highly contorted. At some value of the energy, different for each surface, they disappear. Above that energy there is no closed curve with the given irrational winding number that is invariant under the mapping T. Pictorially, the chaotic orbit is the wrack of broken KAM surfaces, washed out of the central island structure into a stochastic sea. This sea grows with increasing energy.

Now turn back to the neighborhood of Π_8. The nearby stable and unstable periodic orbits share the characteristics of Π_8 and Π_4. In particular, there will be little stochastic lakes in the vicinity of each unstable orbit, and they will grow with increasing energy. KAM surfaces that are close to these periodic orbits are perturbed by them, and break up before those that are farther from periodic orbits. Those that are infinitesimally close break up with an infinitesimal energy increment. Thus at every energy there is a residue of broken KAM surfaces in the neighborhood of every unstable orbit. These most fragile KAM surfaces are excluded from the existence proofs[1] by some number-theoretic measures of the distance from periodic orbits, so perhaps they shouldn't be called KAM surfaces.

In this four-dimensional phase space, orbits in the small stochastic lakes cannot communicate with the larger sea near Π_4. This follows from the fact that the KAM surfaces, together with the energy invariant, completely enclose a volume of phase space. From uniqueness and continuity, no orbits can pass from the inside to the outside of these surfaces.

Major events in the Hénon-Heiles system occur at the energy of disappearance of the last KAM surface between a major lake and the stochastic sea. At that point the volume available to a given orbit increases dramatically, though there may be only a slight increase in the total stochastic volume. An object of the calculations of the next section is to find the energy for one of these major events.

The central idea behind these calculations is that there is a close relation between KAM surfaces and the periodic orbits that bifurcate out of a given orbit. Earlier we saw that these two types of orbits could coexist together through a continuum of bifurcations because the residues of the periodic orbits vanish at the point of bifurcation. These residues increase with increasing energy, and thus make it more difficult for the nearby KAM surfaces to coexist. From a variety of considerations, both numerical and analytic, it appears that these residues have an exponential dependence on orbit length.[2,3] Thus the very long orbits that are closest to the KAM orbits have a sharp break near which the residues jump from a small value to a large one with a corresponding change in stability. Breakup of the KAM surfaces appears to be correlated with this jump.

IV RESULTS

The numerical results reported in this section were designed to further test some of the hypotheses that were formulated in Ref. 3. To some extent these hypotheses were specific for the abstract mapping considered there. The way they work out for this system will now be considered, one by one.

The essence of the first hypothesis of Ref. 3 was that the mean residues, f, calculated from Eq. (23), are approximately similar functions of the energy for all periodic orbits. The simplicity of the mapping utilized there permitted certain bounds to be placed on the behavior of the mean residues as functions of energy. The behavior of the Hénon-Heiles system is a bit more complicated. Mean residue as a function of energy are given in Fig. 2 for three periodic orbits, with $\ell/2\pi = P/Q = 1/5$, $1/6$, and $1/7$. These curves each reach the $f = 0$ axis at the point at which the residue of Π_8 is $\sin^2 \pi P/Q$, according to Eq. (29). It appears that orbits with $P + Q = 0 \pmod 3$ have mean residues with an energy dependence similar to that of the $1/5$ curve, while the others are similar to the $1/6$ and $1/7$ curves. Perhaps perturbation theory could illuminate this dichotomy. In any event, it appears that only a limited variety of energy dependence is displayed by the various periodic orbits of this system.

The mean residue of a number of orbits with energy $E = 0.1145$ in Fig. 3. According to the second hypothesis of Ref. 3, these should show a tendency to be larger for long periodic orbits that are close to short periodic orbits. The short periodic orbits are a strong source of perturbations. This is most visible in the vicinity of the orbit with $Q/P = 6/1$.

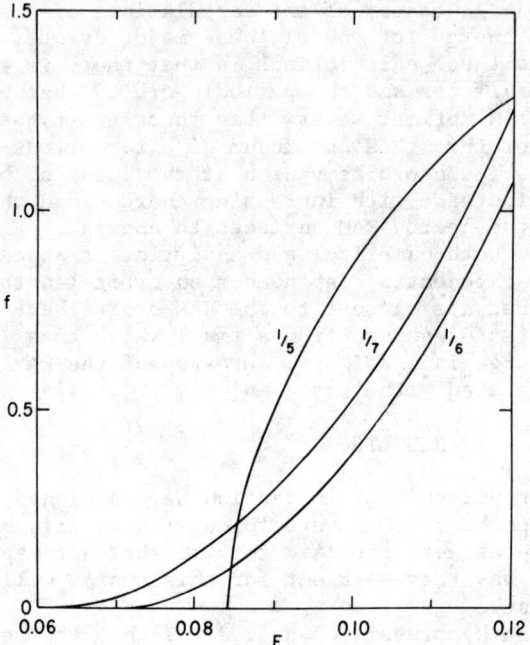

Fig. 2. Mean residue as a function of energy for three periodic orbits. They are labeled by their winding number P/Q.

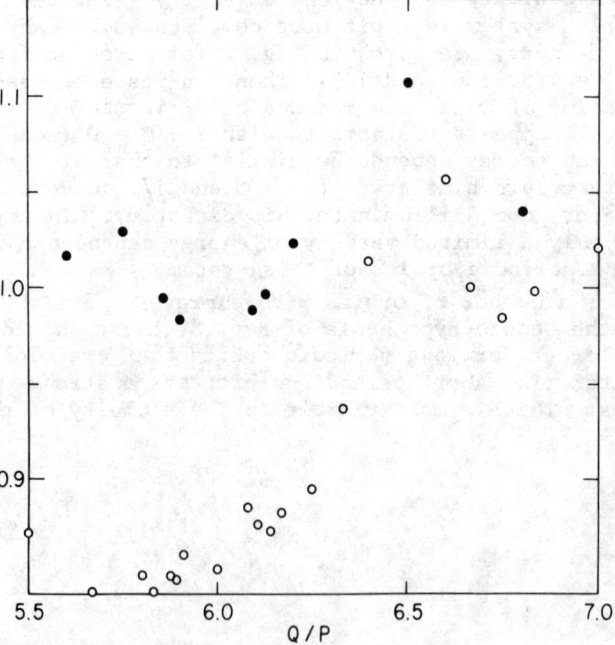

Fig. 3. Mean residue as a function of winding number for a few selected orbits of energy E = 0.1148. The closed circles represent orbits with P + Q = (mod 3).

In the present system this tendency must compete with two other tendencies that were absent from the abstract mapping considered previously. First, mean residues tend to be smaller for orbits closer to Π_8, until they vanish in its immediate vicinity. This accounts for the generally rising trend with Q/P in Fig. 3. Further, orbits with $P + Q = 0$ (mod 3) have anomolously large values of the mean residue, as was noted in Fig. 2. It appears that the mean residue of any particular orbit is a compromise amongst these three tendencies.

The crucial hypothesis of Ref. 3 concerned the convergence of the mean residues for a set of periodic orbits selected to optimally approach a KAM surface. The selection procedure consisted of employing an infinite continued fraction to express the (irrational) winding number of a chosen KAM surface, and then using successive truncations of this continued fraction to define the winding numbers of the desired periodic orbits. This procedure has been followed in Table I for the winding number

$$\ell/2\pi = (7 + 5\sqrt{5})/(41 + 29\sqrt{5}) \simeq 1/5.82200.$$

The convergence with increasing length of the mean residues, f, in Table I is somewhat more irregular than was true of the equivalent calculations of Ref. 3. A major problem is the idiosyncracy of the orbit with $Q/P = 64/11$, so that $P + Q = 0$ (mod 3). It seems possible from other calculations that for very long orbits the value of $P + Q$ (mod 3) becomes less important. There is a tendency of this sort near $Q/P = 6/1$ in Fig. 3. Nevertheless, regardless of these caveats, there is a tendency for convergence of f in Table I.

TABLE I

Mean Residues and Residues for Selected Orbits with Energy

E = 0.1178

Q/P	f	R
29/5	0.9918	0.20
35/6	0.9808	0.13
64/11	1.0204	0.91
99/17	0.9896	0.09
163/28	1.0037	0.46
262/45	0.9972	0.12

The fifth hypothesis of Ref. 3 was that, for a value of the energy chosen so that the mean residue converged to unity, as was approximately done for Table I, the residue R converged to 0.25. This may be valid to within a factor of two in Table I, but it was not possible to carry the calculation to sufficiently long orbits to thoroughly test this hypothesis on this case.

Since the mean residue, f, appears to be converging to a value near unity in Table I, E = 0.1178 must be close to the critical energy for the KAM surface with winding number

$$\ell/2\pi = (7 + 5\sqrt{5})/(41 + 29\sqrt{5}) \simeq 1/5.82$$

It can be seen from Fig. 3 that this winding number is near a minimum of the mean residue. Thus this surface must be close to, or perhaps is precisely, the last KAM surface to break up in the region between Q/P = 6/1 and 11/2. At some energy close to E = 0.1178 the large stochastic sea that includes the orbit Π_4, and that was discussed in the previous section, will have a significant increase in volume. Since

$$df/dE \simeq 43$$

in this vicinity, small errors in estimating the converged mean residues have a very small effect on the estimation of the critical energy.

A calculation in Ref. 4 showing a stochastic orbit approaching the Q/P = 5/1 islands at an energy of E = 0.125 is consistent with the critical energy calculated here.

V CONCLUSIONS

It is straightforward to apply the method of residues, previously developed for abstract mappings, to Hamiltonian systems with two degrees of freedom, and in which the Hamiltonian is independent of time. The case of one degree of freedom, but with the Hamiltonian containing an explicit, periodic time dependence would also be straightforward, and similar to that described here.

The major problem with the Hénon-Heiles Hamiltonian, with respect to the method of residues, appears to be its threefold symmetry. Thus periodic orbits with P + Q = 0 (mod 3) are more strongly perturbed than orbits that do not satisfy this condition. This produces complications that were difficult to explore thoroughly.

Nevertheless, the Hénon-Heiles system appears to exhibit behavior qualitatively similar to that of the analytic mappings considered previously. In particular, it appears to be possible to closely estimate the critical energy for the breakup of any given KAM surface.

ACKNOWLEDGMENTS

Conversations with Dr. R. C. Churchill have been very helpful. This work was supported by United States Department of Energy, Contract No. EY-76-C-02-3073.

1. V. I. Arnol'd, Usp. Mat. Nauk. $\underline{18}$, 13 (1963) [Russian Mathematical Surveys $\underline{18}$, No. 5, 9 (1963)]; J. Moser, Nachr. Akad. Wiss. Göttingen II Math.-Phys. Kl. No. 1 (1962).
2. J. M. Greene, J. Math. Phys. $\underline{9}$, 760 (1968).
3. J. M. Greene, J. Math. Phys. (to be published).
4. M. Hénon and C. Heiles, Astron. J. $\underline{69}$, 73 (1964).
5. R. C. Churchill, G. Pecelli, and D. L. Rod, in Stochastic Behavior in Classical and Quantum Systems, edited by G. Casati and J. Ford, (Springer Verlag, N. Y., 1979) p. 76.
6. W. Magnus and S. Winkler, Hill's Equation, (Interscience, N. Y., 1966).
7. R. deVogelaere, in Contributions to the Theory of Non-linear Oscillations, edited by S. Lefschetz, (Princeton University Press, Princeton, 1958), Vol. IV, p. 53.
8. J. Finn, thesis, physics, University of Maryland, 1974.
9. R. H. G. Helleman, in Topics in Nonlinear Dynamics, edited by S. Jorna, Am. Inst. Phys. Conf. Proc., (AIP, N. Y., 1978) Vol. 46, p. 264.
10. R. H. G. Helleman, and T. Bountis, in Stochastic Behavior in Classical and Quantum Systems, edited by G. Casati and J. Ford, (Springer Verlag, N. Y., 1979) p. 353; T. Bountis, thesis, physics, University of Rochester, 1978.

DIFFUSION IN NEAR-INTEGRABLE HAMILTONIAN SYSTEMS WITH THREE DEGREES OF FREEDOM

by

J. L. Tennyson, M. A. Lieberman and A. J. Lichtenberg
Department of Electrical Engineering and Computer Sciences
&
The Electronics Research Laboratory
University of California
Berkeley, California 94720

ABSTRACT

We review, using a few simple examples, the mechanism for a very general stochastic motion—the Arnold diffusion—which occurs in near-integrable Hamiltonian systems with more than two degrees of freedom. The examples chosen describe the effect of periodic perturbations on a free particle system in three dimensions. The calculated Arnold diffusion rates are in good agreement with the results of simulation studies.

I. INTRODUCTION

One hundred and forty five years after W. R. Hamilton formulated his famous equations, the long term motion they describe is still not completely understood. The nature of the motion is firstly determined by the degree of symmetry of the Hamiltonian system according to the rule[1] that "the less symmetrical the Hamiltonian system, the more intricate is the motion." For each symmetry, there exists an independent global invariant of the motion. Symmetries may be "visible" or "hidden", and often it is not easy to guess whether or not they exist. A notable example is the Toda lattice Hamiltonian,[2] which apart from the energy has one "visible" and one "hidden" symmetry. There is no known analytic procedure for obtaining the global invariants of the motion for a given Hamiltonian, or even for determining their total number. As stated in a recent review,[1] "a theory that could be used to test the presence of one symmetry or another of a given Hamiltonian, and in particular to calculate its degree of symmetry, is still waiting to grow out of 'alchemy' into 'chemistry'. Therefore, eschewing alchemy, we pass this problem by in silence."

In the discussion that follows, we consider only Hamiltonians that are autonomous (the quantity represented by the Hamiltonian function is conserved). Any Hamiltonian that is not autonomous may be made so by introducing an extended phase space.[3]

There are two distinct classes of Hamiltonians, integrable and nonintegrable. For integrable Hamiltonians with N degrees of freedom, N independent symmetries exist (including energy). These allow for the isolation and independent solution of the motion in each degree of freedom. All trajectories are regular and are confined to an N dimensional surface in the phase space. According to a theorem of Siegal,[4] such Hamiltonians are rare. In the generic case, the Hamiltonian is not completely integrable. Hamiltonians in this second class have M (less than N) independent symmetries. These symmetries may be used to reduce the Hamiltonian to a system with (N - M + 1) degrees of freedom which has no global invariants and always some stochastic motion. It has been shown[5,6] that in general, these nonintegrable Hamiltonians generate a finite proportion of trajectories which are stochastic, and a finite proportion which are integrable. The integrable trajectories do not reflect global invariants of the system since their existence depends discontinuously on the initial conditions. Stochastic and integrable trajectories are intimately comingled, with a stochastic trajectory lying arbitrarily close to every point in the phase space.

For two degrees of freedom, stochastic trajectories may be isolated from one another by integrable (KAM) surfaces. For three degrees of freedom (with only one global invariant), the integrable trajectories still exist, but do not isolate the stochastic trajectories. All stochastic regions of the phase space are connected into a single complex network—the Arnold web. The web permeates the entire phase space, intersecting or lying infinitesimally close to every point. For an initial condition on the web, the subsequent stochastic motion will eventually intersect every finite region of the phase space (or energy surface, for an autonomous system)—this is the Arnold diffusion.[7,8]

Although the merging of stochastic trajectories into a single web seems to be a general characteristic of $N > 2$ nonintegrable systems, there is some controversy, and the question is not entirely settled. Contopoulos et al[9] have performed simulation studies on a particular model with $N = 3$, and have found results which they interpret to indicate the existence of segregated stochastic regions. One of these regions exhibits the usual single invariant (energy), while the others seem to be characterized by a second invariant in addition to the energy. Whether this additional invariant is truly a constant, or whether the slow rate of Arnold diffusion has caused it to appear as a constant, is not yet known.

II. FREE PARTICLE HAMILTONIAN SYSTEMS

We consider Hamiltonians comprised of an integrable part, $H_N^{(0)}$, and a small perturbation, $\epsilon H_N^{(1)}$,

$$H_N = H_N^{(0)}(\underline{I}) + \epsilon H_N^{(1)}(\underline{I},\underline{\theta})$$

where $\underline{I}$ and $\underline{\theta}$ are the N-dimensional action and angle vectors of $H_N^{(0)}$, and $H_N^{(1)}$ is periodic in $\underline{\theta}$ with period 2π. The stochastic web appears in the neighborhood of the resonances of the unperturbed Hamiltonian. The resonances are the closed, periodic trajectories of $H_N^{(0)}$.

As a specific example, we examine in detail the system shown in Fig. 1. A free particle moving in three dimensions is subject to the influence of a small, spatially periodic perturbation,

$$H = \frac{p^2}{2m} + \epsilon \sum_{\underline{m}} V_{\underline{m}} e^{i\underline{m} \cdot \underline{\theta}} + c.c. \tag{1}$$

where m is called the <u>resonance vector</u> and has three integer components. We also have

$$\theta_j \equiv K_j x_j$$

$$K_j \equiv \frac{2\pi}{\lambda_j} .$$

For simplicity, we assume the components of K to be equal, $K_j = K$. The case of unequal K_j differs trivially from the one considered here.

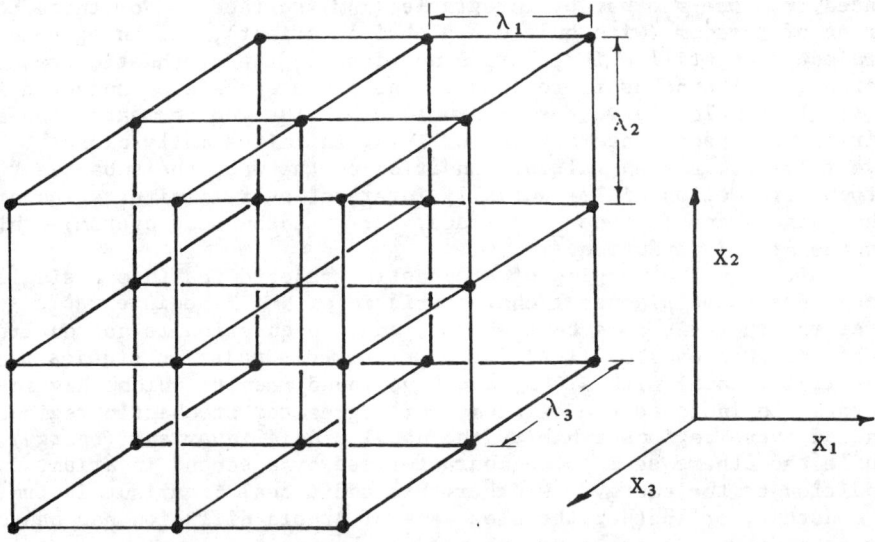

Fig. 1. The free particle is perturbed by a periodic perturbation in three dimensions.

The action-angle variables for the unperturbed motion are

$$\underline{I} \equiv \frac{m}{K} \underline{v} \tag{2}$$

$$\underline{\theta} \equiv K \underline{x} \tag{3}$$

$$\underline{\omega}(\underline{I}) \equiv \frac{K^2}{m} \underline{I}. \qquad (4)$$

Note that the system is nonlinear in the sense that $\underline{\omega}$ is proportional to I (and also to v). The Hamiltonian in the new variables is

$$H = \tfrac{1}{2} \frac{K^2}{m} I^2 + \epsilon \sum_{\underline{m}} V_{\underline{m}} e^{i\underline{m}\cdot\underline{\theta}} + c.c. \;. \qquad (5)$$

When $\epsilon = 0$, the projection of the motion into the three dimensional frequency space (which is also the action and velocity spaces) is a single point which may or may not represent a resonance. The set of resonance frequencies is defined by

$$\underline{m}\cdot\underline{\omega} = 0 \qquad \text{(for all } \underline{m}\text{)}.$$

Thus, for a given m, the loci of resonance points lie in a <u>resonance plane</u> which is perpendicular to m and passes through the origin. The resonance vectors m are shown in Fig. 2 (for m_i = - 2, - 1, 0, + 1, + 2). The resonance planes are perpendicular to these vectors and the entire set intersects every finite region of the action space. The intersection of the resonance planes with the unperturbed energy surface

$$I_1^2 + I_2^2 + I_3^2 = \text{constant}$$

is shown in Fig. 3. The intersection forms a network of resonance lines called the "Arnold web".

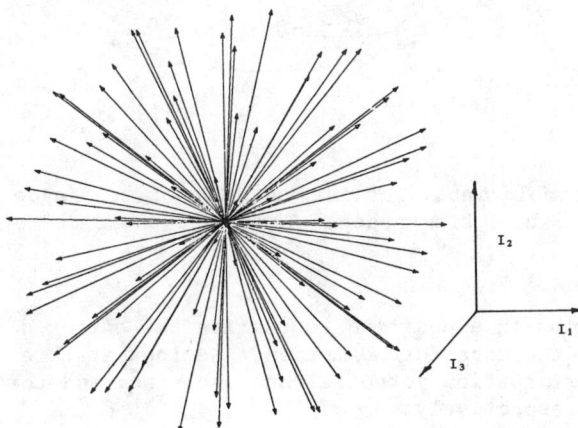

Fig. 2. Resonance vectors m in action space (also frequency and velocity space). The vectors are normalized to unity and projected at an oblique angle. All vectors with $|m_i| \leq 2$ are shown.

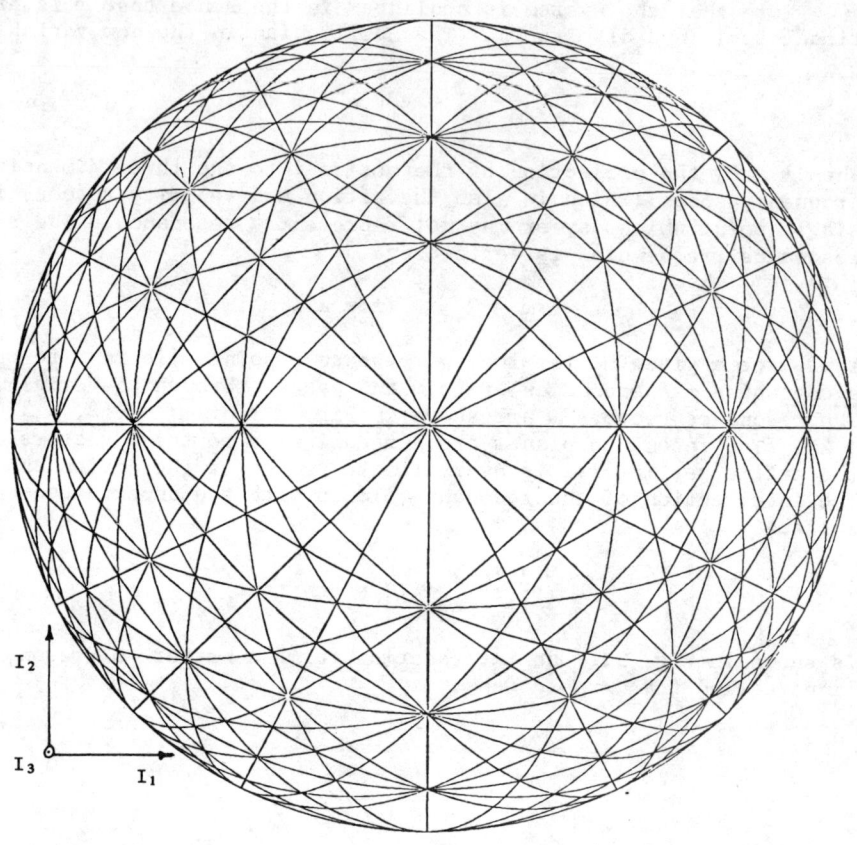

Fig. 3. The Arnold web. The intersection of resonance planes with energy surface in action space. The lines shown correspond to $|m_i| \leq 2$.

We are now in a position to examine the effect of a finite perturbation on the unperturbed motion. We look at three cases in which the perturbation potential has one, two, and three Fourier components, respectively.

We consider first a potential with a single Fourier component n (for V_n independent of $\underline{I}$, this might represent an electrostatic plane wave). Leaving the c.c. notation as understood, the Hamiltonian can be written

$$H = \frac{1}{2}\underline{I}\cdot\underline{\omega} + \epsilon V_{\underline{n}} e^{i\underline{n}\cdot\underline{\theta}} \tag{7}$$

and

$$\frac{d\underline{I}}{dt} = -\frac{\partial H}{\partial \underline{\theta}} = - i\underline{n}\epsilon V_{\underline{n}} e^{i\underline{n}\cdot\underline{\theta}} . \tag{8}$$

Thus, the change in the action must always lie in the direction of the resonance vector n, as shown in Fig. 4. The motion is completely integrable, but the energy surface is thickened by $2\epsilon V_{\underline{n}}$ and a separatrix forms about the resonance. The phase space trajectories are those of a simple pendulum. Typical libration, rotation, and separatrix motion is shown in Fig. 5. We note that since the motion is always perpendicular to the resonance line, any displacement along the resonance line is forbidden.

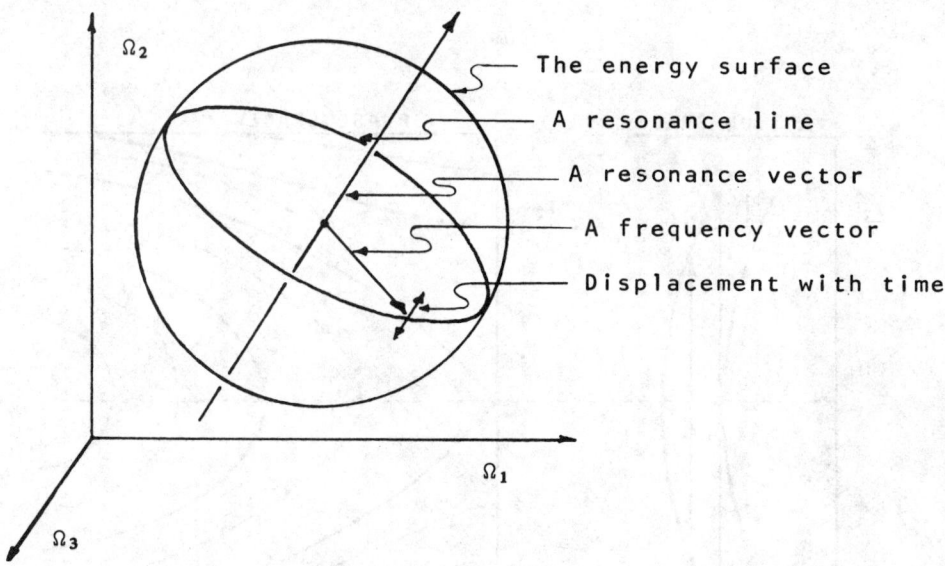

Fig. 4. The energy surface, resonance vector, resonance line, frequency vector and change in action for a free particle perturbed by a single wave.

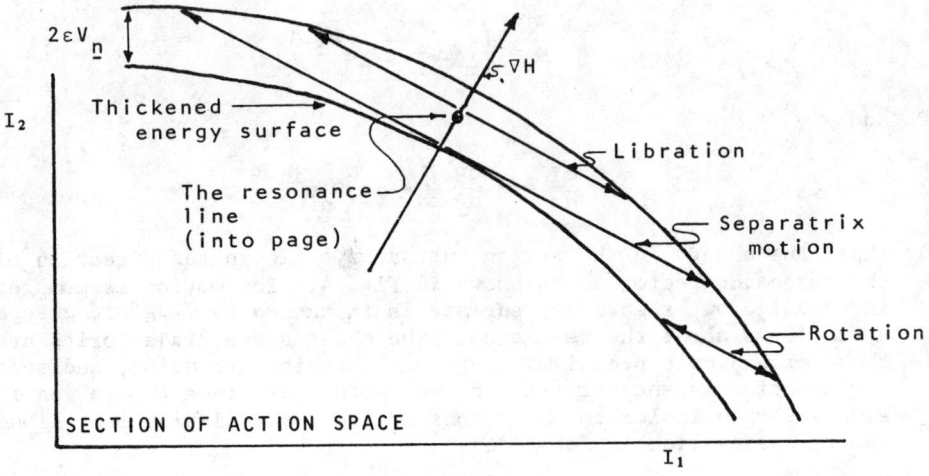
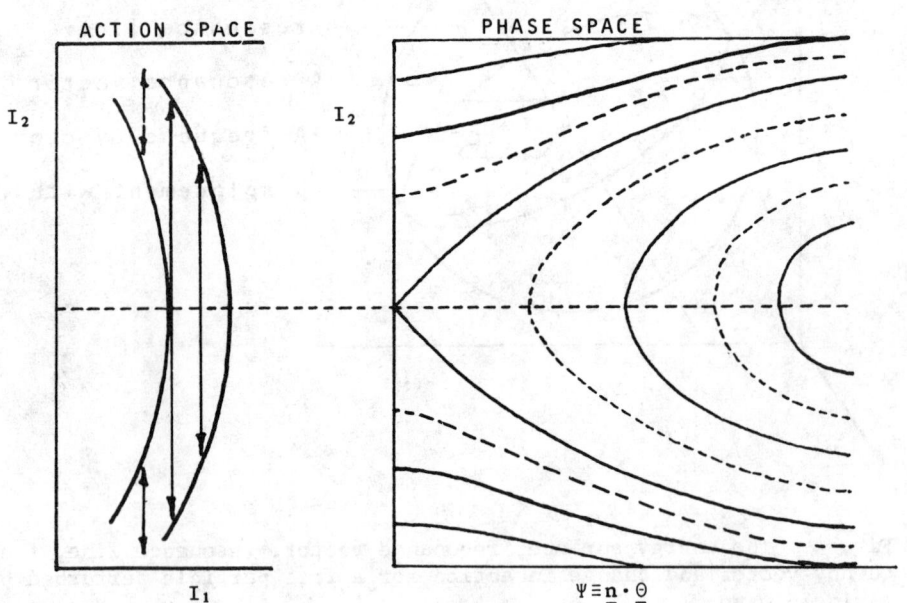

Fig. 5. Thickening of the action surface, libration, separatrix, and rotation motion for a single wave perturbation.

Consider now the case of two nonparallel, nonperpendicular Fourier components (two waves),

$$H = \frac{K^2}{2m} \underline{I} \cdot \underline{I} + \epsilon V_{\underline{n}} e^{i\underline{n} \cdot \underline{\theta}} + \epsilon V_{\underline{\ell}} e^{i\underline{\ell} \cdot \underline{\theta}} \qquad (9)$$

$$-\frac{d\underline{I}}{dt} = i\underline{n} \, \epsilon V_{\underline{n}} e^{i\underline{n} \cdot \underline{\theta}} + i\underline{\ell} \, \epsilon V_{\underline{\ell}} e^{i\underline{\ell} \cdot \underline{\theta}} \qquad (10)$$

$$\frac{d\underline{\theta}}{dt} = \frac{K^2}{m} \underline{I} + \frac{\epsilon \partial V_{\underline{n}}}{\partial \underline{I}} e^{i\underline{n} \cdot \underline{\theta}} + \frac{\epsilon \partial V_{\underline{\ell}}}{\partial \underline{I}} e^{i\underline{\ell} \cdot \underline{\theta}} . \qquad (11)$$

The two primary resonance lines are shown as heavy curves in Fig. 6. In addition to the primary resonances, the nonlinearity of the motion excites an infinite set of secondary resonances. This can be seen by iteratively solving Eqs. (10) and (11) for $\underline{I}$ and $\underline{\theta}$. We choose the special case where the V_m are not functions of $\underline{I}$ (though the results are similar if we include this dependence). Approximating $\underline{\theta} = \underline{\omega}t$ and integrating Eq. (10), we generate the primary resonances

$$\underline{I}(t) \sim A\underline{n} \cos (\underline{n} \cdot \underline{\omega}t) + B\underline{\ell} \cos (\underline{\ell} \cdot \underline{\omega}t). \qquad (12)$$

Using this result in Eq. (11) and integrating yields

$$\underline{\theta}(t) \sim C\underline{n} \cos (\underline{n} \cdot \underline{\omega}t) + D\underline{\ell} \cos (\underline{\ell} \cdot \underline{\omega}t) \qquad (13)$$

where A,B,C, & D are constants. Using Eq. (13), a second iteration of Eq. (10) now generates an infinite set of second order resonances. For example, in Eq. (10) the term $\underline{n} \exp[i(\underline{n} \cdot \underline{\theta})]$ results in an infinite set of harmonics through the expansion

$$\exp[i\underline{n} \cdot \underline{\theta}] \sim \sum_{m_1 m_2} J_{m_1}(C) J_{m_2}(D\underline{n} \cdot \underline{\ell}) \exp[i(m_1 \underline{n} + m_2 \underline{\ell}) \cdot \underline{\omega}t]$$

where the J's are Bessel functions. As can be seen from Eq. (14), all secondary resonances have the form

$$(m_1 \underline{n} + m_2 \underline{\ell}) \cdot \underline{\omega} = 0$$

so the secondary resonance vectors also lie in the ℓ-n plane. These are shown as thin lines in Fig. 6.

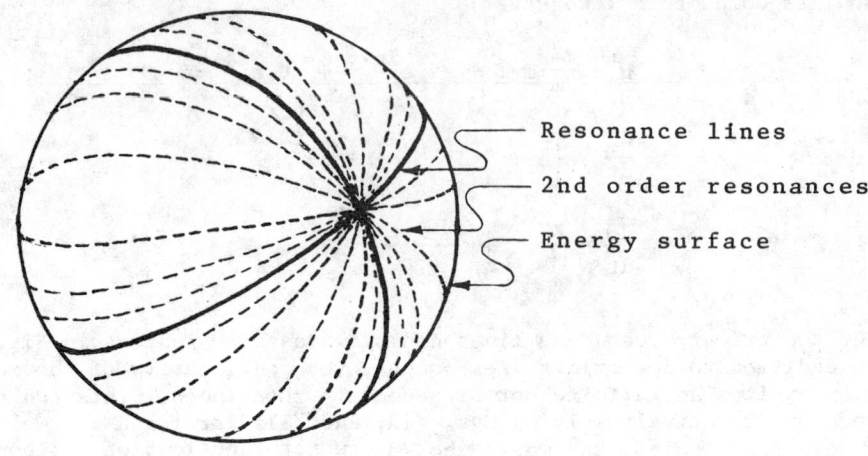

Fig. 6. Action (frequency, velocity) space for the two wave system showing the primary resonances and the network of secondary resonances.

For small perturbations, thin stochastic layers are formed about each resonance line. The formation of stochastic layers occurs when the separatrices of neighboring secondary resonances overlap.[8] Although the motion is now stochastic near a resonance, the system is still confined to a small region of the energy surface. As shown by Eq. (10) (and as illustrated by Fig. 7), all trajectories are restricted to planes parallel to the ℓ-n plane. Motion along a resonance line (the Arnold diffusion) is not allowed. Note that for a small enough total energy (or a large enough perturbation) the secondary stochastic layers may overlap, leading to diffusion around the intersection of the n-ℓ plane and the energy surface. The two wave problem can easily be reduced to two degrees of freedom by separating out the motion in the direction perpendicular to the ℓ-n plane.

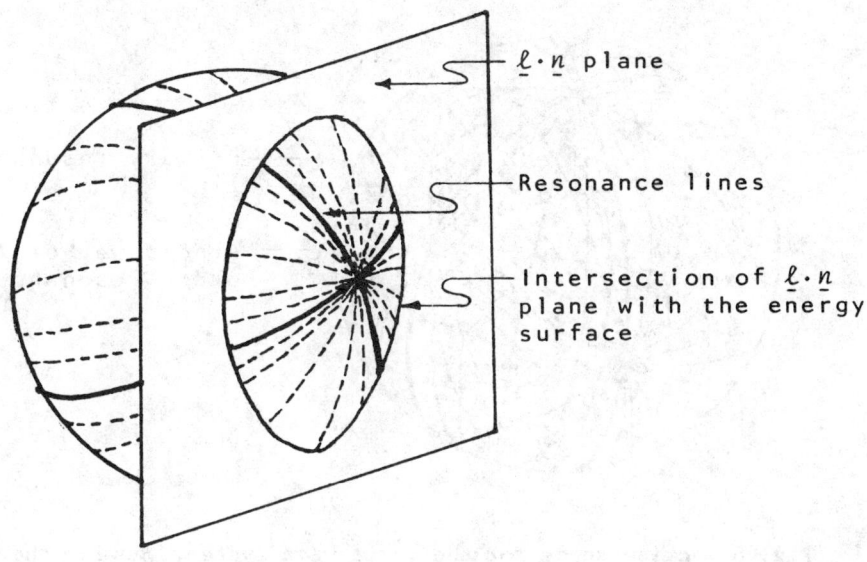

Fig. 7. Confinement of stochastic trajectories for the two wave system. Motion is confined to the intersection of the $\underline{n}$-$\underline{\ell}$ plane with the energy surface.

In our final example, the free particle interacts with three noncoplanar, nonperpendicular Fourier components (or waves). We have

$$\frac{d\underline{I}}{dt} = i\underline{n} \, \epsilon \, V_{\underline{n}} e^{i\underline{n}\cdot\underline{\theta}} + i\underline{\ell} \, \epsilon \, V_{\underline{\ell}} e^{i\underline{\ell}\cdot\underline{\theta}} + i\underline{p} \, \epsilon \, V_{\underline{p}} e^{i\underline{p}\cdot\underline{\theta}} . \qquad (15)$$

The three primary resonance lines are shown in Fig. 8. As before, the nonlinearity of the motion generates a web of second order resonances, with

$$(m_1\underline{\ell} + m_2\underline{n} + m_3\underline{p})\cdot\underline{\omega} = 0. \qquad (16)$$

The system may now move in any direction along the energy surface in I space. Again, thin stochastic layers form near each resonance line. In contrast to the previous examples, stochastic trajectories may now exhibit Arnold diffusion, moving along the resonance lines and exploring every finite region of the energy surface.

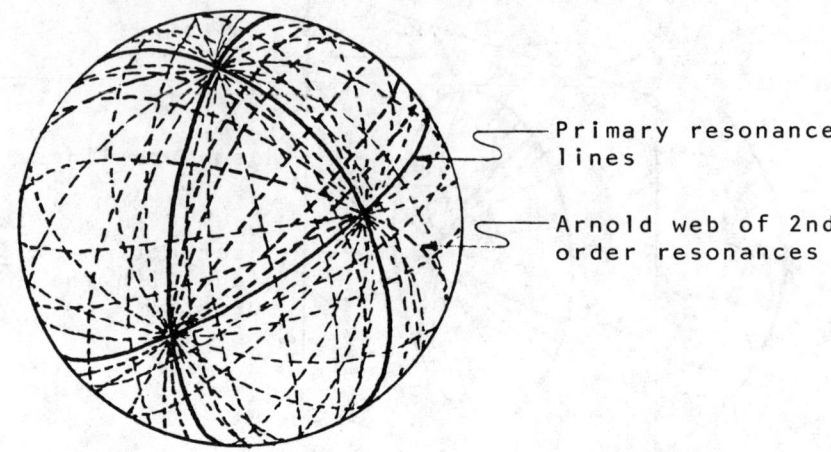

Fig. 8. Action space for the three wave system, showing the three primary resonances and the network of secondary resonances—the Arnold web.

From a practical point of view, there are two major questions concerning Arnold diffusion in a particular system:

(a) What is the relative measure of stochastic trajectories in the phase space region of interest?
(b) How fast will the system diffuse along the thin threads of the Arnold web?

III. THE BILLIARDS PROBLEM

We have studied in detail the motion of a simple physical system with three degrees of freedom. As shown in Fig. 9, a point particle bounces back and forth between a smooth wall located at $z = h$ and a periodically rippled wall located at $z = 0$. The ripple is defined by the equation

$$z = -a_x \cos k_x x - a_y \cos k_y y \qquad (17)$$

which is just a superposition of two perpendicular waves. This "hard wall" system is very similar to the free particle model with a perturbation containing more than three noncoplanar Fourier components. It exhibits both the stable KAM trajectories and Arnold diffusion.

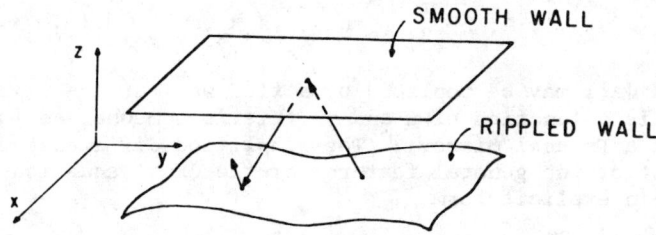

Fig. 9. The three dimensional billiards problem. A point particle bounces back and forth between a smooth and a periodically rippled wall.

It is possible to describe the motion of the particle in terms of four difference equations. These give the evolution of the trajectory angles (α_n, β_n) and position (x_n, y_n) as defined just before the nth bounce. The system is illustrated in Fig. 10 for $a_y = 0$. In this case the y motion is independent of the x-z motion and the system reduces to two dimensions (stochasticity but no diffusion).

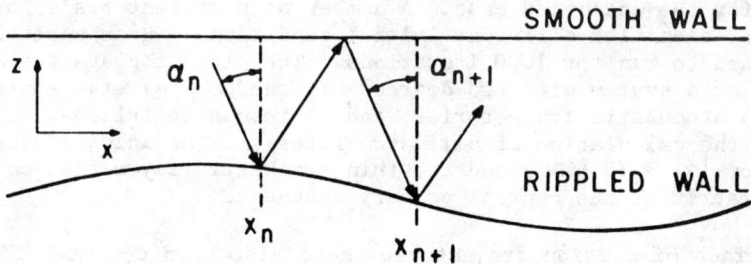

Fig. 10. Motion in two degrees of freedom, illustrating the definition of the trajectory angle α_n, and the bounce position x_n just before the nth collision with the wall. $\alpha = \arctan(v_x/v_z)$, $\beta = \arctan(v_y/v_z)$.

The exact difference equations for this system cannot be written in explicit form, so it is of practical interest (both for analytic calculations and computational speed) to make some simplifying approximations.

If we assume both

$$|\alpha|, |\beta| \ll \frac{\pi}{2} \quad \text{and} \quad a_x k_x, a_y k_y \ll 1$$

the rippled wall may be replaced by a flat wall at $z = 0$ whose normal vector is a function of x and y (this is somewhat analogous to the idea of a Fresnel mirror). The simplified difference equations exhibit most of the general features of the exact equations and may be written in explicit form

$$\alpha_{n+1} = \alpha_n + 2\gamma_x (x_n, y_n) \tag{18}$$

$$x_{n+1} = x_n + 2h \tan(\alpha_{n+1}) \tag{19}$$

$$\beta_{n+1} = \beta_n + 2\gamma_y (x_n, y_n) \tag{20}$$

$$y_{n+1} = y_n + 2h \tan(\beta_{n+1}). \tag{21}$$

These equations may be interpreted as a mapping on the Poincaré surface of section at $z = 0$. A similar set of equations has been studied via computer simulations by Froeschlé and Scheidecker.[10]

If γ_x is not a function of y, and γ_y is not a function of x, the system breaks into two uncoupled parts describing motion in x-z and y-z separately. Figure 11 shows the motion in the α-x surface of section for the uncoupled case. A number of different trajectories are shown, each with different initial conditions. Each particle was allowed to run for 1000 iterations. The plot displays the usual features of a system with two degrees of freedom: a) stable trajectories b) stochastic trajectories, and c) resonance islands. Of concern for the calculation of diffusion rates are the thick stochastic layers for $|\alpha| \gtrsim (0.6)\pi/2$ and the thin stochastic layer that covers the separatrix of the central primary resonance.

Surface of section trajectories have also been computed for the exact problem (a real rippled wall). The motion in the α-x plane is shown in Fig. 12 for the same parameters and initial conditions used in Fig. 11. The basic features of the exact problem are apparently well represented by the approximate equations.

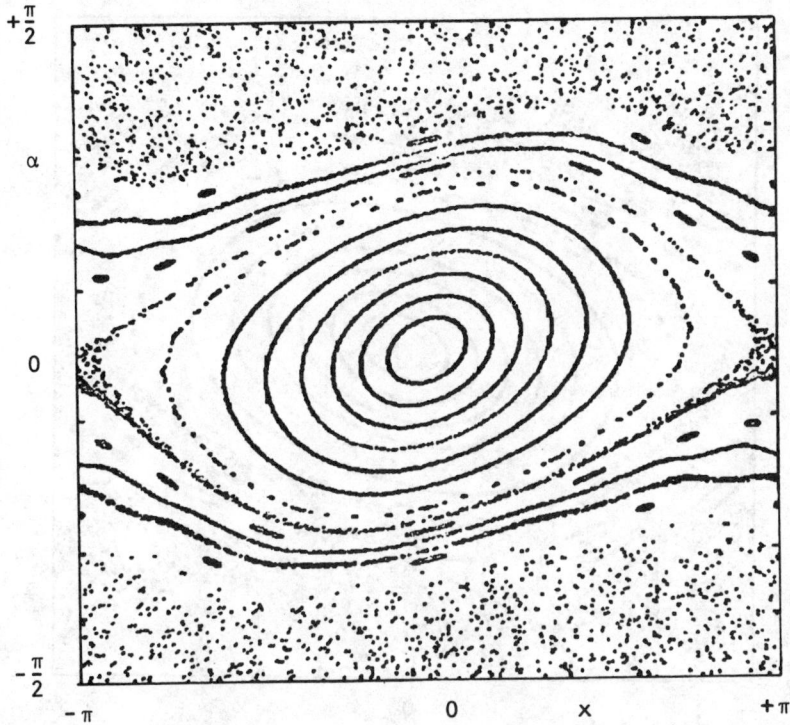

Fig. 11. Motion in the α-x surface of section for the uncoupled billiards problem described by the simplified Eqs. (18)-(24). The parameters are $\epsilon = 0$; λ_x:h:a_x as 100:10:2; $\lambda_x = 2\pi/k_x$. 15 particles are started at $x = 0$ and allowed to run for 1000 iterations each.

The difference equations, Eqs. (18)-(21), will be truly three dimensional when γ_x includes a dependence on y and γ_y a dependence on x. This can be accomplished by adding a small diagonal ripple to the bottom wall. The (virtual) normal to the surface $z = 0$ at (x,y) is then described by the angles

$$\gamma_x = a_x k_x \sin k_x x + \epsilon k_x \gamma_c \tag{22}$$

$$\gamma_y = a_y k_y \sin k_y y + \epsilon k_y \gamma_c \tag{23}$$

where

$$\gamma_c = \sin(k_x x + k_y y). \tag{24}$$

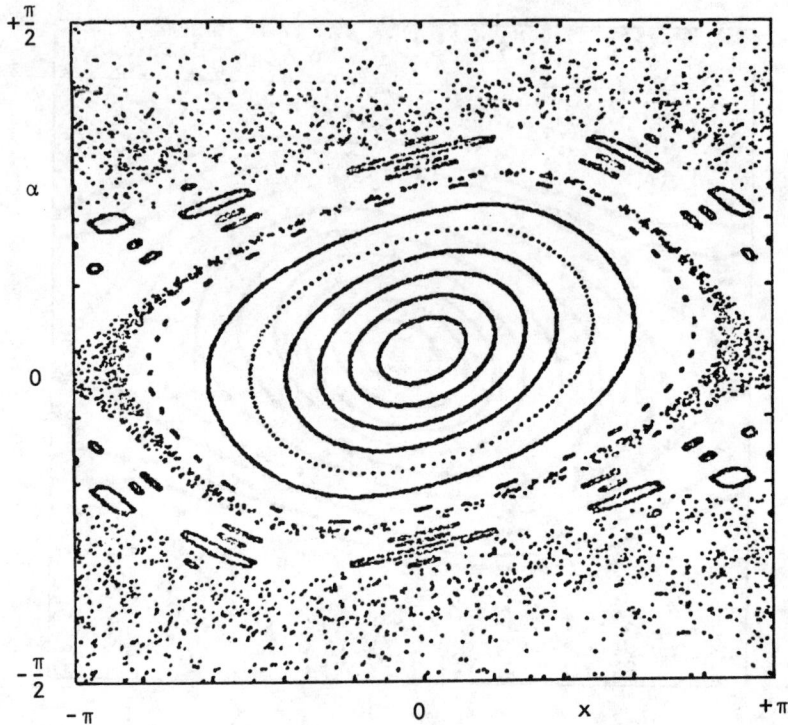

Fig. 12. Exact motion for the uncoupled billiards problem. Same parameters as Fig. 11.

The parameter ϵ is the amplitude of the diagonal ripple and indicates the magnitude of the coupling between the x motion and the y motion. As in the case of two degrees of freedom, there are both integrable and nonintegrable trajectories. Unlike the previous case, however, stochastic trajectories are not confined to a particular locality in the phase space.

An example of integrable motion is shown in Fig. 13. Three invariants of the motion exist. The motion is confined to a two dimensional surface in the four dimensional surface of section. The projection of this motion onto the α-x plane yields an annulus of finite area. Figures 13a and 13b show the motion of a single particle after 5,000 and 50,000 iterations, respectively. The initial conditions were chosen close to the center of the primary resonance. For these initial conditions, the fraction of the phase space area occupied by the Arnold web is very small. It is therefore highly likely, but not absolutely certain, that the motion shown in Fig. 13 is truly integrable. (Computer roundoff errors may destroy this integrability on a very slow time scale.)

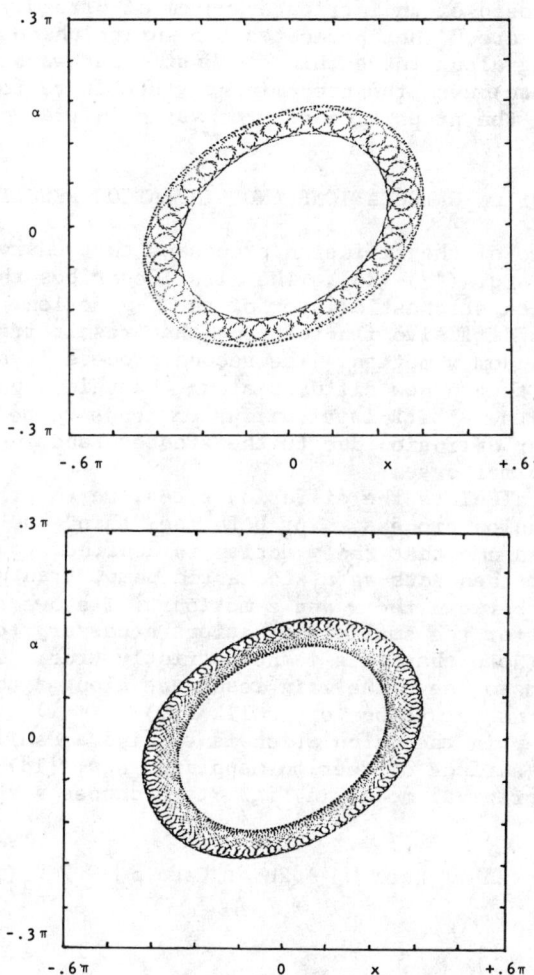

Fig. 13. Integrable motion in the coupled billiards problem. Motion is projected onto the α-x surface of section. The projection onto the β-y surface is qualitatively the same. Parameters are $\epsilon/h = .004$; $\lambda_x:h:a_x$ and $\lambda_y:h:a_y$ as 100:10:2.

We turn, finally, to the featured topic of this paper, stochastic motion in the Arnold web. The stochastic motion for the coupled equations, Eqs. (18)-(24), is characterized by only one (global) invariant. It fills a four dimensional volume in the four dimensional surface of section, and diffuses slowly along the thin stochastic layers which comprise the threads of the pervasive Arnold web. The web itself is composed of an intricate system of "freeways, streets, sidewalks, cracks, etc." that permeates the entire phase space. Particles diffusing along these thin stochastic pathways are able to leave (and penetrate) even the predominantly stable regions of the phase space, where the proportion of stochastic initial conditions is small.

IV. DIFFUSION CALCULATIONS AND SIMULATION RESULTS

We examine two of the diffusion processes that characterize the simplified system, Eq. (18)-(24). The first describes the diffusion of α along the thick stochastic layer of the β-y motion. The quantity α experiences diffusive fluctuations that result from the small coupling to the random y motion. The second process is similar to the first, except that α now diffuses along the thin separatrix layer of the β-y motion. Thick layer diffusion tends to be much faster than thin layer diffusion due to the greater randomness of the y motion in the former case.

In order to calculate the diffusion rates, we adopt a simple model of the diffusion process. For both the "thin" and "thick" layer processes, we assume that the y motion is confined to the stochastic layer. It then acts as a stochastic pump, transporting energy back and forth between the x and z motions. Its own energy may not change except for the small fluctuations necessary to affect the pumping action. (Note that this is not strictly true. It is possible for the system to leave the main resonance along a thin "alleyway", but this turns out to be very unlikely.)

The first step in the calculation is to find a Hamiltonian that will generate the surface of section mappings, Eqs. (18)-(24). In deference to our original model in Fig. 9, we choose a "kicked" Hamiltonian

$$H(\alpha,x,\beta,y,n) = 2h \ln [\sec \alpha] + 2h \ln [\sec \beta] - 2 \delta_1(n) \, C(x,y) \quad (25)$$

where

$$C(x,y) \equiv a_x \cos k_x x + a_y \cos k_y y + \epsilon \cos (k_x x + k_y y) \quad (26)$$

and

$$\delta_1(n) \equiv \sum_{m=-\infty}^{+\infty} \delta(n-m)$$

$$= 1 + 2 \sum_{q=1}^{\infty} \cos (2\pi n q). \quad (27)$$

Equations (18)-(24) may be derived from Eq. (25) by a simple integration of Hamilton's equations. Note that H in Eq. (25) is a nonautonomous Hamiltonian in two degrees of freedom. It is related to the net energy in the x and y motion, and is not conserved.

For a small enough coupling ϵ the diffusion is limited by the rate at which energy can be passed back and forth between the x and y motion. Therefore we neglect the explicit coupling to the z motion and take only the q = 0 term in Eq. (27). (The z coupling is recognized implicitly in the random phase assumptions that are used later.) This approximation makes Eq. (25) autonomous

$$H(\alpha,x,\beta,y) = 2h \ln [\sec \alpha] + 2h \ln [\sec \beta] - 2 C(x,y). \quad (28)$$

If we assume α to be small we may simplify the first term

$$\ln [\sec \alpha] \sim \frac{\alpha^2}{2} \quad (29)$$

Equation (28) may be directly reduced to a nonautonomous Hamiltonian for the x motion only

$$H_x = h\alpha^2 - 2a_x \cos k_x x - 2\epsilon \cos [k_x x + k_y y(n)], \quad (30)$$

where y is now considered to be an explicit function of n. Defining $\theta \equiv k_x x$, $\varphi(n) \equiv k_y y(n)$ and $\mu \equiv -2\epsilon$ we have

$$H_x = h\alpha^2 - 2a_x \cos \theta + \mu \cos [\theta + \varphi(n)]. \quad (31)$$

The type of diffusion observed depends upon the initial conditions. For thick layer diffusion, the initial conditions are chosen close to the center of the primary resonance in the α-x plane, and within the thick stochastic layer of the β-y plane. In the absence of coupling, $\epsilon = 0$, the motion in the α-x plane is confined to a smooth closed curve (like those seen close to the center of Fig. 11). The number of bounces required to go exactly once around the curve is the α-period T_α. For a finite coupling, H_x diffuses slowly due to the small randomizing influence of the stochastic β-y motion. The diffusion of H_x is shown in Fig. 14 for 2,000, 10,000; and 30,000 iterations. The motion eventually explores all of the α-x plane. The corresponding motion in the β-y plane is restricted to the thick stochastic layer, at least until the α-x motion reaches its own thick layer.

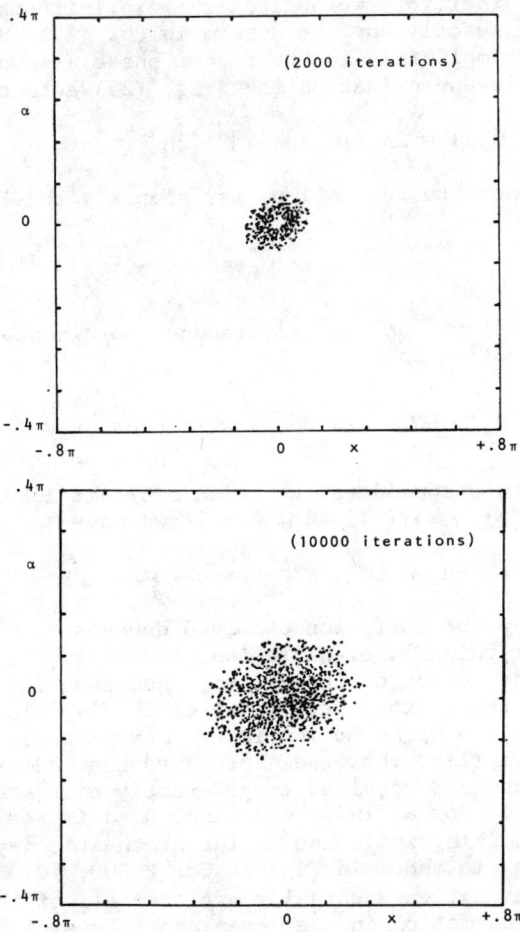

Fig. 14. Thick layer diffusion for the coupled billiards problem. Initial conditions are close to the central "fixed point" in the α-x space and within the thick stochastic layer (near $|\beta| = \pi/2$) of the β-y space. Parameters are the same as Fig. 13.

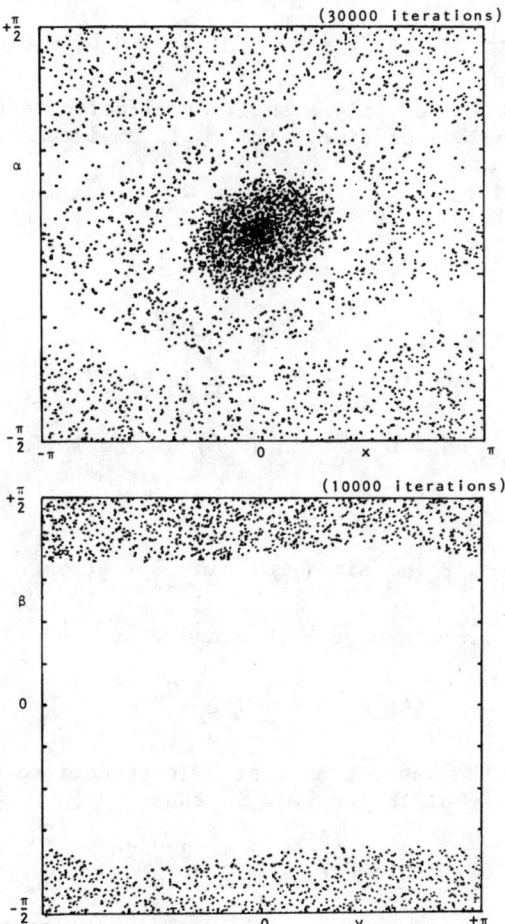

Fig. 14. cont.

The diffusion coefficient D_1 for thick layer diffusion has been calculated from Eq. (31), under the assumption that the n dependent variable $\varphi(n)$ makes a sudden random jump to a new value whenever n = integer.

The evolution of H_x, from Eq. (31), is

$$\frac{dH_x}{dn} = \frac{\partial H_x}{\partial n} = \mu \sin[\theta + \varphi(n)] \frac{d\varphi}{dn} \qquad (32)$$

$$= \frac{d}{dn}[\mu \cos(\theta + \varphi)] - \mu \frac{d\theta}{dn} \sin[\theta + \varphi(n)]. \qquad (33)$$

The first term contributes only a small oscillation with no net change over long periods of time. For small amplitude libration in the x-α plane, we have

$$\theta \sim \theta_o \cos \omega_o n \qquad (34)$$

where

$$\omega_o = \frac{2\pi}{T_\alpha} = 2 k_x (a_x h)^{\frac{1}{2}}$$

Using this, we integrate the second term in Eq. (33)

$$\Delta H_x = \int_m^{m+1} dn\, \mu\, \theta_o\, \omega_o \sin(\omega_o n) \sin[\theta + \varphi(n)]. \qquad (35)$$

For $\omega_o \ll 1$, this is

$$\Delta H_x \sim \mu\, \theta_o\, \omega_o \sin(\omega_o m) \sin[\theta + \varphi(m)]. \qquad (36)$$

We square this and average over both m and φ to get

$$\langle \Delta H_x^2 \rangle_{t,\varphi} = \frac{1}{4} \mu^2 \theta_o^2 \omega_o^2 \qquad (37)$$

where we have used the assumption that φ is randomized at m = integer. The thick layer diffusion rate is then

$$D_1 = \frac{1}{2} \langle \Delta H^2 \rangle_{t,\varphi} = \frac{1}{8} \mu^2 \theta_o^2 \omega_o^2. \qquad (38)$$

The parameters μ and ω_o will remain fairly constant as H_x diffuses. The quantity θ_o, however, increases with H_x, resulting in an increase in the diffusion rate as the x oscillations grow.

In Fig. 15, the theoretical value of D_1 is compared with measurements obtained from the direct iteration of the difference equations. For each experiment, 100 particles were started with identical initial conditions on a libration curve of the α-x plane, and with random initial conditions in the thick stochastic layer of the β-y plane. The motion was followed for 500 collisions, and the rms value of the energy $h(\alpha^2)_{rms}$ was calculated and compared with the

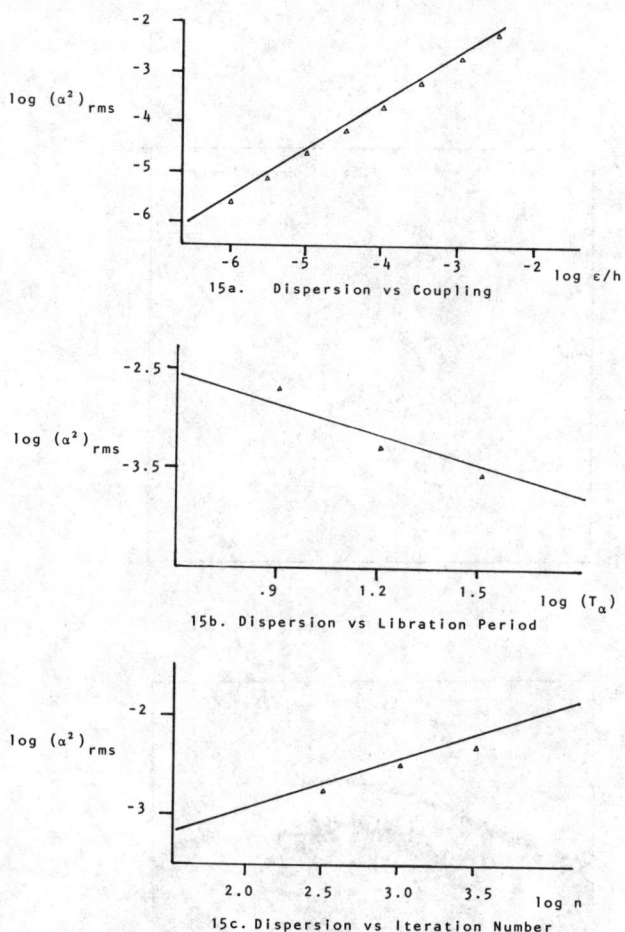

Fig. 15. Thick layer diffusion. Comparison of the theoretical diffusion, Eq. (38) with the results of simulation experiments. In a), the dispersion is plotted vs the coupling amplitude, ϵ. In b), dispersion vs the libration period, T_a. In c), dispersion vs the number of iterations, n. Parameters (except for those varied) are $\epsilon/h = .0001$; $n = 500$; $\lambda_x:h:a_x$ as 10:10:1; $\lambda_y:h:a_y$ as 100:10:1.7. Each triangle gives the spread of 100 particles.

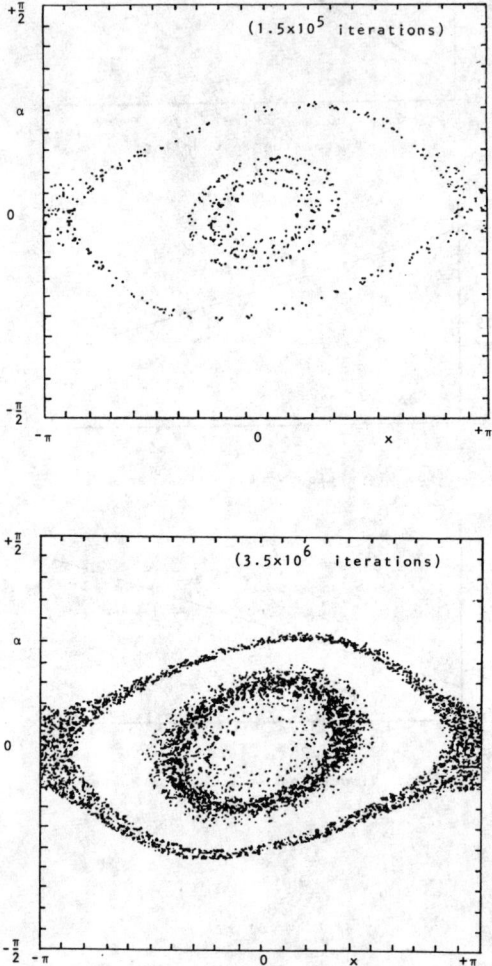

Fig. 16. Thin layer diffusion. Initial conditions are close to the central fixed point in the α-x space and within the separatrix stochastic layer in the β-y space. Parameters are the same as Fig. 13.

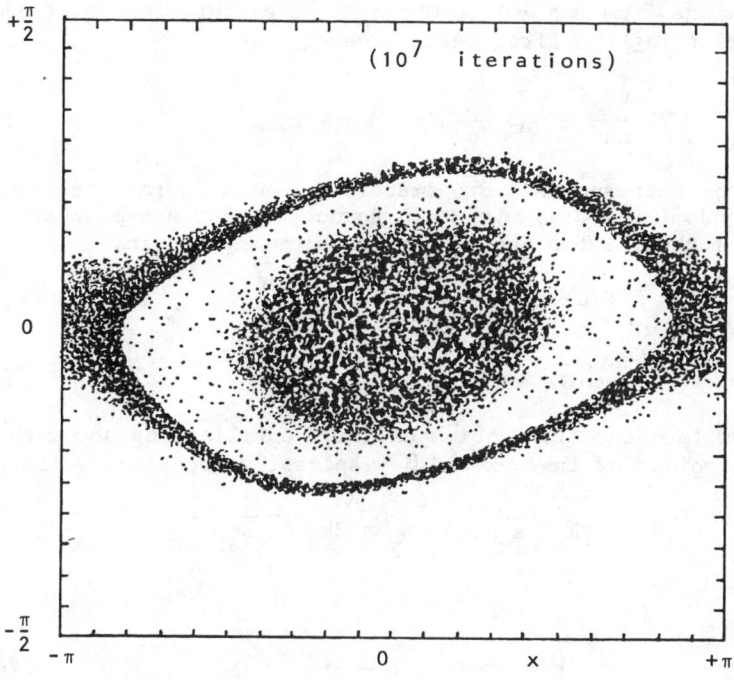

Fig. 16. cont.

theory. Figure 15a shows the variation with coupling strength ε, Fig. 15b the variation with α period T_α, and Fig. 15c the variation with the number of iterations n. The solid lines show the theoretical predictions and the triangles the experimental measurements. Each triangle represents the average of four separate runs. The theoretical predictions, although consistently a little high, are quite good. The discrepancy probably reflects an expected small deviation from true random phase

We turn now to the thin layer diffusion. Although the initial conditions remain close to the central fixed point of the α-x space, they are now chosen inside the thin stochastic layer surrounding the primary separatrix of the β-y space. The diffusion of H_x is again caused by the small coupling to the stochastic y motion, but since thin layer trajectories are considerably less "random" than thick layer trajectories, the diffusion is significantly weaker.

An example of thin layer diffusion is shown in Fig. 16 where both the y and x motions are displayed on the same plot. The y motion is confined to its separatrix until the x motion reaches its own separatrix.

To calculate the diffusion rate, we find the energy change ΔH_x as $\varphi \equiv k_y y$ swings from $\varphi = -\pi$ to $\varphi = +\pi$. Starting with Eq. (33) and again neglecting the first term we have

$$\frac{dH_x}{dn} = -\mu \frac{d\theta}{dn} \sin[\theta + \varphi(n)]. \tag{39}$$

As before, $\theta(n)$ corresponds approximately to small librations. But instead of randomizing $\varphi(n)$ with each bounce, we now assume that it evolves very much like the phase on a pendulum separatrix.

$$\theta(n) = \theta_o \sin[\omega_x(n + n_o)] \tag{40}$$

$$\varphi(n) = 4 \tan^{-1}\left(e^{\omega_y n}\right) - \pi \tag{41}$$

where ω_x and ω_y are the frequencies of small oscillations about the central fixed points of the α-x and β-y spaces, respectively. Let

$$\omega_x = 2k_x \sqrt{a_x h}, \quad \omega_y = 2k_y \sqrt{a_y h}$$

and defining

$$s \equiv \omega_y n, \quad r \equiv \frac{\omega_x}{\omega_y} \tag{42}$$

then

$$\Delta H_x = \mu \theta_o r \int_{-\infty}^{+\infty} ds\, I(s)$$

where

$$I = \cos[r(s + s_o)] \sin\{\theta_o \sin[r(s + s_o) + \varphi]\}. \tag{43}$$

Using $\theta_o \ll 1$

$$I = \cos[r(s + s_o)] \sin\varphi. \tag{44}$$

Only the symmetric part contributes to the integral.

$$I_{sym} = \frac{1}{2} \sin(rs_o)[\cos(\varphi + rs) - \cos(\varphi - rs)]. \tag{45}$$

Thus,

$$\Delta H_x = \frac{1}{2} \mu \theta_o r \sin(rs_o)[A_2(-r) - A_2(r)] \tag{46}$$

where $A_m(\lambda)$ is the Melnikov-Arnold integral[7]

$$A_2(\pm r) = 4\pi r e^{\pm r/2}/\sinh(\pi r). \tag{47}$$

We have finally

$$\Delta H_x = 4\pi\mu\theta_o \, r^2 \sin(rs_o) \sinh(\pi r/2)/\sinh(\pi r). \tag{48}$$

If we assume that $rs_o = \omega_x n_o$ is randomized after every half period of $\varphi(n)$, then we can average ΔH_x^2 to get

$$\langle \Delta H_x^2 \rangle_{s_o} = 8\pi^2\mu^2\theta_o^2 \, F(r) \tag{49}$$

where

$$F(r) = r^4 \sinh^2(\pi r/2)/\sinh^2(\pi r). \tag{50}$$

A plot of $F(r)$ is shown in Fig. 17. It is sharply peaked close to $r = 1$, suggesting that if the characteristic frequencies of the separatrix and libration motion differ by as much as a factor of four, the diffusion will be reduced by two orders of magnitude.

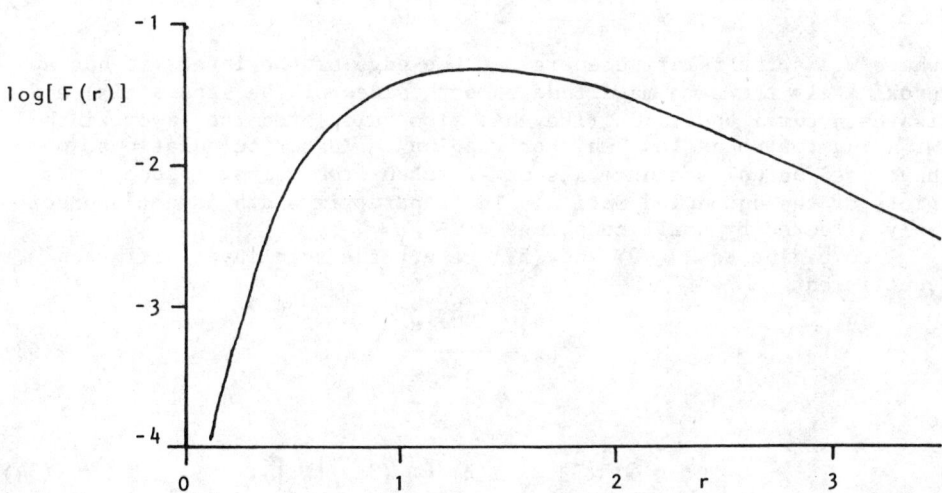

Fig. 17. Plot of the function $F(r) = [r^4 \sinh^2(\pi r/2)]/[\sinh^2 \pi r]$ for the dependence of thin layer diffusion on $r = \omega_x/\omega_y$.

To obtain the diffusion coefficient, we need to know the mean half period of the motion in the thin stochastic layer $\overline{T}_\beta$. The half period of a true pendulum that follows a trajectory very close to the separatrix is approximately

$$T_\beta = \frac{1}{\omega_y} \ln \frac{32}{|W|} \tag{51}$$

where

$$W \equiv \frac{H_y - H_s}{H_s} \ll 1,$$

and

$$H_s = \frac{\omega_y^2}{h}$$

is the separatrix energy. Chirikov[8] has shown that the average half period inside the stochastic layer may be computed by simply integrating the half period over the energy interval of the layer. The result is

$$\overline{T}_\beta = \frac{1}{\omega_y} \ln \frac{32e}{|W_o|} \tag{52}$$

where W_o is the relative energy at the edge of the layer (it has approximately the same magnitude on both sides of the separatrix) and e is the natural base. Chirikov has also calculated the layer width W_o using the so-called "whisker mapping". In our calculations, we have used actual measurements of W_o taken from computer generated plots of the uncoupled motion. The separatrix width is not appreciably affected by small couplings $\epsilon \ll a_y$.

Combining Eqs. (49) and (52) we get the thin layer diffusion coefficient

$$D_2 = \frac{\langle \Delta H_x^2 \rangle_{s_o}}{2\overline{T}_\beta} \tag{53}$$

or

$$D_2 = 4\pi^2 \mu^2 \theta_o^2 \omega_y \, F(r)/\ln(32e/|W_o|). \tag{54}$$

In Fig. 18, the theoretical thin layer diffusion is compared with experimental measurements. Each triangle represents the final spread of 100 particles that have been started with identical initial conditions in the α-x space and slightly different initial conditions in the thin stochastic layer of the β-y space. The motion

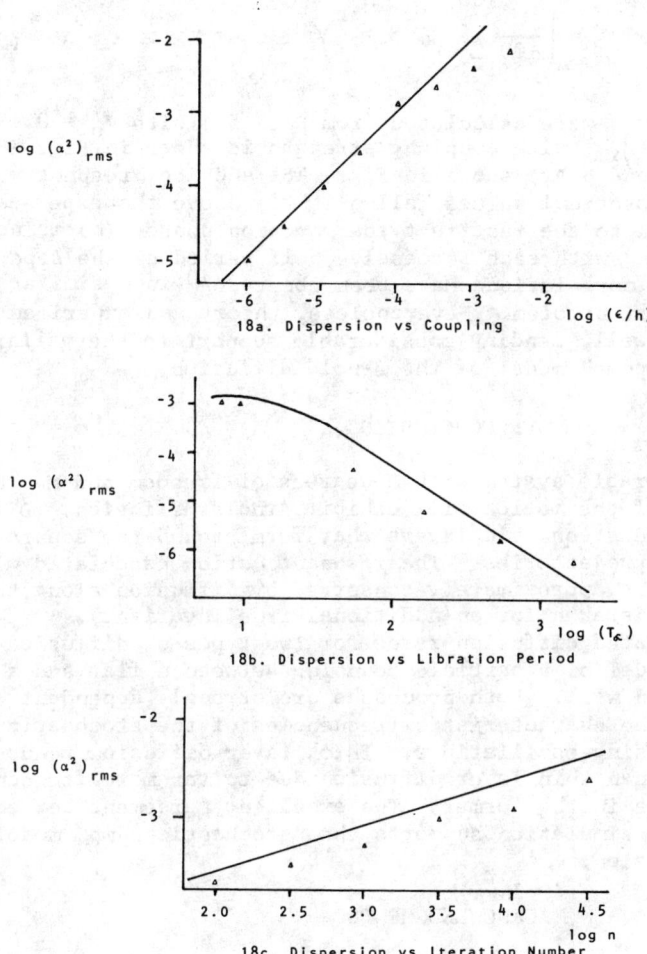

Fig. 18. Thin layer diffusion. Comparison of the theoretical diffusion, Eq. (54) with the results of simulation experiments. The three graphs are the same relations as those shown in Fig. 15. Parameters (except for those varied) are $\epsilon/h = .0001$; $n = 2000$; $\lambda_x : h : a_x$ as 100:10:1; $\lambda_y : h : a_y$ as 100:10:1.8. Each triangle gives the spread of 100 particles.

was followed for 2000 iterations and the rms spread was computed using

$$\alpha^2_{rms} = \left[\frac{1}{100}\sum_{i=1}^{100}\left(\bar{\alpha}^2 - \alpha_i^2\right)^2\right]^{\frac{1}{2}}. \qquad (55)$$

The theoretical curves were calculated from Eq. (54) with $W_o = 0.191$. The variation of $(\alpha^2)_{rms}$ with coupling strength is shown in Fig. 18a. Variations with ω_x and n are shown in Figs. 18b and 18c, respectively. Again, the theoretical values fall slightly above the exper-nmental, probably due to the fact that the y motion phase $\varphi(n)$ is not completely randomized with each successive half period of the separatrix motion. Phase correlations have been observed[10] in a similar mapping for the Fermi problem. Nevertheless, theory and experiment agree surprisingly well, lending considerable support to the validity of the "stochastic pump" model of the Arnold diffusion.

V. CONCLUSION

Any near-integrable system with N degrees of freedom and $M < N - 1$ global invariants of the motion will exhibit Arnold diffusion. Diffusion occurs on the stochastic layers that form around the separatrices of resonant trajectories. The resonant action associated with a particular layer is approximately conserved by diffusion along that layer (and may be mistaken for an additional true invariant).

We have calculated diffusion rates for two types of diffusion using the simple model of a particle bouncing between a flat and a periodically rippled wall. Both processes are strongly dependent upon the ratio of the characteristic frequencies of the stochastic layer and the diffusing oscillations. Thick layer diffusion tends to be much faster than thin layer diffusion due to the more frequent phase randomizations in the former. The excellent agreement between theory and computer simulation supports the "stochastic pump" model of the Arnold diffusion.

ACKNOWLEDGMENTS

We are especially in debt to B. V. Chirikov for providing us with many of the concepts and techniques used in this paper. This work was supported under National Science Foundation Grant ENG78-09424 and Department of Energy Contract EY-76-S-03-0034-PA215.

REFERENCES

1. A. M. Vinogradov and B. A. Kupershmidt, Russ. Math. Surveys 32, 177 (1977).
2. M. Toda, "Studies of a Nonlinear Lattice," Phys. Reports 18c, 3 (1975). See also, M. V. Berry, "Regular and Irregular Motion" in Nonlinear Dynamics, Am. Inst. Phys. Conf. Proc. Series Vol. 46, A.I.P. New York (1978).

3. E. T. Whittaker, A Treatise on the Analytical Dynamics of Particles and Rigid Bodies, (Cambridge University Press, fourth edition, London 1937), p 313-314.
4. C. L. Siegal, Ann. Math. 42, 806 (1941) Math. Anal. 128, 144 (1954).
5. A. N. Kolmogorov, Dokl. Akad. Nauk SSSR 98 527 (1965).
6. V. I. Arnold, A. Avez, Ergotic Problems of Classical Mechanics (Benjamin, Inc., New York, 1968).
7. V.I. Arnold, Dokl. Akad. Nauk SSSR 156, 9 (1964).
8. B. V. Chirikov, "A Universal Instability of Many Dimensional Oscillator Systems", Institute of Nuclear Physics, 630090 Novosibirsk, USSR.
9. G. Contopoulos, L. Galgani and A. Giorgilli, "On the Number of Isolating Integrals in Hamiltonian Systems" ESO Preprint n. 26, submitted to Phys. Rev. A (1978).
10. C. Froeschlé, Astrophs. Space Sci. 14, 110 (1971); C. Froeschlé and J. P. Scheidecker, Astrophys. Space Sci. 25, 373 (1973).
11. V. K. Melnikov, Dokl. Akad. Nauk SSSR 144, 747 (1962); 148, 1257 (1963); Trudy Moskovskova Mat. Obschestra 12, 3 (1963). See also Ref. 7 and Ref. 8 (Appendix).
12. See Ref. 8, Sections 4.4 and 6.1.
13. M. A. Lieberman and A. J. Lichtenberg, Phys. Rev. 45, 1852 (1972).

VARIATIONAL PRINCIPLES FOR INVARIANT TORI AND CANTORI

I. C. Percival

Queen Mary College (University of London)

ABSTRACT

A variational principle for invariant tori with desirable properties is described, and used for the standard mapping. A static analogue of dynamical systems is introduced in order to provide insight. This leads to cantori, which are invariant Cantor sets in the irregular or stochastic region of phase space.

I. INTRODUCTION

Minimum principles have been of great assistance to the solution of linear problems, particularly in the theory of linear vibrations and in the quantum theory. Variational principles are of recognized importance to the foundations of dynamics, whether linear or nonlinear, but they have not been used extensively to solve specific problems in nonlinear dynamics for the following reasons:

R1 We need variational principles for invariant tori, not for the orbits.
R2 Small divisors cause difficulties for variational principles, as they do for perturbation theory.
R3 Unlike the Rayleigh-Ritz principle for linear systems, the variational principles of classical dynamics do not usually provide useful bounds.

The general theory of invariant tori and of the KAM theorem, which clarifies the role of small divisors is reviewed by Whiteman[1] and by Berry.[2] R1 is no longer a difficulty as variational principles for invariant tori have been formulated by Percival[3] and applied to nonlinear problems in molecular dynamics (Percival and Pomphrey,[4] Percival[5]). More recently (Percival[6]) a fixed-frequency Lagrangian variational principle has been formulated for invariant tori in which the small divisor problems are not so severe. Furthermore in some important cases the principle provides a bound for the Lagrangian functional.

The method is applied to the standard mapping in Section 3 and results are presented in Section 4.

A static analogue of particle dynamics is discussed in Section 5, where it is applied to the vertical pendulum and the standard mapping.

In Section 6 it is shown how this analogue leads to the proposal that there are invariant Cantor sets of fixed frequency, which we name "cantori", in the irregular or stochastic regions of phase space. The sawtooth mapping is introduced and it is shown analytically that there are cantori for that mapping.

II. VARIATIONAL PRINCIPLE

Consider a system of n freedoms, with Lagrangian $L(\underline{q}, \underline{\dot{q}})$. We define an invariant torus Σ by a configuration function

$$\underline{q}_\Sigma(\underline{\theta}) \tag{2.1}$$

of the vector angle variable

$$\underline{\theta} = (\theta_1, \theta_2, \ldots, \theta_n), \tag{2.2}$$

and by an angular frequency vector

$$\underline{\omega} = (\omega_1, \omega_2, \ldots, \omega_n), \tag{2.3}$$

where ω_j is the angular frequency of the change with respect to time of the angle variable θ_j. The function $\underline{q}_\Sigma(\underline{\theta})$ is periodic of period 2π in each of the angle variables θ_j.

The gradient operator in $\underline{\theta}$ space is

$$\underline{\nabla}_\theta = (\frac{\partial}{\partial \theta_1}, \frac{\partial}{\partial \theta_2}, \ldots, \frac{\partial}{\partial \theta_n}) . \tag{2.4}$$

There is a dynamical differential operator $D_{\underline{\omega}}$ for a torus which is equivalent to the operator d/dt for the orbit. $D_{\underline{\omega}}$ is defined by the

$$D_{\underline{\omega}} f(\underline{\theta}) = \underline{\omega} \cdot \underline{\nabla}_\theta f(\underline{\theta}) \tag{2.5}$$

for any differentiable function $f(\underline{\theta})$.

The Lagrangian representation of a torus in position-velocity space is given by the function

$$(\underline{q}_\Sigma(\underline{\theta}), D_{\underline{\omega}} \underline{q}_\Sigma(\underline{\theta})). \tag{2.6}$$

The more usual Hamiltonian representation of a torus in phase space can be obtained from $\underline{q}_\Sigma(\underline{\theta})$ and $\underline{\omega}$ through the definition of the momentum function

$$\underline{p}_\Sigma(\underline{\theta}) = \underline{\nabla}_{\dot{q}} L(\underline{q}, D_{\underline{\omega}} \underline{q}), \tag{2.7}$$

which together with the coordinate function $\underline{q}_\Sigma(\underline{\theta})$ provides a parametric definition of the torus. We use $\underline{\nabla}_q L$ and $\underline{\nabla}_{\dot{q}} L$ to represent gradients with respect of the first and second arguments of L.

The mean of a function $f(\underline{\theta})$ over the torus is defined to be

$$\langle f(\underline{\theta}) \rangle = \frac{1}{(2\pi)^n} \oint d^n\theta \cdot f(\underline{\theta}). \tag{2.8}$$

The variational principle for the torus states that the Lagrangian functional

$$\Phi = \langle L(\underline{q}(\underline{\theta}), D_{\underline{\omega}}\underline{q}(\underline{\theta})) \rangle \tag{2.9}$$

is stationary with respect to arbitrary variations in the differentiable function $\underline{q}(\underline{\theta})$, for a fixed value of the frequency vector $\underline{\omega}$. The Euler-Lagrange equations for Φ are then the angle Lagrange equations

$$D_{\underline{\omega}}[\underline{\nabla}\cdot_{\underline{q}} L(\underline{q}, D_{\underline{\omega}}\underline{q})] = \underline{\nabla}_{\underline{q}} L(\underline{q}, D_{\underline{\omega}}\underline{q}) \tag{2.10}$$

for the torus.

If $\underline{q}_{\Sigma}(\underline{\theta})$ is a solution of equation (2.10) then $\underline{q}_{\Sigma}(\underline{\omega}t + \underline{\theta}^o)$ with $-\infty < t < \infty$ defines an orbit which satisfies Lagrange's equations of motion with initial vector coordinate and velocity given by

$$\underline{q}_{\Sigma}(\underline{\theta}^o), \quad D_{\underline{\omega}}\underline{q}_{\Sigma}(\underline{\theta}^o). \tag{2.11}$$

III. THE STANDARD MAPPING

We use a method based on the variational principle to determine invariant tori.

In order to evaluate this method we apply it to the so-called standard area-preserving mapping

$$p_{t+1} = p_t - (\alpha/2\pi)\sin(2\pi q_t) \quad \text{(modulo 1)},$$

$$q_{t+1} = q_t + p_{t+1} \quad \text{(modulo 1)},$$

$$(0 \leq q < 1, \; 0 \leq p < 1, \; t \text{ integral}), \tag{3.1}$$

which has a two-dimensional toric phase space. This system has already been extensively studied by Taylor,[7] Chirikov[8] and Greene.[9]

Poincaré introduced area-preserving mappings as a means of studying dynamical systems under perturbation. The orbits of the mappings are sequences of points (q_t, p_t) which are much easier to compute than the orbits of conservative dynamical systems.

For $\alpha = 0$ the standard mapping has invariant tori which are closed curves in the two-dimensional phase space. As α is increased, more and more of the invariant tori are destroyed, and apparently replaced by irregular or stochastic orbits which are dense in a two-dimensional region of the phase space.

The standard mapping is the discrete analogue of the vertical pendulum, with derivatives replaced by differences. It describes the motion of a one-dimensional rotor with orientation $2\pi q$ under the action of a uniform succession of impulses, each with the same sinusoidal dependence on q. The position q_t is recorded at the time of each impulse and the momentum p_t just beforehand.

The usual form (3.1) for the standard mapping is Hamiltonian. We need the Lagrangian form which is

$$q_{t+1} - 2q_t + q_{t-1} = -(\alpha/2\pi)\sin(2\pi q_t). \quad \text{(modulo 1)} \quad (3.2)$$

This difference equation is obtained on applying the stationary condition to the functional

$$S(q^1, t^1; q^0, t^0) = \sum_{t=t^0}^{t^1-1} L(q_t, q_{t+1} - q_t), \quad (3.3)$$

where the Lagrangian L is given by

$$L(q_t, q_{t+1} - q_t) = \frac{(q_{t+1} - q_t)^2}{2} + \frac{\alpha}{(2\pi)^2}\cos(2\pi q_t). \quad (3.4)$$

The analogy with Hamilton's principle for the vertical pendulum is obvious.

An invariant torus for the mapping is represented in the Lagrangian theory by the continuous function $q(\hat{\theta})$ where

$$q(\hat{\theta}) = \hat{\theta} + x(\hat{\theta}), \quad (3.5)$$

The function $x(\hat{\theta})$ is periodic in $\hat{\theta}$ of period 1, and has zero mean.

The normalized angle variable

$$\hat{\theta} = \theta/(2\pi) \quad (3.6)$$

depends linearly on t (modulo 1) with normalized frequency coefficient

$$\nu = \omega/(2\pi). \quad (3.7)$$

Thus we have

$$\theta = \nu t + \hat{\theta}^o \quad \text{(modulo 1)} \quad (3.8)$$

and the mean period of the motion is $\tau = \nu^{-1}$, as illustrated in Fig. 1.

The Lagrangian functional for the torus of a mapping is given by

$$\Phi = \langle L[q(\hat{\theta}), q(\hat{\theta} + \nu) - q(\hat{\theta})]\rangle, \quad (3.9)$$

where the mean is taken over the normalized angle variable $\hat{\theta}$. For

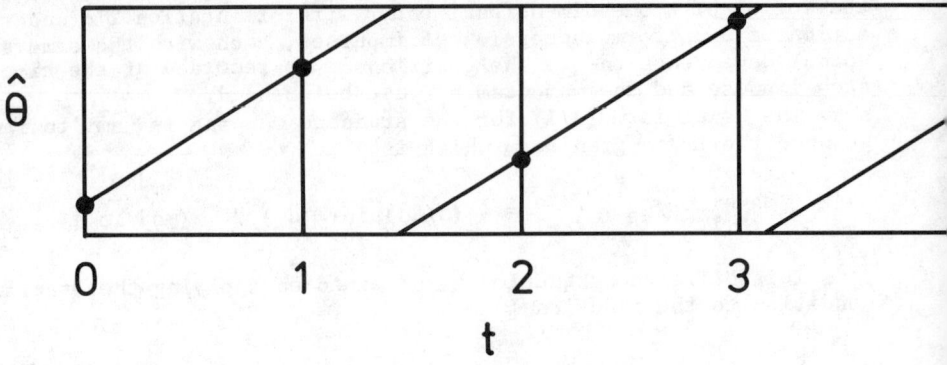

Fig. 1.

the standard mapping we have

$$\Phi = \langle \frac{[q(\hat{\theta} + \nu) - q(\hat{\theta})]^2}{2} + \frac{\alpha}{(2\pi)^2} \cos[2\pi q(\hat{\theta})] \rangle. \quad (3.1)$$

On applying the condition that Φ should be stationary we obtain the angle Lagrange equation for the standard mapping, which is

$$q(\hat{\theta} + \nu) - 2q(\hat{\theta}) + q(\hat{\theta} - \nu) = -(\alpha/2\pi)\sin[2\pi q(\hat{\theta})]. \quad (3.1)$$

Clearly the functional Φ is bounded below, and for $\alpha = 0$ the low bound is attained by that solution of equation (3.11) which satisfies the condition (3.5). We do not know that this is still true when $\alpha \neq 0$, but numerical experiments suggests that it is.

IV. METHOD OF SOLUTION

A numerical method was used for minimizing the Lagrangian. It is based on an iterative solution of Eq. (3.11) subject to the condition (3.5). For $\alpha = 0$ the solution is given by (3.5) with $x(\hat{\theta}) = 0$. The value of α was increased in steps $\Delta\alpha$. At each such step the numerical solution for the previous value of α was used as a starting solution $q^o(\hat{\theta})$ for the sequence of solutions $q^i(\hat{\theta})$ obtained by solving the equation

$$q^i(\hat{\theta} + \nu) - 2q^i(\hat{\theta}) + q^i(\hat{\theta} - \nu) = F^i(\nu), \quad (4.1)$$

with

$$F^i(\hat{\theta}) = -(\alpha/2\pi)\sin[2\pi q^{i-1}(\hat{\theta})] \quad (4.2)$$

subject to condition (3.5). The Eq. (4.1) was solved by Fourier analysis using a fast Fourier transform program. The Fourier coefficients in the expansions

$$F^i(\hat{\theta}) = \sum_{s=-\infty}^{\infty} F^i_s \exp 2\pi i \hat{\theta} \qquad (4.3)$$

$$x^i(\hat{\theta}) = \sum_{s=-\infty}^{\infty} x^i_s \exp 2\pi i \hat{\theta} \qquad (4.4)$$

are related by the equation

$$x^i_s = F^i_s / (4 \sin^2(s\nu/2)). \qquad (4.5)$$

In practice it was only possible to use a finite Fourier sum. At each step the Lagrangian was calculated to ensure that its value was reduced. Results are presented in Table I for 15 Fourier components.

TABLE I

α	$\Delta\Phi$	$\Delta\Phi$	$\Delta\Phi$	$\Delta\Phi$	$\Delta\Phi$	Φ
0.0						0
0.09716	-2(-5)	-6(-9)	-1(-11)			-1.720720(-5)
0.19432	-2(-5)	-3(-8)	-2(-10)			-6.886596(-5)
0.29148	-2(-5)	-6(-8)	-1(-9)			-1.550918(-4)
.	.	.	.			
.	.	.	.			
.	.	.	.			
.	.	.	.			
0.77728	-2(-5)	-6(-7)	-8(-8)			-1.116506(-3)
0.87444	-2(-5)	-9(-7)	-2(-7)			-1.41962(-3)
0.97160	-2(-5)	-1(-6)	-3(-7)			-1.762635(-3)
0.97161	-1(-7)	-4(-8)	+2(-7)	+3(-6)	+5(-5)	

The starting torus for each value α was the converged torus for the previous value of α. $\Delta\Phi$ is the change in the Lagrangian functional Φ at each step of the iteration.

For sufficiently large perturbation $\Delta\Phi$ becomes positive and the procedure diverges. In the illustrated case this takes place near the value of a α for which Greene[9] finds that the tori break up. However, the geometric iteration procedure is subject to small divisor problems, and it <u>cannot</u> be said that this is a reliable test of the break-up value of α.

V. THE STATIC ANALOGUE

It may be necessary to use a Newton iteration method, as in the proof of the KAM theorem and as used by Helleman[10] in his procedure for determining periodic orbits.

Problems in dynamics have their analogues in statics, which are sometimes easier to think about. A particle moving along its orbit in configuration space under the action of a potential V(q) has its analogue in a stretched uniform elastic string in equilibrium in the <u>negative</u> of the potential -V(q). This follows from the correspondence between the variational principles for the orbit and for the string.

In each case $\delta S = 0$, where for the orbit of the particle, mass m, moving in potential V(q)

$$S(q_1, q_0) = \int_{q_0}^{q_1} dt\, L(q, dq/dt) = \int_{q_0}^{q} dt\left[\frac{1}{2m}(dq/dt)^2 - V(q)\right] \quad (5.1)$$

and for the string, of stiffness A, in equilibrium in potential -V(q),

$$S(q_1, q_0) = \int_{q_0}^{q_1} ds\, E(q, dq/ds) = \int_{q_0}^{q_1} ds\left[\frac{A}{2}(dq/ds)^2 - V(q)\right], \quad (5.2)$$

where E(q, dq/ds) is the energy density along the string. The variable s labels points on the string by the "natural" distance along the string in its neutral state, in the absence of potentials. The string is assumed to obey Hooke's Law at each point. Clearly for fixed end points (q_0, q_1) the string takes up an equilibrium configuration identical to the orbit of the particle. We can think of an orbit as an elastic object under the influence of a potential -V(q). Notice that a center of repulsion <u>attracts</u> that part of an orbit between two fixed points. The <u>continuity</u> of the orbit corresponds to the unlimited tensile strength of an elastic string that obeys Hooke's Law.

Now we consider the static analogue of the one-dimensional torus of a rotating vertical pendulum. It is shown in Fig. 2. The natural distance along the string is $2\pi\hat{\theta}$. The energy is

$$E = \left\langle \frac{A}{2}\left(\frac{dq}{d\hat{\theta}}\right)^2 + B\cos 2\pi q \right\rangle. \quad (5.3)$$

The elastic string droops to a configuration $q(\hat{\theta})$ in equilibrium which is the <u>same</u> function as the one-dimensional invariant torus of the vertical pendulum in a gravitational field.

There is also an analogue of the standard mapping, consisting of a <u>light</u> string wrapped helically around a horizontal cylinder with weights which are equally spaced at interval ν on the band in its natural state. $\hat{\theta}$ labels the weights by their orientation in their natural position in the absence of the gravitational field. The function q($\hat{\theta}$) represents the position of the weights in the presence of the field.

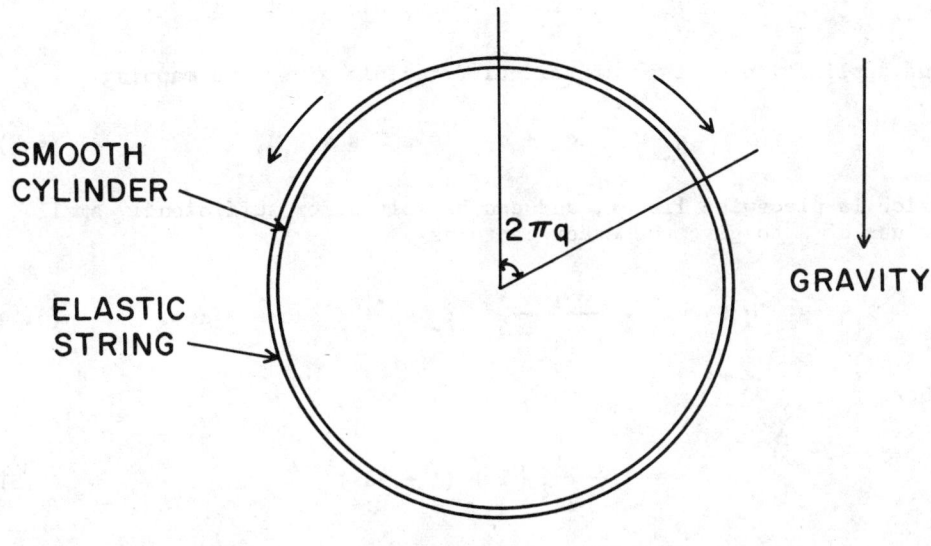

FIG. 2.

The energy functional is the same as the Lagrangian functional (3.10) of the standard mapping. According to the KAM theorem the invariant torus q($\hat{\theta}$) will persist for sufficiently small values of the gravitational field, or sufficiently light weights, and furthermore it will be a continuous function of $\hat{\theta}$.

VI. CANTORI AND THE SAWTOOTH MAPPING

But what happens if the field is strong? Clearly for a sufficiently strong field the elastic string will not be strong enough to maintain the weights near the top of the cylinder. Those on the left will slide to the left, and those on the right will slide to the right, leaving a gap at the top where q($\hat{\theta}$) is discontinuous. We can say that the torus has a finite tensile strength. Furthermore this gap will propagate around the cylinder at intervals of $\Delta\theta = \nu$, becoming smaller the further it goes. There will be a countable infinity of such gaps, so that q($\hat{\theta}$) has an infinite number of discontinuities, which sum to a finite value. The invariant torus breaks up into a Cantor set, or cantorus.

We can obtain analytic cantori for a discontinuous mapping, the <u>sawtooth</u> mapping. Suppose we define

$$\text{saw}(x) = x - \tfrac{1}{2} - \text{ent}(x) \qquad (6.1)$$

where ent(x) is the largest integer greater or equal to x. Clearly saw(x) is periodic of period 1.

The action functional for an orbit of the sawtooth mapping has the form

$$S = a^2 \sum_t \left[\frac{(q_{t+1}-q_t)^2}{2} \right] + \sum_t \frac{1}{2} [\text{saw}(q_t)]^2, \qquad (6.2)$$

and application of the variational principle gives the mapping

$$q_{t+1} - 2q_t + q_{t-1} = a^{-2} \text{saw}(q_t), \qquad (6.3)$$

which is piecewise linear, and can be solved for sufficiently small values of a to give the exact cantorus

$$q(\hat{\theta}) = \hat{\theta} - \frac{1}{(1+4a^2)^{\frac{1}{2}}} \sum_{s=-\infty}^{\infty} \rho^{-|s|} \text{saw}(\hat{\theta} + s\nu), \qquad (6.4)$$

where

$$\rho = 1 + \frac{1}{2a^2}\left[1 + (1+4a^2)^{\frac{1}{2}} \right]. \qquad (6.5)$$

The motion on the cantorus is almost periodic, but is unstable.

ACKNOWLEDGMENTS AND ADDENDUM

I thank J. M. Greene and D. Richards for stimulating discussions. The Lagrangian variational principle has been obtained independently by Klein and Lee.[11]

REFERENCES

1. K. J. Whiteman, Rep. Prog. Phys., **40**, 1033 (1977).
2. M. V. Berry, in *Topics in Nonlinear Dynamics*, ed. S. Jorna, AIP Conf. Proc. Series **46**, 16-120 (New York: American Institute of Physics) (1978).
3. I. C. Percival, J. Phys. A: Math. Nucl. Gen., **7**, 794 (1974).
4. I. C. Percival and N. Pomphrey, Molec. Phys., **31**, 97 (1976).
5. I. C. Percival, Adv. Chem. Phys. 36 (1977).
6. I. C. Percival, J. Phys. A: Math. Nucl. Gen., **12**, 157 (1979).
7. J. B. Taylor, Private communication (1968).
8. B. V. Chirikov, *Universal Instability of Many-Dimensional Oscillator Systems*, Report of Institute of Nuclear Physics, Novosibirsk, submitted to Physics Reports (Phys. Lett. Sec. C.).
9. J. M. Greene, J. Math. Phys., in press.
10. R. H. G. Helleman, in *Topics in Nonlinear Dynamics*, ed. S. Jorna, AIP Conf. Proc. Series **46**, 264-285 (New York: American Institute of Physics) (1978).
11. A. Klein and Li Ching-Toh, J. Math. Phys. **20**, 572 (1979).

ON AN APPLICATION OF PERTURBATION METHODS

TO THE BEAM-BEAM INTERACTION

Tassos Bountis

Dept. of Math and Computer Science

Clarkson College, Potsdam, N. Y. 13676

and

Evangelos Coutsias

Department of Mathematics

University of New Mexico

Albuquerque, N. M. 87131

I. INTRODUCTION

Recently,[1,2] a model equation has been proposed to describe the interaction between colliding beams in intersecting storage rings:

$$\frac{d^2y}{d\theta^2} + ay = \underline{P}(\theta) \, F(y). \tag{1.1}$$

In Eq. (1.1) y is the displacement transverse to the ideal (circular) orbit of the particle, θ is the angle of revolution around the center of the ring, and $\sqrt{a}$ (after denoted by Q in the Accelerator literature) is the "tune" or frequency of the "betatron" oscillations due to the presence of magnetic fields.[5] F(y) represents the electromagnetic interaction between the two beams upon collision and $\underline{P}(\theta)$ is a periodic "pulse" which is nonzero once per revolution during very short $\Delta\theta$ collision intervals.

In this paper we apply the "multiple-scaling" methods of Asymptotic Perturbation Theory[6-8] and obtain analytical results (valid to order ϵ, ϵ^2, etc.) exhibiting the main behavior of the solutions of some linear and nonlinear equations of the type (1.1).

In section 2 we consider <u>linear</u> equations of the form,

$$\frac{d^2y}{d\theta^2} + ay + \epsilon \, (\cos 2\theta + \cos 4\theta + \ldots) \, y = 0, \tag{1.2}$$

for ϵ sufficiently small. In particular, we derive expressions valid to order ϵ^2 for the <u>boundary curves</u> separating the regions of bounded and unbounded solutions of (1.2) in the ϵ,a plane.

The sum of cosines in (1.2) is meant to approximate a "δ-function" type, 2π-periodic $\underline{P}(\theta)$, while F(y), in (1.1), is often taken to have the form of an error function,[4] i.e. F(y) $\propto$ erf(y). Keeping only one cosine and one nonlinear term in an approximation of F(y) we consider[1,2]

$$\frac{d^2y}{d\theta^2} + ay + \epsilon\cos 2\theta (y - by^3) = 0, \quad b > 0 \tag{1.3}$$

[For additional remarks concerning the relevance of Eqs. (1.2), (1.3) to Accelerators see Ref. 1-5].

In Section 3 we obtain zeroth order expressions for the solutions of (1.3) in the neighborhood of a stable solution of the corresponding linear (b = 0) equation. We discuss the (un)boundedness properties of such solutions which, of course, depend strongly on the initial conditions. Moreover, we find that the region of initial conditions leading to unbounded motion dramatically increase with increasing b.

II. INSTABILITY REGIONS FOR A LINEAR, MATHIEU-LIKE EQUATION

In this section we obtain by second order perturbation methods the boundaries of the first two instability regions of the Mathieu-like equation

$$\frac{d^2y}{d\theta^2} + ay + \epsilon(\cos 2\theta + \cos 4\theta)y = 0 \qquad (2.1)$$

in the (ϵ, a) plane, for $0 < \epsilon \leqslant 1$ i.e. near the a-axis. Our results are in agreement with the "exact" boundaries, obtained recently by novel variational methods.[1,2]

Equation (2.1) is of the general type of Hill's equation[9,10] and according to the general theory it has bounded as well as unbounded solutions. These solutions are, in general, given as complex exponentials multiplied by 2π (or 4π) periodic functions. Along the boundary curves in the (ϵ, a) plane, the exponentials are equal to one and the solutions of (2.1) are either 2π-periodic or grow linearly with θ.[9,10] These boundary curves between the bounded and unbounded solutions of (2.1) intersect the ($\epsilon = 0$) a-axis at the points

$$a_c = n^2, \qquad n = 1, 2, 3, 4, \ldots \qquad (2.2)$$

We consider here the cases n = 1 and n = 2 (other values of n can be similarly treated) and look for uniformly valid (in θ) solutions of (2.1) as $\epsilon \to 0$. Close to the <u>first</u> instability region we write

$$a = 1 + \epsilon a_1 + \epsilon^2 a_2 + \ldots \qquad (2.3)$$

Using the standard "multiple-scaling" techniques of perturbation theory[6-8] we write the solution of (2.1) as a series expansion in ϵ:

$$y(\theta, \epsilon) = F_0(\theta, \tilde{\theta}) + \epsilon F_1(\theta, \tilde{\theta}) + \epsilon^2 F_2(\theta, \tilde{\theta}) + \ldots \qquad (2.4)$$

where

$$\tilde{\theta} \equiv \epsilon\theta \qquad (2.5)$$

is the "slow" variable associated with the small fluctuations of the "spring force" in Eq. (2.1). Inserting (2.3), (2.4) in (2.1) and equating like powers of ϵ we obtain a hierarchy of equations[6-8]

$$\frac{\partial^2 F_0}{\partial \theta^2} + F_0 = 0, \qquad (2.6)$$

$$\frac{\partial^2 F_1}{\partial \theta^2} + F_1 = -2 \frac{\partial^2 F_0}{\partial \theta \partial \tilde{\theta}} - (a_1 + \cos 2\theta + \cos 4\theta) F_0, \qquad (2.7)$$

$$\frac{\partial^2 F_2}{\partial \theta^2} + F_2 = -2 \frac{\partial^2 F_1}{\partial \theta \partial \tilde{\theta}} - \frac{\partial^2 F_0}{\partial \tilde{\theta}^2} - (a_1 + \cos 2\theta + \cos 4\theta) F_1 - a_2 F_0, \qquad (2.8)$$

etc.

The solution of (2.6) is

$$F_0(\theta, \tilde{\theta}) = A_0(\tilde{\theta}) \cos\theta + B_0(\tilde{\theta}) \sin\theta \qquad (2.9)$$

where $A_0(\tilde{\theta})$, $B_0(\tilde{\theta})$ are two as yet undetermined functions. Substituting (2.9) in (2.7) yields

$$\frac{\partial^2 F_1}{\partial \theta^2} + F_1 = 2 \frac{dA_0}{d\tilde{\theta}} \sin\theta - 2 \frac{dB_0}{d\tilde{\theta}} \cos\theta - a_1(A_0 \cos\theta + B_0 \sin\theta) -$$

$$- A_0 \cos 3\theta - \tfrac{1}{2} A_0 \cos\theta - \tfrac{1}{2} A_0 \cos 5\theta + \tfrac{1}{2} B_0 \sin\theta \qquad (2.10)$$

$$- \tfrac{1}{2} B_0 \sin 5\theta$$

Since we are seeking <u>uniformly</u> valid (i.e. valid for all θ) solutions we cannot allow terms proportional to $\cos\theta$, $\sin\theta$ on the rhs of (2.10). This requires that the A_0, B_0 satisfy:

$$2 \frac{dA_0}{d\tilde{\theta}} + (\tfrac{1}{2} - a_1) B_0 = 0$$

$$2 \frac{dB_0}{d\tilde{\theta}} + (a_1 + \tfrac{1}{2}) A_0 = 0 \qquad (2.11)$$

It can be easily verified that the solutions of (2.8) grow exponentially if $a_1^2 < \tfrac{1}{4}$ and oscillate if $a_1^2 > \tfrac{1}{4}$. Therefore, on the boundary of the instability region

$$a_1 = \pm \tfrac{1}{2} \qquad (2.12)$$

So far, our results are identical with the corresponding ones for the pure Mathieu equation[6,11] (2.6) without the $\cos 4\theta$ term. The boundaries of the instability regions in the (ε, a) plane are -to this order- given by

$$a_- \simeq 1 - \frac{\varepsilon}{2} : \quad \text{Left Branch} \tag{2.13a}$$

$$a_+ \simeq 1 + \frac{\varepsilon}{2} : \quad \text{Right Branch} \tag{2.13b}$$

In order to carry this procedure to second order we first determine the form of $F_1(\theta, \tilde{\theta})$ by solving (2.10):

$$F_1(\theta, \tilde{\theta}) = A_1(\tilde{\theta})\cos\theta + B_1(\tilde{\theta})\sin\theta + \frac{A_0}{8}\cos 3\theta + \frac{A_0}{48}\cos 5\theta + \frac{B_0}{48}\sin 5\theta \tag{2.14}$$

making use of (2.11). Inserting $F_1(\theta, \tilde{\theta})$ in the rhs of (2.8) and demanding again that all resonance producing terms cancel gives

$$2\frac{dA_1}{d\tilde{\theta}} + (\tfrac{1}{2} - a_1)B_1 = \frac{d^2 B_0}{d\tilde{\theta}^2} + \frac{1}{96}B_0 + a_2 B_0$$

$$2\frac{dB_1}{d\tilde{\theta}} + (a_1 + \tfrac{1}{2})A_1 = -\frac{d^2 A_0}{d\tilde{\theta}^2} - \frac{13}{96}A_0 - a_2 A_0 \tag{2.15}$$

The functions $A_0(\tilde{\theta})$, $B_0(\tilde{\theta})$ are readily obtained from (2.11):

$$A_0(\tilde{\theta}) = a_0 e^{\lambda\tilde{\theta}} + a_0^* e^{-\lambda\tilde{\theta}}, \quad B_0(\tilde{\theta}) = b_0 e^{\lambda\tilde{\theta}} + b_0^* e^{-\lambda\tilde{\theta}}$$

where, $\tag{2.16}$

$$\lambda = \tfrac{1}{2}\sqrt{\tfrac{1}{4} - a_1^2}$$

and a_0, b_0 and their complex conjugates a_0^*, b_0^* are specified by the initial conditions [the solutions of the homogeneous equations (2.15) are of the same type as (2.16)].

Looking for particular solutions of (2.15), of the form[13]

$$\vec{X}_p(\tilde{\theta}) \equiv \sum_{i=1}^{2} c_i(\tilde{\theta}) \vec{X}_i(\tilde{\theta}) \quad [\vec{X} \equiv (A_1, B_1)],$$

where $X_i(\tilde{\theta})$, $i = 1, 2$, are a basis set of solutions of the homogeneous

Eq. (2.15), we arrive at the following equations for $c_1(\tilde{\theta})$, $c_2(\tilde{\theta})$:

$$-\sqrt{\tfrac{1}{2} - a_1}\, e^{\lambda\tilde{\theta}} \frac{dc_1}{d\tilde{\theta}} + e^{-\lambda\tilde{\theta}} \sqrt{\tfrac{1}{2} - a_1}\, \frac{dc_2}{d\tilde{\theta}} = (\lambda^2 + \tfrac{1}{96} + a_2) B_0 \quad (2.17a)$$

$$\sqrt{\tfrac{1}{2} + a_1}\, e^{\lambda\tilde{\theta}} \frac{dc_1}{d\tilde{\theta}} + e^{-\lambda\tilde{\theta}} \sqrt{\tfrac{1}{2} + a_1}\, \frac{dc_2}{d\tilde{\theta}} = -(\lambda^2 + \tfrac{13}{96} + a_2) B_0 \quad (2.17b)$$

of (2.15), (2.16). Clearly, on the left boundary of the instability region, where $a_1 = -\tfrac{1}{2}$, (2.17b) yields

$$a_2 = -\frac{13}{96}$$

and, hence, (2.13a) becomes to second order in ϵ:

$$a_- \cong 1 - \frac{\epsilon}{2} - \frac{13}{96}\epsilon^2 : \quad \text{Left Branch} \quad (2.18a)$$

On the other boundary $a_1 = \tfrac{1}{2}$ and (2.17a) gives

$$a_2 = -\frac{1}{96}$$

and (2.13b(becomes

$$a_+ \cong 1 + \frac{\epsilon}{2} - \frac{1}{96}\epsilon^2 : \quad \text{Right Branch} \quad (2.18b)$$

Expressions (2.18a,b) are already in very good agreement with the "exactly" computed boundary curves in Fig. 12 of Ref. 1 up to $\epsilon \cong 1.2$. We have plotted them below in Fig. 1 (solid curves) together with the corresponding results of Ref. 1 (dashed curves) for purposes of comparison. In fact the "dashed" curves in Fig. 1 were obtained for a Mathieu-like Eq. (2.1) with <u>five</u> cosines.[1] However, if one includes in (2.1) the terms $\cos 6\theta$, $\cos 8\theta$, $\cos 10\theta$ and repeats the above analysis one finds that the resulting expressions for a_+ and a_- differ only slightly from (2.18a,b) by an amount that is not distinguishable in the scale of Fig. 1.

In the remainder of this section we carry out the above procedure for the <u>second</u> instability region of Eq. (2.1) which touches the a-axis at $a_c = 4$, cf. Fig. 1. Near that point, a is given approximately by

$$a = 4 + \epsilon a_1 + \epsilon^2 a_2 + \ldots \quad (2.19)$$

cf. (2.3).

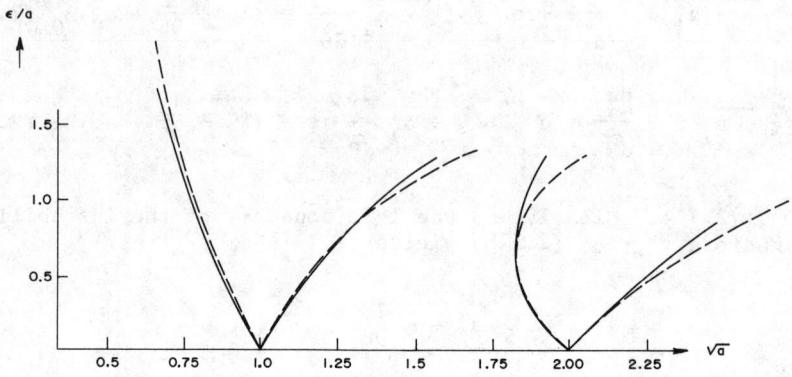

Figure 1. Instability regions for the Mathieu-like Eq. (2.1).

Attempting again a series solution of the type (2.4) we find that its zeroth order term satisfies

$$\frac{\partial^2 F_0}{\partial \tilde{\theta}^2} + 4F_0 = 0$$

$$F_0(\theta,\tilde{\theta}) = A_0(\tilde{\theta})\cos 2\theta + B_0(\tilde{\theta})\sin 2\theta \qquad (2.20)$$

Substituting (2.20) on the rhs of the first order equation corresponding to (2.7) and setting the resonance producing terms equal to zero, we arrive at equations similar to (2.11):

$$4\frac{dA_0}{d\tilde{\theta}} + (\tfrac{1}{2} - a_1)B_0 = 0$$

$$4\frac{dB_0}{d\tilde{\theta}} + (\tfrac{1}{2} + a_1)A_0 = 0 \qquad (2.21)$$

Had we not included $\cos 4\theta$ in (2.1), the terms $\tfrac{1}{2}A_0$ and $\tfrac{1}{2}B_0$ in (2.21) would not have been present. This is the case with the pure Mathieu equation in which $a_1 = 0$ and the boundary curves of the <u>second</u> instability region are of <u>second order</u> in ε.[6,11]

With the term $\cos 4\theta$, however, in Eq. (2.1) we conclude from (2.21) that the boundaries of the second instability region are given to order ϵ by the expressions

$$a_- \cong 4 - \frac{\epsilon}{2} : \quad \text{Left Branch}$$
$$a_+ \cong 4 + \frac{\epsilon}{2} : \quad \text{Right Branch,} \qquad (2.22)$$

cf. (2.19) and Fig. 1. This explains why the second instability region, obtained "exactly" for Eq. (2.1) with five cosines in Ref. 1, is so much larger than the corresponding one of the pure Mathieu equation with only the $\cos 2\theta$ term.[6,11]

Proceeding exactly as before we find the second order corrections to the second pair of boundary curves

$$a_- \cong 4 - \frac{\epsilon}{2} + \frac{37}{384} \epsilon^2 : \quad \text{Left Branch}$$
$$a_+ \cong 4 + \frac{\epsilon}{2} - \frac{11}{384} \epsilon^2 : \quad \text{Right Branch.} \qquad (2.23)$$

Comparing (2.23) with the dashed curves (of Ref. 1) coming down to $\sqrt{a} = 2$ in Fig. 1, we find very good agreement even up to $\epsilon \approx 3.5$.

III. THE NONLINEAR MATHIEU EQUATION

Using the perturbation methods of the previous sections we discuss here the boundedness properties of the solutions of the __nonlinear__ Mathieu equation

$$\frac{d^2 y}{d\theta^2} + ay + \epsilon \cos 2\theta (y - by^3) = 0, \quad b \geq 0, \qquad (3.1)$$

"near" the points where the corresponding __linear__ (i.e. $b = 0$) equation has bounded solutions __for all__ ϵ!

Defining the new variable

$$Y \equiv y\sqrt{b} \qquad (3.2)$$

we rewrite (3.1) as

$$\frac{d^2 Y}{d\theta^2} + aY + \epsilon \cos 2\theta (Y - Y^3) = 0. \qquad (3.3)$$

Here again, as in the linear Mathieu equation the $\cos 2\theta$ is the major resonance producing term to lowest order in the perturbation. Near these lowest order resonances the solutions oscillate with __unperturbed__ frequency squared.

$$a = n^2 + \epsilon a_1 + \epsilon^2 a_2 + \ldots \qquad (3.4)$$

where

$$n = 0, \tfrac{1}{2}, 1, 3/2, 2, \ldots \tag{3.4a}$$

For convenience we consider here only the half integer values of n since for all other n values-with the exception of the integers-resonances appear at higher order and the calculations become more cumbersome. In particular we examine below only the case $n = \tfrac{1}{2}$ since all other half-integer cases can be similarly treated.

We remark that the case of <u>integer</u> n cannot be studied by the methods of this paper or most other perturbation methods for that matter. The reason is that for $b = 0$ and $n = 1, 2, 3, \ldots$ the solution of the $\epsilon = 0$ (linear, "integrable"[14,15], i.e. solvable ....) problem is <u>unstable</u>, i.e. corresponds to a <u>saddle point</u> in the y, $dy/d\theta$ phase plane. Near such saddle points, when $b \neq 0$ (hence the problem becomes nonlinear, "non-integrable",[14,15] etc.) asymptotic analysis breaks down and is seen to give inaccurate results at zeroth order already!

Introducing again the "slow" variable

$$\tilde{\theta} \equiv \epsilon\theta \tag{3.5}$$

we expand the solution of (3.3) as before

$$Y(\theta,\epsilon) = F_0(\theta,\tilde{\theta}) + \epsilon F_1(\theta,\tilde{\theta}) + \epsilon^2 F_2(\theta,\tilde{\theta}) + \ldots \tag{3.6}$$

Treating $\theta,\tilde{\theta}$ as independent variables we substitute (3.6) and (3.4), with $n = \tfrac{1}{2}$, in (3.3) and equate like powers of ϵ to find the hierarchy of equations

$$\frac{\partial^2 F_1}{\partial \theta^2} + \frac{1}{4} F_0 = 0, \tag{3.7a}$$

$$\frac{\partial^2 F_1}{\partial \theta^2} + \frac{1}{4} F_1 = -2\frac{\partial^2 F_0}{\partial \theta \partial \tilde{\theta}} - \cos 2\theta (F_0 - F_0^3) - a_1 F_0, \tag{3.7b}$$

$$\frac{\partial^2 F_2}{\partial \theta^2} + \tfrac{1}{4} F_2 = -2\frac{\partial^2 F_1}{\partial \theta \partial \tilde{\theta}} - \frac{\partial^2 F_0}{\partial \tilde{\theta}^2} - \cos 2\theta (F_1 - 3F_0^2 F_1) - a_1 F_0 - a_2 F_0, \tag{3.7c}$$

etc. The solution of (3.7a) is

$$F_0(\theta,\tilde{\theta}) = A_0(\tilde{\theta})\cos\frac{\theta}{2} + B_0(\tilde{\theta})\sin\frac{\theta}{2} \tag{3.8}$$

Inserting (3.8) on the rhs of (3.7b) and setting all terms proportional to $\cos(\theta/2)$, $\sin(\theta/2)$, respectively, equal to zero we find

$$\frac{dA_0}{d\tilde{\theta}} - a_1 B_0 = \frac{1}{8}(-B_0^3 + 3A_0^2 B_0) , \qquad (3.9a)$$

$$\frac{dB_0}{d\tilde{\theta}} + a_1 A_0 = \frac{1}{8}(A_0^3 - 3A_0 B_0^2) . \qquad (3.9b)$$

In order to determine the (un) boundedness of the zeroth order solution (3.8), it suffices to plot the solution curves[12,13] of the system (3.9) in the A_0, B_0 plane. This may be done by observing that (3.9) possesses the integral

$$A_0^4 + B_0^4 - 6A_0^2 B_0^2 - 16a_1(A_0^2 + B_0^2) + C = 0 \qquad (3.10)$$

C being an arbitrary constant. Using elementary methods[12,13] we find the equilibrium points of (3.9) in the A_0, B_0 plane and determine their stability. We then make use of (3.10) and draw schematically the solution curves of (3.9) in Fig. 2 below.

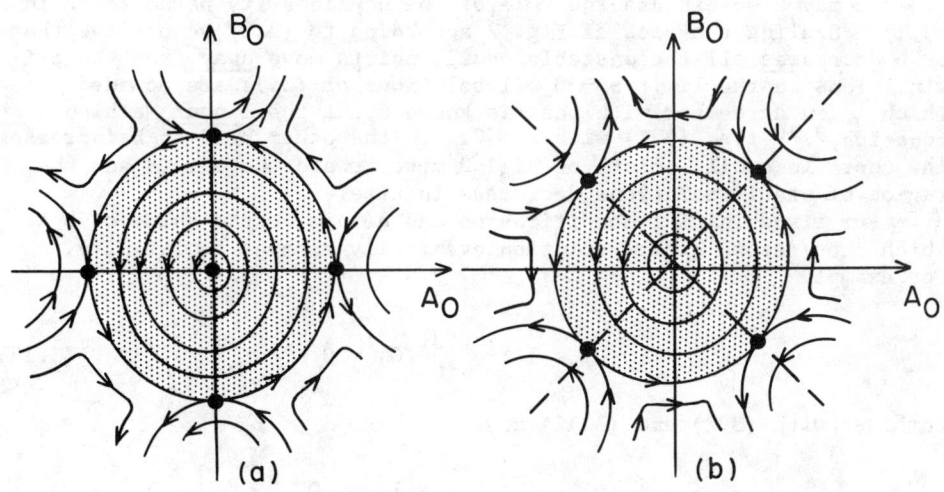

Fig. 2. The solution curves (3.10) for (a) $a_1 > 0$, (b) $a_1 < 0$ [The equil. pts. in (a) are $(0,0)$, $(\pm\sqrt{8a_1}, 0)$, $(0, \pm\sqrt{8a_1})$ and in (b) they are $(0,0)$ and $(\pm\sqrt{-4a_1}, \pm\sqrt{-4a_1})$].

We wish to point out here the analogy between Fig. 2 above and the pictures one gets when one approximates a <u>non-integrable</u> dynamical system by an <u>integrable</u> one (see e.g. Ref. 16). The exact solution of (3.1), plotted on a suitable "phase plane" (y, dy/dθ) are <u>not</u> expected to generate <u>everywhere</u> continuous curves as one finds in Fig. 2. According to KAM theory,[14,15] even though "most" of these curves are still present, there are regions between them where the solutions behave in a "wild" and "chaotic" manner! For small enough ε these regions are practically indistinguishable but as ε increases they grow dramatically in size and eventually render the approximate solutions obsolete.[16]

It is also evident from Fig. 2 that no matter what a_1 is we cannot draw here "global" stability charts for <u>all</u> initial conditions, as we did in Fig. 1. The reason is that the (un) boundedness of the solutions of (3.3) depends strongly on the (zeroth order) initial conditions

$$Y(0) = A_0(0), \quad \frac{dY}{d\theta}(0) = B_0(0). \tag{3.11}$$

Thus if one starts with $A_0(0)$, $B_0(0)$ in any of the shaded regions of Fig. 2, the solutions (to lowest order in ε) are bounded, whereas if $A_0(0)$, $B_0(0)$ lie in the unshaded regions of Fig. 2, the motion is unbounded.

We may also discuss the role of the nonlinearity parameter b in (3.1): Scaling the axes of Fig. 2 according to (3.2) we observe that as b <u>decreases</u> all the unstable equil. points move <u>away</u> from the origin. Thus in the limit b → 0 all solutions of (3.1) are bounded which is in agreement with what is known about the <u>linear</u> Mathieu equation,[9,10] i.e. (3.1) with b = 0. On the other hand as b increases the unstable equil. points of Fig. 2 move toward the origin and the region of stable solutions decreases in size.

For fixed initial conditions we can determine the value of b at which the (zeroth order) solution eventually escapes to infinity. For example

$$y(0) = 1; \quad \frac{dy}{d\theta}(0) = 0 \tag{3.12}$$

combined with (3.2) and (3.11) give

$$A_0(0) = \sqrt{b}; \quad B_0(0) = 0$$

Thus, for $a_1 > 0$ our (zeroth order) solution corresponding to (3.12) becomes unbounded when

$$b \geq 8a_1,$$

cf. Fig. 2(a).

IV. CONCLUSION

We derived approximate expressions for the boundaries between bounded and unbounded solutions of certain <u>linear</u> Mathieu-like equations arising in problems of accelerator dynamics. In most cases first order estimates are sufficient to determine the main behavior of the solutions. In the cases where we went to second order in ϵ, it was found that our expressions agreed very well with the "exact" results of Refs. 1,2 up to $\epsilon \leqslant 1$.

Applying our perturbation techniques to a <u>nonlinear</u> Mathieu equation we determined the zeroth order behavior of the solutions near a stable periodic solution of the corresponding linear equation. We found, as expected, that this behavior depends strongly on the initial conditions and sketched the regions of initial conditions leading to bounded and unbounded motion.

We emphasize that the results reported here are incomplete and great care must be exercised before one uses them to draw any general conclusions. They do suggest, however, that when studying such equations of accelerator dynamics it may be instructive to use simple perturbation techniques to obtain, for small enough ϵ, some information about the general behavior of the solutions.

ACKNOWLEDGMENTS

We wish to thank Dr. Mel Month for organizing a stimulating conference and spreading to all of us his contageous enthusiasm and interest in the Beam-Beam Interaction. During the research one of us (T.B.) was partially supported by a DOE grant under contract EG-77-C-03-1538. Part of this work was carried out under the auspices of the Applied Mathematics Department of Cal Tech, whose hospitality is gratefully acknowedged here.

REFERENCES

1. C. R. Eminhizer, P. Vuillermot, T. Bountis and R. H. G. Helleman in "Review of the Beam-Beam Phenomena," BNL-25703, Accelerator Department, Brookhaven Nat. Lab., Upton, N. Y. (1979).
2. See the paper by R. H. G. Helleman in this volume.
3. M. Month, IEEE Trans. Nucl. Sci. <u>NS-22</u>, 1376 and 1443 (1975).
4. J. C. Herrera in this volume as well as in the volume of Ref. 1.
5. E. D. Courant and H. S. Snyder, Am. Phys. <u>3</u>, 1-42 (1958).
6. J. D. Cole, "Perturbation Meth. in Appl. Math., Blaisdell, Waltham, Mass. (1968).
7. J. Kevorkian and J. D. Cole in <u>Nonlinear Differential Equations and Nonlinear Mechanics</u>, Academic Press, New York (1965).
8. A. Nayfeh, <u>Perturbation Methods</u>, Wiley, New York (1973) and his "Nonlinear Oscillations", Wiley, N. Y. (1979).
9. W. Magnus and S. Winkler, <u>Hill's Equation</u>, Intersci. Publ., New York (1966).
10. M. Abramowitz and I. Stegun, <u>Handbook of Mathematical Functions</u>, Dover, New York (1965).
11. J. J. Stoker, "Nonlinear Vibrations", Intersci. Publ. New York (1950).

12. G. Birkhoff and G. Rota, <u>Ordinary Differential Equations</u>, J. Wiley & Sons, New York (1962), 3d. ed. (1978).
13. W. Hurewicz, "Lectures on Ordinary Differential Equations", Cambridge Tech. Press MIT (1968).
14. J. Moser, <u>Stable and Random Motions in Dynamical Systems</u>, Princeton Univ. Press (1973) as well as his article in this volume.
15. M. V. Berry in "Topics in Nonlinear Dynamics", A.I.P. Conf. Proc. Vol. 46, Ed. S. Torna, A.I.P., New York (1978).
16. F. G. Gustavson, Astron. J. <u>71</u>, No. 8, 670 (1966).

SOME NUMERICAL STUDIES OF ARNOLD DIFFUSION IN A SIMPLE MODEL

Boris V. Chirikov

Insitiute of Nuclear Physics, Novosikirsk 630090, USSR

and

Joseph Ford and Franco Vivaldi

School of Physics, Georgia Institute of Technology,
Atlanta, Georgia, 30332, U.S.A.

INTRODUCTION

One of the most powerful experimental techniques for studies of the deepest laws of nature related to the world of elementary particles is the so-called colliding beam devices, or storage rings, which provide the highest center-of-mass energy for a pair of accelerated particles. An important characteristic of such devices is the so-called luminosity, i.e. the rate of collisions between particles of the two beams which determines the efficiency (the rate) as well as the range of experiments available. The luminosity depends, in turn, on the intensity (the current) of the colliding beams. Besides some technical difficulties in the storage of a large number of particles, especially heavy ones, e.g. antiprotons, the principal limitations of beam intensity is due to the beam-beam interaction via their collective electromagnetic fields. This interaction is known to disrupt the beam completely above a certain critical beam current (see, e.g. Ref. 1 and papers in these Proceedings). Even much below this critical intensity the transverse dimensions of the beams are generally enlarged appreciably which results in a troublesome drop of luminosity. For the new generation of heavy particle colliding beam machines, now under design or even construction, the beam-beam interaction is expected to be an even more crucial phenomenon since there is no radiation damping in this case to stop the beam vlow-up, and also in view of the enormously long life time of heavy particles in a storage ring which is required (of the order of a few days, or $\sim 10^{11}$ interactions, a truly cosmic time scale!) Thus extensive studies of the beam-beam phenomena seem to be highly in order, and even somewhat urgent we would say.

Generally, the collision of two intense bunches of particles is a very complicated dynamical process with an infinite number of degrees of freedom. There seems to be not much hope in any near future to study it either analytically or numerically. Thus, the importance of experimental work on existing, smaller storage rings must be emphasized, including model experiments with electrons to simulate the heavy particle behavior in future storage rings as, for example, recent experiments with low-energy electrons on SPEAR.[2]

A common analytical, as well as numerical, approach to the problem under consideration is the so-called weak-strong approximation of the beam-beam interaction. What is actually hidden behind this terminology is a dramatic simplification of the original problem by a cut in the number of degrees of freedom from infinity down to at most three. Namely, the influence of the weaker beam on the stronger one is completely ignored, and the motion of a single particle in a given field of the strong beam is studied. Within the framework of such a weak-strong approximation a number of studies has been done (with some further simplifications) attempting to understand this simplified beam-beam interaction and to obtain some estimates for the critical intensity of the beam (the strong one in this approximation) (see these Proceedings and also Ref. 3 for example). The importance of those studies is related to the fact that the weak-strong stability of particle motion is certainly a necessary, even though obviously not sufficient, condition for the performance of colliding beam machines.

To the best of our knowledge all earlier studies were concerned with a strong instability of motion related to the so-called overlap of nonlinear resonances (see, e.g. review papers[4,5]). Meanwhile, a much weaker instability - the so-called Arnold diffusion - is knows to occur in such dynamical systems (see, e.g. the review paper 5 and references therein). This latter instability turns out to be a universal one in the sense that there is no critical perturbation strength for this instability. The perturbation influences the rate of instability only. This particular feature of the Arnold diffusion makes it especially dangerous for colliding beams of heavy particles where a very long particle life time is required. The phenomenon of Arnold diffusion has already been studied on simple models both numerically and analytically.[6-7] The theories developed in these works seem to agree with numerical results within a factor of two, provided the perturbation is not too weak. The latter condition is essential because for a relatively strong perturbation, but of course still below the resonance overlap, only a few (minimum 3) resonances (perturbation terms) determine the diffusion rate, and those can be explicitly taken into account when evaluating the diffusion rate analytically (see Ref. 5 for details). For a weaker perturbation many resonances are involved, and analytical evaluation becomes much more complicated. We call this latter region of parameters the Nekhoroshev region (after the name of a Soviet mathematician who first has given rigorous upper estimates for the diffusion rate in this region[8]). Unfortunately due to obvious technical difficulties, his estimates seem to be much above the actual values of the diffusion rate (see Section 6).

The present paper is a brief report on our mainly numerical studies of the Arnold diffusion on a simple model developed in Ref. 6 (see Section 2). This model has no immediate relation to the beam-beam interaction because of the different types of nonlinearity and perturbation chosen, yet the phenomenon of the Arnold diffusion remains the same in both cases due to essentially the same phase space structure of both dynamical systems. Our main objective in these studies is an attempt to find a simple, semi-empirical relation for the diffusion rate in the Nekhoroshev regions. We seem to have found one (Section 5 Eq. 20) and we hope that these results will help other physicists in this interesting and important field of research.

2. DESCRIPTION OF MODEL

The model we have studied in this work had been developed and applied in Refs. 6,5. We have made only minor improvements to facilitate computation. The model is described by four dynamical variables: two coordinates (x_1, x_2) and two canonically conjugated momenta (p_1, p_2).

We specify the equations of motion in the form of a mapping:
$(x_i, p_i) \to (\bar{x}_i, \bar{p}_i)$, $i = 1, 2$ where,

$$\bar{p}_1 = p_1 - x_1^3 + \mu x_2 + \epsilon f(t)$$
$$\bar{p}_2 = p_2 - x_2^3 + \mu x_1$$
$$\bar{x}_1 = x_1 + \bar{p}_1$$
$$\bar{x}_2 = x_2 + \bar{p}_2$$
(1)

Here μ and ϵ are small perturbation parameters, and $f(t)$ is some driving force, typically periodic in integer time t which is actually the serial number of interactions for the Mapping (1).

To simplify understanding the model, we observe that for a sufficiently small difference $|\bar{p}_i - p_i| \ll 1$; $|\bar{x}_i - x_i| \ll 1$, which was actually the case in our studies, the <u>Difference</u> Equations (1) can be changed into the <u>differential</u> equations related to the Hamiltonain:

$$H(x_i, p_i) = \frac{p_1^2 + p_2^2}{2} + \frac{x_1^4 + x_2^4}{4} - \mu x_1 x_2 - \epsilon x_1 f(t) \quad . \tag{2}$$

Alternatively, one may consider the Mapping (1) as a particular numerical procedure to solve the equations of motion for the System (2). The important feature of this procedure is that Mapping (1) conserves phase space volume and, moreover, is a cononical mapping. This prevents a fast accumulation of computational errors. The latter result only in some bounded oscillations, or background, as we call it, which proves to be fairly low (Section 4).

The Hamiltonian (2) describes two nonlinear oscillators with a weak linear coupling (small parameter μ) and a driving force acting upon one of them (small parameter ϵ). It is interesting to mention that, in spite of a strong (quartic) nonlinearity, the anharmonicity, i.e. the amplitude of higher harmonics in the free uncoupled oscillations ($\mu = \epsilon = 0$) is less than 4%. The frequency of free oscillations, however, grows in proportion to the oscillation amplitude a_i. In particular,

where
$$\omega_i \approx \beta a_i \; ; \; i = 1, 2,$$
$$\beta = 0.8472.$$
(3)

Two types of the driving force were used:

$$f_1(t) = \cos(\Omega_1 t) + \cos(\Omega_2 t) \tag{4}$$

and

$$f_2(t) = \frac{\cos(\Omega t)}{1 - A\cos(\Omega t)} \; . \tag{5}$$

For $(1 - A) \ll 1$, the latter force possesses a rich spectrum:

$$f_2(t) \approx \sum_m \frac{2e^{-\sigma m}}{\sigma} \cos(m\Omega t), \qquad (6)$$

where

$$\sigma \approx \sqrt{1 - A^2}.$$

The main coupling resonance $\omega_1 = \omega_2$ ($a_1 = a_2$) has been chosen as the guiding resonance for Arnold diffusion. The term guiding resonance[5] refers to the fact that the diffusion is going on just along this resonance, or to be more precise, along a stochastic layer around the separatrix of this resonance (see Ref. 5). The stochastic layer is formed due to driving resonances, or driving perturbation terms in Hamiltonian (2). The disposition of basic (first order) resonances is outlined in Fig. 1 for the driving force (5). In case of Force (4) only two nearest driving resonances remain.

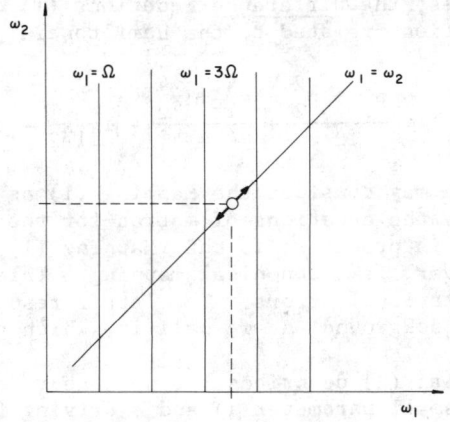

Fig. 1. A set of the first order resonances for the System (2) with driving Force (5): circle indicates location of the initial conditions; arrows show direction of the Arnold diffusion.

The initial conditions were chosen about the midpoint between two driving resonances. Typically the diffusion rate was so slow that the diffusion range, or an average shift of a trajectory was much smaller than the spacing between driving resonances, so the initial disposition of trajectories persisted during a computational run. To provide the initial location of a trajectory within the stochastic layer it was sufficient to set $x_1(0) = -x_2(0)$ ($p_1(0) = p_2(0) = 0$). Under opposite conditions: $x_1(0) = x_2(0)$ the trajectory started near the center of the coupling resonance (see Ref. 5). Both conditions are independent of the perturbation strength which greatly simplified the computation.

The diffusion rate in energy (2) was computed. Since typically the perturbation was very small, it sufficed to take account of the unperturbed energy ($\mu = \epsilon = 0$) only.

3. COMPUTATIONAL TECHNIQUES

The iteration of Mapping (1) for various values of parameters μ, ϵ, and Ω (or Ω_1, Ω_2) and various initial conditions was performed on the CRAY-1 computer. The code has been written in FORTRAN language in such a way as to have the innermost loop vectorized which provides a much higher computation speed. To achieve this, we ran 64 trajectories simultaniously. The computation speed was around 85 MFLOPS (millions of floating point operations per second) that is about half the maximal speed (160 MFLOPS). The computation time for one iteration and one trajectory was about 0.34 μs, or 27 clock periods which corresponded to approximately 3.7 minutes of CPU time for a typical run of 10^7 iterations or to 18.5 minutes for a few of the longest runs of 5.10^7 iterations.

The sixty-four trajectories were distributed in eight groups related to eight different values of the perturbation parameter μ. Typical values of this parameter corresponded to:

$$\mu^{-1/2} = 50, 100, 150, 200, 250, 300, 350, 400;$$

and have covered a fairly wide range of the perturbation strength:

$$\mu = 4 \times 10^{-4} \text{ to } 6.3 \times 10^{-6}.$$

For a stronger perturbation the region of resonance overlap would be entered (see Section 4). The driving force small parameter ϵ was changed in proportion to μ so that the ratio ϵ/μ was constant for a particular run. Typically, $\epsilon/\mu = 0.1$ to 0.001.

The eight trajectories of each group were chosen with slightly different initial conditions to supress big fluctuations in Arnold diffusion by averaging the diffusion rate over all trajectories of each group. Due to the exponential local instability inside a stochastic layer, a very small variation in initial conditions was sufficient for this purpose, typically $|\Delta x_i| \sim 10^{-5}$ ro 10^{-8}.

In the computation of the diffusion rate, we followed the procedure developed in Ref. 16 and described in detail in Ref. 5, namely:

$$D_k = \frac{2}{N_k(N_k-1)} \sum_{m>n} \frac{(\bar{H}_m \bar{H}_n)^2}{(\Delta t)_k (m-n)} \tag{7}$$

To compute this quantity the total motion time t_{max} was subdivided into N_k time intervals of length $(\Delta t)_k$ each. The current value of energy $H(t)$ was then averaged over each of these intervals to give quantities $\bar{H}_m$. The diffusion rate for a given pair $\bar{H}_m$, $\bar{H}_n$ separated by the time interval $(\Delta t)_k (m-n)$ would be $(\bar{H}_m - \bar{H}_n)^2/(\Delta t)_k |m-n|$. This rate was averaged then over all combination $m \neq n$ to give Eq. (7). Two diffusion rates were computed in each run related to two different subinterval $(\Delta t)_k$, namely:

$$(\Delta t)_1 = \frac{t_{max}}{100}; \quad N_1 = 100,$$

$$(\Delta t)_2 = \frac{t_{max}}{10}; \quad N_2 = 10.$$

For a true diffusion process both rates must be close (to the accuracy of statistical fluctuations). If, on the other hand, there are only bounded oscillations one would expect

$$\left(\frac{D_2}{D_1}\right)_{osc} \approx \frac{(\Delta t)_1^3}{(\Delta t)_2^3} \to 10^{-3}, \qquad (8)$$

since each of the differences ($\overline{H}_m - \overline{H}_n$) would decrease in proportion to $(\Delta t)_1/(\Delta t)_2$. For the same reason the computational value of each D in the case of bounded oscillations rapidly decreases with the motion time:

$$D_{osc} \sim t_{max}^{-3}. \qquad (9)$$

Comparing the two values of D_1 and D_2, one can get a rough idea as to the possible portion of side (non-diffusion) processes. It is clear also that the D_2 rate is the much more reliable of the two, and so all the data below correspond just to this diffusion rate D_2.

4. NUMERICAL RESULTS

An example of the dependence of the diffusion rate (decimal logarithm of D_2) on the perturbation ($1/\sqrt{\mu}$) is plotted in Fig. 2 for the following set of parameters: $|x(0)| = 0.225$; $\Omega = 0.03466$; $A = 0.995$ [force (5)]; $\epsilon/\mu = 0.01$ with μ in the range: $1/\sqrt{\mu} = 6$ to 400, and the motion time $t_{max} = 10^3$ to 5.10^7 depending on perturbation. Dependence on $1/\sqrt{\mu}$ has been taken merely in analogy with a simple theory of the Arnold diffusion[5,6], where the quantity $1/\sqrt{\mu}$ enters the exponent and, thus, essentially determines the dependence $D(\mu)$ (see below, Section 5).

For the chosen amplitude $a = |x(0)| = 0.225$ the oscillation frequency $\omega(0) = 0.19$, and the ratio $\omega/\Omega = 5.5$, that is, the system is located between resonances of 5th and 6th harmonics of the driving force. This frequency value is sufficiently small to provide a low background. A rough estimate of the background can be gotten from the 'diffusion rate' near the center of resonance where the motion is perfectly stable for a sufficiently weak perturbation. As is seen from Fig. (triangles), the background level $D_b \approx 10^{-25}$ for the motion time $t_{max} = 10^7$. The rate D_1 in this case is $\sim 10^{-22}$, as one would expect from Estimate (8). If $t_{max}^1 = 10^6$ the background (D_2) grows by about 2 to 4 orders of magnitude, again in rough accordance with Estimate (9). On the contrary, for initial conditions inside the stochastic layers the diffusion rate does not depend on the motion time t_{max} within a factor of 2. This difference seems to be mainly due to still appreciable fluctuations in spite of averaging over 8 trajectories. As explained at length in Ref. 5, the fluctuations are related to the complicated structure of a stochastic layer, especially its peripheral part near the layer edge. A trajectory may 'stick' here for a relatively long period of time which results in a big deviation from the average diffusion rate. This is apparently also the main cause for an always present slight difference between the two rates D_1 and D_2. Typically, inside the stochastic layer $D_2/D_1 \approx 0.8$.

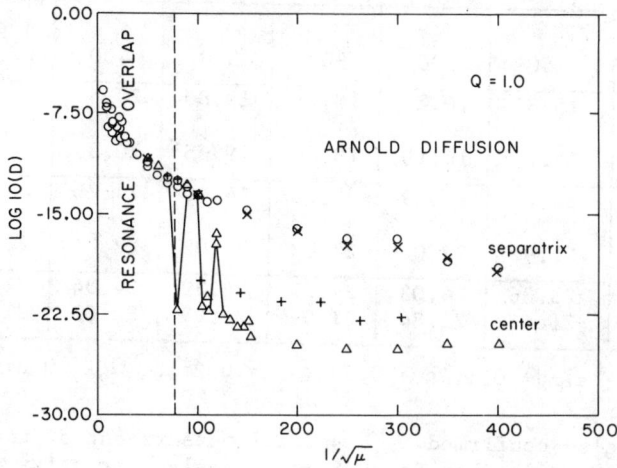

Fig. 2. Diffusion rate D vs. perturbation $1/\sqrt{\mu}$: O-inside the stochastic layer; x-same with a lower accurately reduced by a factor of 500; Δ-at the center of resonance, $t_{max} = 10^7$; +-same for $t_{max} = 10^6$.

Another estimate for the background can be obtained from runs with a single driving resonance [for the force (4)]. As is explained in detail in Ref. 5, no long-range diffusion is possible in this case. The point is that a single driving resonance would provide the diffusion only in a certain particular direction in the action or frequency plane of the system (Fig. 1). This direction does not generally, and particularly for the model under consideration, coincide with the direction of guiding resonance (and its stochastic layer). Since the width of stochastic layer is typically very small for weak perturbation the resulting motion would be of the type of bounded oscillations. In Table 1 we compare the diffusion rates for a single and two driving resonances as well as for trajectories near the resonance center with parameters given below the Table. We see that the background as determined near the resonance center and inside the stochastic layer with a single driving resonance is practically the same and much lower than the rate of Arnold diffusion. The ratio $D_2/D_1 \approx 5 \cdot 10^{-3}$ for two background runs is somewhat larger than expected from Eq. (8) but much lower than for the Arnold diffusion (first run, $D_2/D_1 \approx 0.8$). A low background value for a single driving resonance shows that the Arnold diffusion in the first run is really a long-term one, that is, its range during the motion time $t_{max} = 10^7$ is much larger that the width of stochastic layer.

Table 1. Background.

$1/\sqrt{\mu}$	150	200	300	350	400	Average background	
2 driving resonances	15.31	16.23	18.71	19.80	20.54	-	$-\log D_2$
	15.23	16.15	18.66	19.65	20.39	-	$-\log D_1$
1 driving resonance	24.20	24.29	24.25	24.40	24.26	24.28	$-\log D_2$
	21.92	22.06	22.02	22.02	22.01	22.01	$-\log D_1$
resonance center	23.86	24.03	24.27	24.29	24.34	24.16	$-\log D_2$
	21.66	21.84	21.99	21.99	22.03	21.90	$-\log D_1$

$t_{max} = 10^7$; $\epsilon/\mu = 0.1$; $a = 0.27$; $\Omega_1 = 0.2513$; $\Omega_2 = 0.2167$

It is also confirmed by a special measurement of that width using a procedure described in Refs. 5, 6. Namely, for a faxed perturbation the initial conditions were chosen according to the expression ($p_1(0) = p_2(0) = 0$):

$$x_1(0) = a + d_i; \quad x_2(0) = a - d_i , \quad (10)$$

where i stands for the serial number of a trajectory group. If $d = 0$ the trajectory starts at the resonance center and reveals only a background 'diffusion' as we have seen above. Yet as d increases, a trajectory eventually crosses the stochastic layer that is immediately obvious from numerical data by a 'jump-up' of the diffusion rate. An example of the dependence of diffusion rate on the shift d is plotted in Fig. 3. The stochastic layer is clearly seen at $d \approx 4.5 \times 10^{-3}$, in reasonable agreement with the theoretical prediction[5] $d_\perp = \sqrt{\mu} = 5 \cdot 10^{-3}$. The layer width $\Delta d \approx 4 \cdot 10^{-4}$ is also close to the expected value $\Delta d \approx 4.3 \times 10^{-4}$ according to the expression derived in Ref. 5. Note that the diffusion rate drops by about 8 orders of magnitude (!) at both edges of the layer. According to Ref. 5, the layer width in energy is related to (Δd) by a simple expression:

$$\Delta H = a^2 \cdot d \cdot \Delta d . \quad (11)$$

Whence the time interval required for the diffusion to reach across the layer is

$$T_\ell \sim \frac{(\Delta H)^2}{D} \sim 10^3 . \quad (12)$$

Numerical estimate is given for the parameters used in Fig. 3. Time interval T_ℓ is much less then the motion time $t_{max} = 10^7$, so that the diffusion is spreading along the stochastic layer at a distance 100 times as much as the layer width. However for this run the relative change in energy is only $\sim 1\%$, and even 4 times less in amplitude and in frequency $\omega(a)$. This results in only $\sim 1\%$ relative change in detuning $\delta\omega = \Omega_1 - \omega \approx \omega - \Omega_2$.

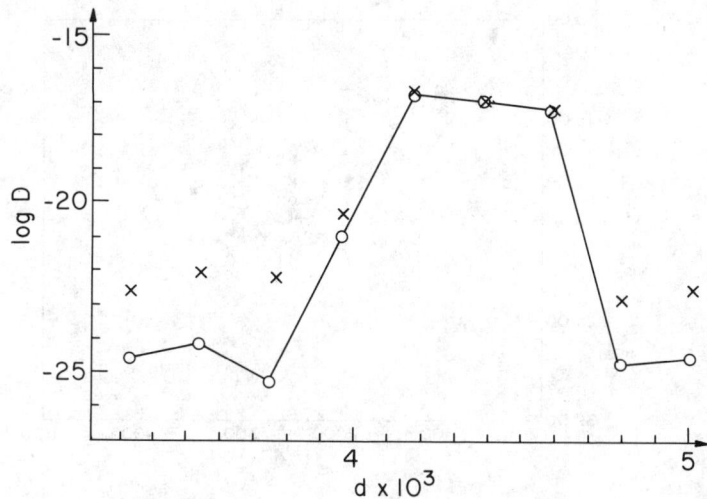

Fig. 3. Diffusion rate vs. initial conditions: $d = [x_1(0) - x_2(0)]/2$; $p_1(0) = p_2(0) = 0$; $1/\sqrt{\mu} = 200$; $\epsilon/\mu = 0.1$; $a = 0.2$; $t_{max} = 10^7$; $\log D_1$ - X; $\log D_2$ - o.

Coming back to Fig. 2 we see that even for the weakest perturbation with $1/\sqrt{\mu} = 400$, the background is more than 5 orders of magnitude lower as compared to the rate of Arnold diffusion inside stochastic layer (Fig. 2). However, for a sufficiently strong perturbation ($1/\sqrt{\mu} \lesssim 80$) the diffusion rate near the resonance center 'jump up' by almost 10 orders of magnitude (!) and attains that level inside the stochastic layer. This simply means that the resonance center has become completely destroyed by a strong resonance overlap whose border is marked in Fig. 2 by the dashed vertical line as evaluated according to Ref. 5 (there, see Section 4.1). This line, thus, divides the whole perturbation range into two regions- the resonance overlap, or a strong stochasticity, and the Arnold diffusion, or a weak but universal instability. Note, that there is no obvious change in the dependence $D(\mu)$ between these two regions in spite of a quite different mechanism of the diffusion. This allows us to describe diffusion in both regions by a single expression except, perhaps, under the very strong perturbation at the leftmost part of the plot in Fig. 2 (see Section 5).

Another set of data is plotted in Fig. 4. It corresponds to 10 runs with parameters given in Table 2. The quantity $\delta\omega$ is the detune between the oscillation frequency $\omega(0)$ and that of the nearest driving resonance. In all runs of Table 2 but the first the two detunes, which are related to the two nearest driving resonances, are equal. If they are not equal, the bigger of the two values should be taken. The point is that, as was explained above in this section (for more detail see Refs. 5, 6), a long-range diffusion can be provided only by at least two driving resonances. Hence, if the influence of two resonances is different (due to a different detune, for example), the diffusion will be determined by the weaker one (with a larger detune).

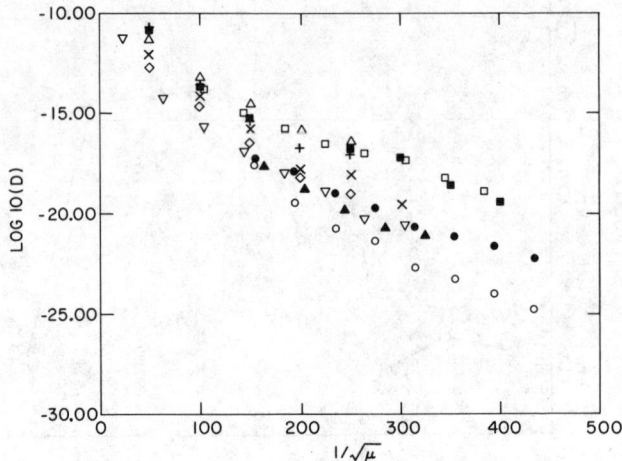

Fig. 4. Diffusion rate D vs. perturbation $1/\sqrt{\mu}$, parameters of trajectories are given in Table 2.

A large dispersion of points between runs in Fig. 4 is obviously due to the variation of the model parameters for these different runs. Since our model has no immediate relation to the problem of the beam-beam interaction in which we are interested, the 'raw' data presented in Fig. 4 are only of a minor importance. We need thus to trace some regularities in this raw material to get rid of the peculiarities of the particular model under consideration.

5. SCALING

We may start with an explicit expression for the rate of Arnold diffussion in our model as derived in Ref. 5:

$$D \approx \frac{(\pi a \omega \epsilon f_m)^2}{3 L \Omega_\mu} \cdot e^{-\frac{\pi |\delta \omega|}{\Omega_\mu}}, \qquad (13)$$

where f_m is the Fourier amplitude for the nearest harmonic of the driving force $f(t)$, $\Omega_\mu \approx 0.85\sqrt{\mu}$ the frequency of small phase oscillations on the guiding resonance[5], and L is some logarithmic factor which for our rough estimates may be considered as a constant. The physical meaning of the first factor if very simple and transparent. Indeed, the quantity $a \omega \epsilon f_m \sim \epsilon f_m \dot{x}$ gives the order of magnitude of the driving force power responsible for the diffusion in energy. Further, since an elementary change ΔH happens each halfperiod of the phase oscillations (see Ref. 5), that is on time scale $T_\mu \sim 1/\mu$, one would expect $\Delta H \propto \epsilon f_m a \omega \cdot T_\mu$, or for the diffusion rate

$$D \sim \frac{(\Delta H)^2}{T_\mu} \propto \frac{(\epsilon f_m a \omega)^2}{\Omega_\mu}, \qquad (14)$$

Table 2. Parameters of trajectories related to Fig. 4.

No.	Range, $1/\sqrt{\bar{\mu}}$	Range, $-\log D$	Backgrd*, $-\log D_\ell$	t_{max}	$\frac{\epsilon}{\bar{\mu}}$	a	A	$\frac{\omega}{\Omega}$	$\frac{\Omega_1}{\Omega_2}$	$\|\delta\omega\|$	Symbol in Fig. 4	Comments
1	154-434	17.6-24.7	26.3	$5 \cdot 10^7$	0.01	0.27	-	-	29/25	0.0226	●	Force (4)
2	50-400	11.4-18.4	24.9	10^7	0.01	0.225	0.995	6.5	-	0.0147	△	Force (5)
3	50-400	10.8-20.3	24.9	10^7	0.01	0.225	0.995	4.5	-	0.0212	+	"
4	50-400	12.0-20.7	24.9	10^7	0.01	0.295	0.995	3.5	-	0.0272	×	"
5	50-400	12.7-21.6	24.9	10^7	0.01	0.295	0.98	5.5	-	0.0173	◆	"
6	164-444	17.7-23.2	27.0	$5 \cdot 10^7$	0.001	0.225	0.995	5.5	-	0.0173	⊕	"
7	24-304	11.3-20.7	24.8	10^7	0.001	0.225	0.995	5.5	-	0.0173	⊠	"
8	154-434	17.3-22.3	24.8	10^7	0.001	0.265	0.995	5.5	-	0.0177	⊠	"
9	50-400	10.9-19.5	27.0	$5 \cdot 10^7$	0.01	0.225	0.995	5.5	-	0.0173	Y	"
10	104-384	13.8-18.9	27.0	$5 \cdot 10^7$	0.01	0.225	0.995	5.5	-	0.0173	⋈	"

*For $1/\sqrt{\bar{\mu}} > 150$, see Fig. 2

in accordance with explicit Relation (13). Note, that the symbol $\propto$ here means proportionality rather than order of magnitude (symbol $\sim$), since obviously an exponential factor is present. Nevertheless, this estimate for the factor is part of the exponential term seems to be fairly general, and we may try to rescale the numerical data making use of the dependence in Relationship (14). Namely, let us introduce the reduced diffusion rate D* according to the relation

$$D^* = \frac{D/\mu}{\varepsilon^2 f_m^2 a^4} = D/\varepsilon^2 f_m^2 a^4 \mu^{3/2} \left(\frac{\varepsilon}{\mu}\right)^2 . \tag{15}$$

The latter expression takes into account the fact that in a single run ε/μ = constant, and so besides a constant factor only dependence on μ is to be introduced in rescaling.

As was observed in Refs. 5, 6, relation (13) does not hold for a sufficiently weak perturbation (see Fig. 5 below), so the exponent in

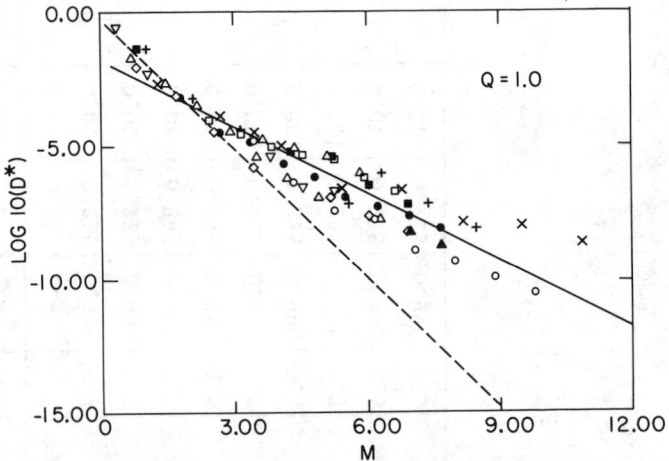

Fig. 5. Dimensionless diffusion rate D* vs. rescaled perturbation $M = |\delta\omega|/\sqrt{\mu}$. Dashed line shows dependence given in Eq. (13), and the solid line is the least square fit of the data to the Relation (16), using Q = 1.

Eq. (13) has to be changed somehow. At this point we may try to make use of some rough estimates derived in Ref. 5 (see also Ref. 9 for a simplified version) which may be represented in our case as follows:

$$D^* = A \cdot \exp\left[- B \cdot \left(\frac{F}{\Omega_\mu}\right)^{1/Q}\right] \tag{16}$$

where A, B. f and Q are some constants independent of the perturbation. Since D* is a rescaled diffusion rate, the factor A is dimensionless and would be expected to be of the order unity. Parameter F may be chosen to have the dimensions of frequency in order to make factor B dimensionless. Note that the analytical estimate (13) is also of the type of Eq. (16) with

$$A = \frac{\pi^2 \beta}{3L} \approx 0.56 \ ; \ B - \pi \ ; \ F = |\delta\omega| \ ; \ Q = 1 \ ,$$

where $L \approx 5$ is assumed for $1/\sqrt{\mu} \sim 100$, and relations $\omega = \beta a$ and $\Omega_\mu = \beta/\sqrt{\mu}$ are taken into account.[5] Relation of the type (16) follows also from Nekboroskev's estimates in Ref. 8. Since the Relation (13) describes the Arnold diffusion fairly well for a relatively large perturbations[5,6], it seems plausible to assume $F = |\delta\omega|$ also for a weaker perturbation. In other words, we may rescale the perturbation as well by introducing into Eq. (16) a dimensionless quantity:

$$M = \frac{|\delta\omega|}{\Omega_\mu} = \frac{|\delta\omega|}{\beta/\sqrt{\mu}}$$

Now we may rescale numerical data to the variables D* and M and try to fit them to Eq. (16). An example of rescaled data for the 10 runs listed in Table 2 is given in Fig. 5 using coordinates logD* and M for Q = 1. The dashed line represents Relation (13) which is valid only for $M \leq 3$.

From comparison of Fig. 4 with Fig. 5, it is clearly seen that the dispersion of numerical points has been considerably reduced by rescaling. Thus, Eq. (16) at least represents some partial dependence of the diffusion rate on the model parameters. (Note the difference in scale of the vertical axis on the two Figs.). To find the optimal values of the parameters A, B and Q in Relation (16), a least squares fit of the numerical data was done in the following way. For a given value of Q the pairs of quantities logD* and $M^{1/Q}$ for all the 10 runs in Table 2 have been fitted to a straight line corresponding to the Relation (16), each fit giving two parameters A and B in dependence on Q value. The quality of a fit is characterized by the root-mean-square deviation

$$S = \sqrt{\langle(\log D^* - \log D_F^*)^2\rangle} \ , \tag{18}$$

where D* are computational data and where D_F^* is related to Eq. (16). It is convenient to introduce another characteristic of dispersion

$$R = 10^S = 10^{\sqrt{\langle(\lg\frac{D^*}{D_F^*})^2\rangle}} \ , \tag{19}$$

which represent a certain average ratio $\langle D^*/D_F^*\rangle$ (or vice versa $\langle D_F^*/D^*\rangle$, of course). The results of the fit are summarized in Table 3.

Table 3. The least square fit of data in Table 2 to the Relation (16).

Q	1	2	3	4	5	6	7	8
A	0.0155	25.8	2.94×10^4	3.06×10^7	3.06×10^{10}	3.01×10^{13}	2.94×10^{16}	2.84×10^{19}
B	1.92	7.86	14.5	21.2	28.1	34.9	41.8	48.6
R	4.97	3.87	3.95	4.09	4.21	4.31	4.39	4.45

The ratio R characterizing the dispersion of numerical points (see Fig. 5) is not very sensitive to the value of Q, especially for a large Q, but factor changes drastically with Q. Since $A \sim 1$ is expected, as mentioned above, we have only to choose between $Q = 1$ and $Q = 2$. The value $Q = 2$ seems to be better in regard to a lower dispersion R and a reasonable value of A. Thus, we would suggest the following estimate for the rate of Arnold diffusion

$$D^* \approx 26 \cdot \exp(-7.9 \times \sqrt{M}) . \qquad (20)$$

In Fig. 6 numerical data are plotted in coordinates $\log D^*$ and $\sqrt{M}$, the straight line representing Relation (20). Surprisingly, this relation describes satisfactorily also the region $M < 3$ where more accurate estimate (13) is applicable (comp. Fig. 5), and even the resonance overlap region. The latter would correspond roughly to $\sqrt{M} \lesssim 1.3$ except trajectories No. 3 and 4 (see Table 2 and Fig. 2).

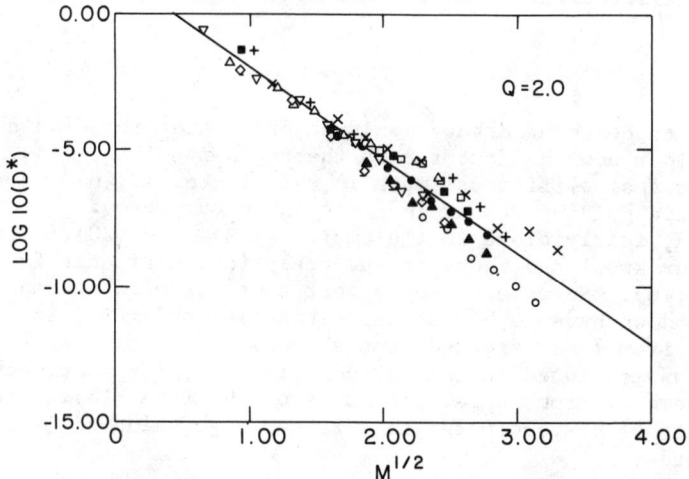

Fig. 6. Same as in Fig. 5 for $Q = 2$.

For another set of 10 runs with different values of the parameters the least dispersion was obtained also for $Q = 2$ with $A = 42.1$ and $B = 8.2$ which is rather close to the rough estimate given by Eq. (20). The minimal dispersion $R = 3.21$ was still less in the latter case (comp. Table 3). If one should try to fit original numerical data (in coordinates $\log D$ and $1/\sqrt{\mu}$) as plotted in Fig. 4, one would obtain the minimal $R \approx 30$ for both sets of runs. On the other hand, the minimal dispersion for a single run fit would be $R \approx 1.13$ only. So, the above mentioned dispersion $R \approx 3$ characterizes the accuracy of a simplified Estimate (20).

6. DISCUSSION

The main question to be discussed is whether Estimate (20) describes Arnold diffusion only for the model under consideration, and moreover, only in the restricted range of parameters actually studied numerically, or can one hope to apply Eq. (20) in a more general situation. This question may be answered partly via a comparison of the empirical parameters in Eq. (20) with the analytical consideration of Ref. 5 (see also Ref. 9). Let us consider first the most important parameter Q. According to Ref. 5, it is equal to the number N of basic frequencies of the unperturbed system whose combinations give rise to a dense set of driving resonances. It is precisely these higher order resonances which increase the diffusion rate as compared to the simple Estimate (13) (see Fig. 5). Even though their amplitudes are much smaller than for the first order resonances (Fig. 1) on which Eq. (13) is based, they are much closer so that the corresponding detune $|\delta\omega|$ is much smaller. Eventually, as the perturbation decreases, just those high order resonances determine the diffusion rate (for more details see Refs. 5, 9). Since in the model under consideration there are 3 basic frequencies $[\omega_1, \omega_2, \Omega$, Force (5)], Relation (20) with Q = 2 seems to contradict at the first glance with the conclusion in Ref. 5, 9. However, the two frequencies in our computation are very close, especially for a weak perturbation $\omega_1 \approx \omega_2 \approx \omega_0$. Hence, the high order resonances are essentially related to the considerations of only two frequencies: ω_0 and Ω, and the general relation

$$Q = N \tag{21}$$

needs not to be changed but should be only interpreted in a proper way.

Consider now the factor B on the exponent of Eq. (16). Again, according to a simplified estimate in Ref. 5 [Eq. (7.46)], $B \approx 2\pi$ for a low harmonic guiding resonance ($\omega_1 = \omega_2$ in our case). The latter value is, indeed, fairly close to the empirical $B \approx 7.9$ (20). Finally, to the best of our knowledge there are no analytical estimates for the factor A in Eq. (16). So we are simply left thus far with an empirical value $A \approx 26$, and we have to be satisfied that its order is, indeed, not much different from 1 as expected (see above).

As was mentioned in the Introduction to this paper, Nekhoroshev obtained[8] some rigorous upper estimates of the Type (16). The main peculiarity of his estimates is a rather high value of Q. This best result reads:

$$A = \frac{3N^2 - N + 8}{4} = 4.5 \tag{22}$$

where numerical value is given for N = 2. This value seems to contradict with our numerical results since the corresponding factor A would be incomprehensibly big (see Table 3).

Summarizing, we have reason to believe that Estimate (20) may have a wider application than to just the simple immediate model from which it has been obtained.

Let us try to apply our estimate to a rather different system, namely a particle in a magnetic trap. The alleged Arnold diffusion of electrons in two different experiments was analysed in Ref. 9 using a rough estimate

$$\tau \nu^2 \sim \exp(b/\Omega_\phi^{1/3}), \tag{23}$$

where τ is electron lifetime; ν dimensionless strength of driving resonances due to the azimuthal variation of the magnetic field; $Q = N = 3$ since there is no special relation between the basic frequencies, Larmor frequency $\omega = 1$, and the frequency of phase oscillations on a guiding resonance

$$\Omega_\phi \sim \exp(1/2\epsilon), \tag{24}$$

with some small adiabaticity parameter ϵ. Now we may rescale the perturbation as was done in Section 5 for our simple model, namely: $\Omega_\phi \rightarrow \Omega_\phi/|\delta\omega|$, where mean detune ($\delta\omega$) is determined now by the lowest drift frequency of an electron $\Omega_g \sim (\rho/\ell)^2$ with ρ equal to the electron Larmor radius, and ℓ the scale of the magnetic field, whence:

$$\tau\nu^2 \sim \exp\left[-B\frac{\rho}{\ell}^{2/3} e^{\frac{1}{2\epsilon}}\right]. \tag{25}$$

A new factor B may be calculated from the data in Ref. 9 using relation: $B = b(\ell/\rho)^{2/3}$ with factor b from the Estimate (23). The results are shown in Table 4. One can see that B values are much closer than those of b. The ratio of mean b values for the two experiments is 2.33, whereas that for B values is 1.35. So the scaling used in Eq. (20) seems to work in this case as well. On the other hand, the mean value $<B> = 12.9$ is a great deal larger than for our numerical model ($B = 7.9$). Is it a real contradiction? Not necessarily at all!

The point is that Eq. (20) describes the rate of Arnold diffusion in relatively very narrow stochastic layers (see Fig. 3), whereas for most initial conditions the motion is perfectly stable. If it would be so in a real dynamical system, the Arnold diffusion would be of no practical importance. However, there is always some additional 'external' (in regard to a dynamical system) diffusion, or noise, which brings system into all of stochastic layers. In the case of electrons in a magnetic trap it is the gas scattering. But then the average diffusion inside a stochastic layer, since the sojourn time inside a layer amounts to only a small fraction of the motion time. The corresponding reduction of the diffusion rate may be very roughly estimated as follows. From a simple theory of the Arnold diffusion resulting in a relation like Eq. (13), it is known[5] that the width of a stochastic layer (w_s) is proportional to an exponential factor similar to one in the diffusion rate.

$$w_s \propto \exp\left(-\frac{\pi|\delta\omega|}{2\Omega_\mu}\right).$$

The exponent in the latter expression differs from that in the diffusion rate by a factor of (1/2) only. One can draw from this comparison a very rough conclusion that the reduction of the diffusion rate due to some external noise, being proportional to the layer width, would result in an increase of the factor B in Eq. (16) by a factor 1.5. In other words the reduced diffusion rate would be roughly a product of the rate inside a layer and the layer width. The actual ratio of B values for a electron in magnetic traps (Table 4) and for our purely dynamical model [Eq. (20)] is 1.63. Since estimate (25) is very rough and is based upon only a few experimental points (see Table 4), any recomendation of this estimate seems premature. Instead, additional studies of the Arnold diffusion in more realistic conditions are very much in order.

Table 4. Arnold diffusion in magnetic traps.

	$\frac{1}{\epsilon}$	$\frac{\ell \ell}{\rho \rho}$	b	B
1st. experiment	7.89	8.96	3.66	15.8
	6.68	6.61	4.38	15.4
	8.25	6.61	3.37	11.9
2nd. experiment	13.6	17.6	1.75	11.8
	14.5	15.5	1.51	9.4
Mean Value				12.9

ACKNOWLEDGMENT

The authors express their sincere appreciation for the assistance, stimulating conversations, and useful comments which were provided by G. Casati, G. Contopoulos, E. Courant, F. Izraelev, E. Keil, A. Lichtenlerg, M. Lieberman, O. Manley, M. Month. R. Peierls, and J. Tennyson. This work was performed under the U.S.A. - U.S.S.R. Cooperative Program on Research in the Fundamental Properties of Matter. We appreciate the assistance of the U.S. Department of Energy and the U.S.S.R. Academy of Science in making this joint research possible.

REFERENCES

1. E. Keil, "Experimental Data on the Beam-Beam Limit and Their Consequences on the Design of LEP," CERN/ISR/TH/79-32, 1979.
2. M. Wiedemann, private communication.
3. F. M. Izrailev, S. I. Mishnev, G. M. Tunaikin, "Numerical Studies of Stochasticity Limits in Colliding Beams (One-Deimensional Model)," Preprint 77-43, Institute of Nuclear Physics, Novosbirsk, 1977.
4. J. Ford, in Fundamental Problems in Statistical Mechanics, Vol. 3, Edited by E.G.D. Cohen (North-Holland, Amsterdam, 1975).
5. B. V. Chirikov, "A Universal Instability of Many-Dimensional Oscillator Systems," Physics Reports 52:5 (1979) 265.
6. G. V. Gadiyak, F. M. Izraelev, B. V. Chirikov, "Numerical Experiments on a Universal Instability in Nonlinear Oscillator Systems (The Arnold Diffusion)," Proc. 7th Int. Conf. on Nonlinear Oscillations (Berlin, 1975), Vol. II, 1, 315; "The Arnold Diffusion in a System of Three Resonances" (to be published).
7. J. L. Tennyson, M. A. Liebermann, A. J. Lichtenberg, "Diffusion in Near-Integrable Hamiltonian Systems With Three Degrees of Freedom," Memorandum No. UCB/ERL M79/1, College of Engineering, Berkeley, 1979.
8. N. N. Nekhoroshev, Usp. Mat. Nauk 32:6 (1977).
9. B. V. Chirikov, "The Problems of Motion Stability for a Charged Particle in a Magnetic Trap," Fizika Plasmy 4:3, 527 (1978).

RAYMOND H. FOGLER LIBRARY
DATE DUE

BOOKS ARE SUBJECT TO
RECALL AFTER TWO WEEKS